INTERNATIONAL UNION OF PURE AND APPLIED CHEMISTRY

ANALYTICAL CHEMISTRY DIVISION
COMMISSION ON SOLUBILITY DATA

SOLUBILITY DATA SERIES

Volume 34

4-AMINOBENZENESULFONAMIDES

Part I

Non-cyclic Substituents

A complete list of volumes published in the Solubility Data Series will be found on p. 347.

SOLUBILITY DATA SERIES

Editor-in-Chief
A.S. KERTES

Volume 34

4-AMINOBENZENESULFONAMIDES

Part I

Non-cyclic Substituents

Volume Editors

ANTHONY N. PARUTA
University of Rhode Island
Kingston, RI, USA

RYSZARD PIEKOS
Medical Academy
Gdansk, Poland

PERGAMON PRESS

OXFORD · NEW YORK · BEIJING · FRANKFURT
SÃO PAULO · SYDNEY · TOKYO · TORONTO

U.K.	Pergamon Press plc, Headington Hill Hall, Oxford OX3 0BW, England
U.S.A.	Pergamon Press, Inc., Maxwell House, Fairview Park, Elmsford, New York 10523, U.S.A.
PEOPLE'S REPUBLIC OF CHINA	Pergamon Press, Room 4037, Qianmen Hotel, Beijing, People's Republic of China
FEDERAL REPUBLIC OF GERMANY	Pergamon Press GmbH, Hammerweg 6, D-6242 Kronberg, Federal Republic of Germany
BRAZIL	Pergamon Editora Ltda, Rua Eça de Queiros, 346, CEP 04011, Paraiso, São Paulo, Brazil
AUSTRALIA	Pergamon Press Australia Pty Ltd., P.O. Box 544, Potts Point, N.S.W. 2011, Australia
JAPAN	Pergamon Press, 5th Floor, Matsuoka Central Building, 1-7-1 Nishishinjuku, Shinjuku-ku, Tokyo 160, Japan
CANADA	Pergamon Press Canada Ltd., Suite No. 271, 253 College Street, Toronto, Ontario, Canada M5T 1R5

First edition 1988

The Library of Congress has catalogued this serial title as follows:

Solubility data series. — Vol. 1 — Oxford; New York;
Pergamon, c 1979-
v.; 28 cm.
Separately catalogued and classified in LC before no. 18.
ISSN 0191-5622 = Solubility data series.
1. Solubility — Tables — Collected works.
QD543.S6629 541.3′42′05-dc19 85-641351
AACR 2 MARC-S

British Library Cataloguing in Publication Data

4-Aminobenzenesulfonamides.
Pt. 1: Non-cyclic substituents
1. Sulfonamides. Solubility
I. Paruta, Anthony N. II. Piekos,
Ryszard III. Guillory, J. Keith
IV. Series
547.7′5045′42

ISBN 0-08-030742-6

Printed in Great Britain by A. Wheaton & Co. Ltd., Exeter

CONTENTS

Page Number

FOREWORD

If the knowledge is
undigested or simply wrong,
more is not better

How to communicate and disseminate numerical data effectively in chemical science and technology has been a problem of serious and growing concern to IUPAC, the International Union of Pure and Applied Chemistry, for the last two decades. The steadily expanding volume of numerical information, the formulation of new interdisciplinary areas in which chemistry is a partner, and the links between these and existing traditional subdisciplines in chemistry, along with an increasing number of users, have been considered as urgent aspects of the information problem in general, and of the numerical data problem in particular.

Among the several numerical data projects initiated and operated by various IUPAC commissions, the *Solubility Data Project* is probably one of the most ambitious ones. It is concerned with preparing a comprehensive critical compilation of data on solubilities in all physical systems, of gases, liquids and solids. Both the basic and applied branches of almost all scientific disciplines require a knowledge of solubilities as a function of solvent, temperature and pressure. Solubility data are basic to the fundamental understanding of processes relevant to agronomy, biology, chemistry, geology and oceanography, medicine and pharmacology, and metallurgy and materials science. Knowledge of solubility is very frequently of great importance to such diverse practical applications as drug dosage and drug solubility in biological fluids, anesthesiology, corrosion by dissolution of metals, properties of glasses, ceramics, concretes and coatings, phase relations in the formation of minerals and alloys, the deposits of minerals and radioactive fission products from ocean waters, the composition of ground waters, and the requirements of oxygen and other gases in life support systems.

The widespread relevance of solubility data to many branches and disciplines of science, medicine, technology and engineering, and the difficulty of recovering solubility data from the literature, lead to the proliferation of published data in an ever increasing number of scientific and technical primary sources. The sheer volume of data has overcome the capacity of the classical secondary and tertiary services to respond effectively.

While the proportion of secondary services of the review article type is generally increasing due to the rapid growth of all forms of primary literature, the review articles become more limited in scope, more specialized. The disturbing phenomenon is that in some disciplines, certainly in chemistry, authors are reluctant to treat even those limited-in-scope reviews exhaustively. There is a trend to preselect the literature, sometimes under the pretext of reducing it to manageable size. The crucial problem with such preselection - as far as numerical data are concerned - is that there is no indication as to whether the material was excluded by design or by a less than thorough literature search. We are equally concerned that most current secondary sources, critical in character as they may be, give scant attention to numerical data.

On the other hand, tertiary sources - handbooks, reference books and other tabulated and graphical compilations - as they exist today are comprehensive but, as a rule, uncritical. They usually attempt to cover whole disciplines, and thus obviously are superficial in treatment. Since they command a wide market, we believe that their service to the advancement of science is at least questionable. Additionally, the change which is taking place in the generation of new and diversified numerical data, and the rate at which this is done, is not reflected in an increased third-level service. The emergence of new tertiary literature sources does not parallel the shift that has occurred in the primary literature.

With the status of current secondary and tertiary services being as briefly stated above, the innovative approach of the *Solubility Data Project* is that its compilation and critical evaluation work involve consolidation and reprocessing services when both activities are based on intellectual and scholarly reworking of information from primary sources. It comprises compact compilation, rationalization and simplification, and the fitting of isolated numerical data into a critically evaluated general framework.

The *Solubility Data Project* has developed a mechanism which involves a number of innovations in exploiting the literature fully, and which contains new elements of a more imaginative approach for transfer of reliable information from primary to secondary/tertiary sources. *The fundamental trend of the Solubility Data Project is toward integration of secondary and tertiary services with the objective of producing in-depth critical analysis and evaluation which are characteristic to secondary services, in a scope as broad as conventional tertiary services.*

Fundamental to the philosophy of the project is the recognition that the basic element of strength is the active participation of career scientists in it. Consolidating primary data, producing a truly critically-evaluated set of numerical data, and synthesizing data in a meaningful relationship are demands considered worthy of the efforts of top scientists. Career scientists, who themselves contribute to science by their involvement in active scientific research, are the backbone of the project. The scholarly work is commissioned to recognized authorities, involving a process of careful selection in the best tradition of IUPAC. This selection in turn is the key to the quality of the output. These top experts are expected to view their specific topics dispassionately, paying equal attention to their own contributions and to those of their peers. They digest literature data into a coherent story by weeding out what is wrong from what is believed to be right. To fulfill this task, the evaluator must cover *all* relevant open literature. No reference is excluded by design and every effort is made to detect every bit of relevant primary source. Poor quality or wrong data are mentioned and explicitly disqualified as such. In fact, it is only when the reliable data are presented alongside the unreliable data that proper justice can be done. The user is bound to have incomparably more confidence in a succinct evaluative commentary and a comprehensive review with a complete bibliography to both good and poor data.

It is the standard practice that the treatment of any given solute-solvent system consists of two essential parts: I. Critical Evaluation and Recommended Values, and II. Compiled Data Sheets.

The Critical Evaluation part gives the following information:

(i) a verbal text of evaluation which discusses the numerical solubility information appearing in the primary sources located in the literature. The evaluation text concerns primarily the quality of data after consideration of the purity of the materials and their characterization, the experimental method employed and the uncertainties in control of physical parameters, the reproducibility of the data, the agreement of the worker's results on accepted test systems with standard values, and finally, the fitting of data, with suitable statistical tests, to mathematical functions;

(ii) a set of recommended numerical data. Whenever possible, the set of recommended data includes weighted average and standard deviations, and a set of smoothing equations derived from the experimental data endorsed by the evaluator;

(iii) a graphical plot of recommended data.

The Compilation part consists of data sheets of the best experimental data in the primary literature. Generally speaking, such independent data sheets are given only to the best and endorsed data covering the known range of experimental parameters. Data sheets based on primary sources where the data are of a lower precision are given only when no better data are available. Experimental data with a precision poorer than considered acceptable are reproduced in the form of data sheets when they are the only known data for a particular system. Such data are considered to be still suitable for some applications, and their presence in the compilation should alert researchers to areas that need more work.

The typical data sheet carries the following information:

(i) components - definition of the system - their names, formulas and Chemical Abstracts registry numbers;
(ii) reference to the primary source where the numerical information is reported. In cases when the primary source is a less common periodical or a report document, published though of limited availability, abstract references are also given;
(iii) experimental variables;
(iv) identification of the compiler;
(v) experimental values as they appear in the primary source. Whenever available, the data may be given both in tabular and graphical form. If auxiliary information is available, the experimental data are converted also to SI units by the compiler.

Under the general heading of Auxiliary Information, the essential experimental details are summarized:

(vi) experimental method used for the generation of data;
(vii) type of apparatus and procedure employed;
(viii) source and purity of materials;
(ix) estimated error;
(x) references relevant to the generation of experimental data as cited in the primary source.

This new approach to numerical data presentation, formulated at the initiation of the project and perfected as experience has accumulated, has been strongly influenced by the diversity of background of those whom we are supposed to serve. We thus deemed it right to preface the evaluation/compilation sheets in each volume with a detailed discussion of the principles of the accurate determination of relevant solubility data and related thermodynamic information.

Finally, the role of education is more than corollary to the efforts we are seeking. The scientific standards advocated here are necessary to strengthen science and technology, and should be regarded as a major effort in the training and formation of the next generation of scientists and engineers. Specifically, we believe that there is going to be an impact of our project on scientific-communication practices. The quality of consolidation adopted by this program offers down-to-earth guidelines, concrete examples which are bound to make primary publication services more responsive than ever before to the needs of users. The self-regulatory message to scientists of the early 1970s to refrain from unnecessary publication has not achieved much. A good fraction of the literature is still cluttered with poor-quality articles. The Weinberg report (in 'Reader in Science Information', ed. J. Sherrod and A. Hodina, Microcard Editions Books, Indian Head, Inc., 1973, p. 292) states that 'admonition to authors to restrain themselves from premature, unnecessary publication can have little effect unless the climate of the entire technical and scholarly community encourages restraint...' We think that projects of this kind translate the climate into operational terms by exerting pressure on authors to avoid submitting low-grade material. The type of our output, we hope, will encourage attention to quality as authors will increasingly realize that their work will not be suited for permanent retrievability unless it meets the standards adopted in this project. It should help to dispel confusion in the minds of many authors of what represents a permanently useful bit of information of an archival value, and what does not.

If we succeed in that aim, even partially, we have then done our share in protecting the scientific community from unwanted and irrelevant, wrong numerical information.

A. S. Kertes

PREFACE

With few exceptions, these volumes of the solubility data series deal with solubilities of the derivatives of 4-aminobenzenesulfonamide, usually referred to as "sulfanilamide" (sulfanilic acid amide), a name coined in 1937 (1). The history of sulfanilamide begins in 1906, when Schroeter (2) synthesized the molecule containing a 4-acetylaminosulfanilamide portion.

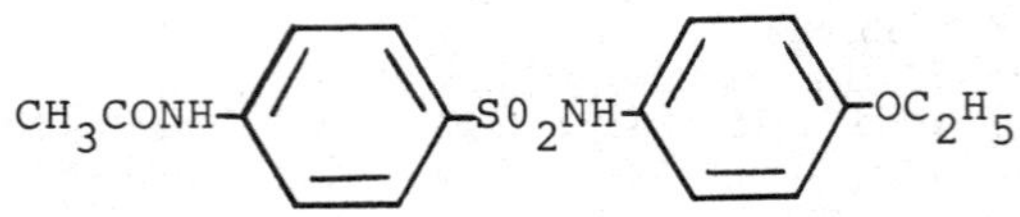

In 1908, Gelmo (3) described sulfanilamide and 13 of its derivatives and gave solubility values for these compounds. In 1935, Domagk (4) detected antibacterial activity of a synthetic azo dye, prontosil, with the structure.

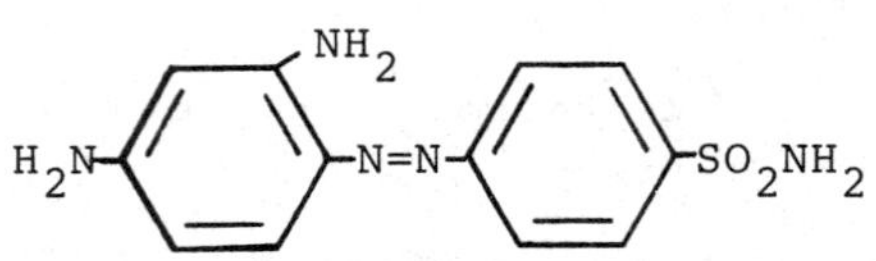

This compound had been tested for antibacterial activity (5), the "sulfanilamide" portion being responsible for its activity. This was confirmed (6) by isolation of sulfanilamide in the urine of patients. Fildes (7) and Wood (8), in 1940, demonstrated that the derivatives of sulfanilamide were antimetabolites of p-aminobenzoic acid (PABA) which is a step in the folic acid synthesis of bacteria. Thus, the structural similarity of PABA and sulfonamides caused interference by competitive antagonism and resulted in a bacteriostatic effect. The discoveries of antibacterial activity led to an exciting flood of research, and thousands of sulfanilamide derivatives have been synthesized. As early as 1948, the number of sulfonamide derivatives (9) was estimated to be several thousand. In the two decades after that, the number of synthesized sulfonamides have gone past 10,000(10)

Clinical trials of these sulfonamides and derivatives have been associated with low solubilities and some renal crystalluria. The low solubility, and its sensitivity to pH, could cause crystalline precipitation in the renal tubules in the filtration of blood into acidic urine. Some of the problems of limited solubility were overcome by complexation or salt formation, and solid state manipulations which in turn have stimulated investigations into solubility of the drugs in water, buffers and some binary solvent system. Analytical methodologies span a wide spectrum of techniques and the relevant references are in pharmaceutical, medical and chemical literature.

In all volumes the chemical structures, registry number and the molecular weight of the compounds considered are collected in the front of each volume. The compounds as they occur on the data sheets are given successively in each volume. In the first volume of this series there are 35 compounds. The second and third volumes have 58 compounds and 108 compounds, respectively.

NOMENCLATURE:

The nomenclature of sulfanilamide derivatives has conventionally been based on the following numbering system: substituents at the nitrogen atom of the amide group ($-SO_2NH_2$) are called $\underline{N}^1$-substituents, whereas substitutents at the 4-amino nitrogen ($4-H_2N-$) are called $\underline{N}^4$-substituents. Substitution in either or both of the two positions lead to compounds referred to as "sulfonamides" (sometimes "sulfanilamides" or even "sulfamides"). Here are illustrative examples of this nomenclature.

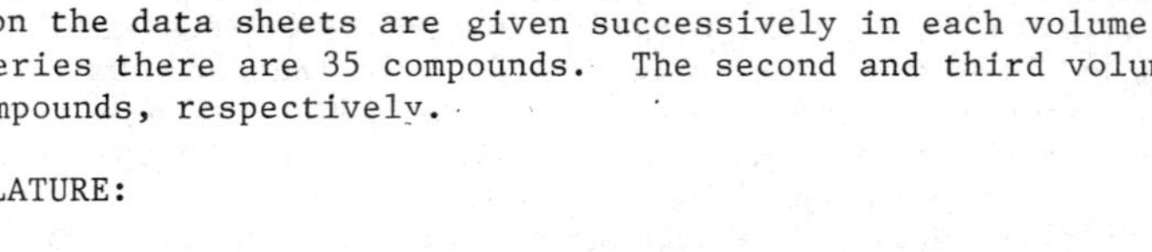

$\underline{N}^1$-substituted sulfonamides

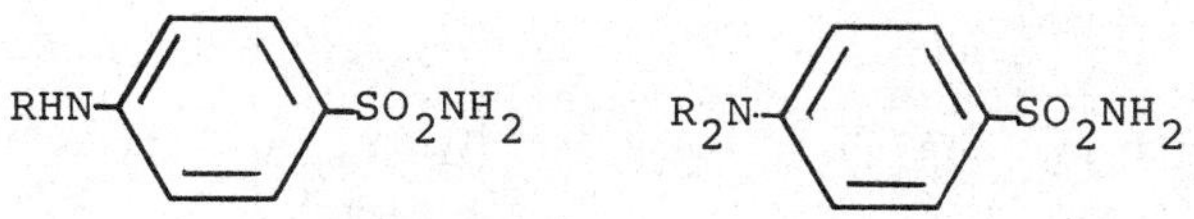

N^4-substituted sulfonamides

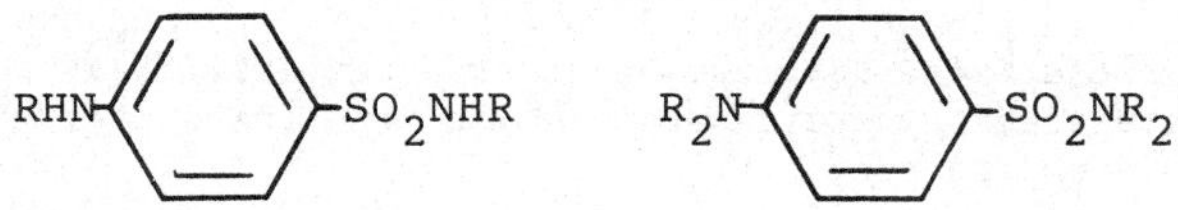

N^1,N^4-substituted sulfonamides

The 4-amino group can be diazotized to give derivatives of the formula

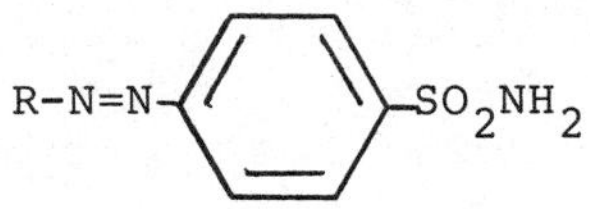

As the sulfonamide molecule carries a basic 4-NH_2 group and an acid -SO_2NH_2 one, it is capable of formation the respective salts or complexes, e.g.

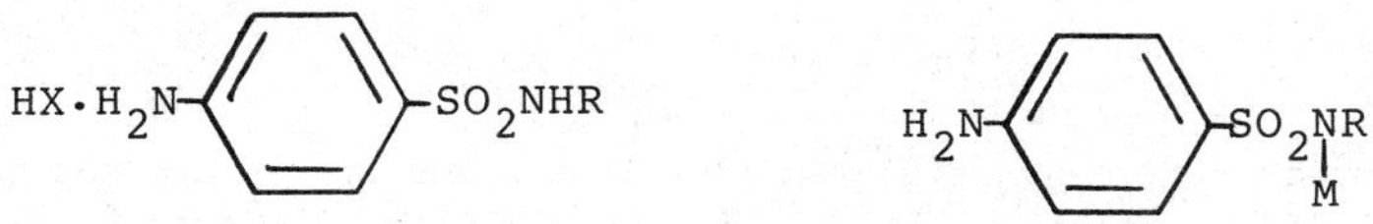

where HX stands for an acid and M is a univalent metal atom.

In common use by health practitioners are nonproprietary names of sulfonamides which are brief and reflect the chemical nature of their molecules. Examples are: sulfacetamide, sulfapyridine, sulfathiazole, sulfadiazine, sulfaguanidine, etc. There are numerous trivial names; for example, sulfanilamide has as many as 140 synonyms, and sulfathiazole has 113. Negwer (11) has compiled an excellent guide to this nomenclature. In chemical literature, systematic names in line either with IUPAC (12) or Chemical Abstract rules (13) are used. The latter has been adopted in these volumes and the systematic name is, where appropriate, followed by the nonproprietary or trivial name.

ORGANIZATION OF THE VOLUMES:

The numerical data on the solubility of 2-aminobenzenesulfonamide, 3-aminobenzenesulfonamide, and 4-aminobenzenesulfonamide and its N^1 and N^4- derivatives, salts and complexes, compiled up to 1985 inclusive, have been divided into three volumes on the basis of chemical structure of the compounds.

The first volume includes the solubility of 2-aminobenzenesulfonamide, 3-aminobenzenesulfonamide, 4-aminobenzenesulfonamide and the derivatives of the last-named compound substituted at either of the nitrogen atoms, or both, with non-cyclic substituents (see System Index at the end of the first volume). The aroyl substituents, -C(:O)aryl, have also been included here. The second volume includes sulfanilamide derivatives substituted with 5-membered heterocyclic rings at either of the nitrogen atoms, and their derivatives. The third volume covers the solubilities of the derivatives substituted with 6-membered rings, mixtures of sulfonamides, and miscellanea. The compilations do not include compounds devoid of the -NH_2, -NHR or -NR_2 group in the benzene ring.

The solvent systems include all solvents with the exception of body fluids. The order of solvents for a particular solute are as follows: water; water-mineral acid; water-mineral base; water-mineral salt; water-miscellaneous mineral components; water-mineral and organic compounds; water-organic components; organic solvents; carboxylic acid and their salts; aliphatic acids; aromatic acids; other acids; alcohols, phenols (mono-, di-, polyhydric); amides; amines; aliphatic amines (primary, secondary, tertiary); aromatic amines (primary, secondary, tertiary); other amines; aminoalcohols; carboxylic acid esters; ethers (excluding tensides); hydrocarbons; aliphatic hydrocarbons; aromatic hydrocarbons; miscellaneous hydrocarbons; halogenated hydrocarbons (flouro-, chloro-, bromo-, iodo-); aliphatic halogenated hydrocarbons; aromatic halogenated hydrocarbons; ketones; tensides (surface-active agents); miscellaneous organic solvents.

SIGNIFICANT FIGURES AND GRAPHICAL DATA:

In most cases, solubility values given in the primary source by various workers are overstated with respect to significant figures. Since the author(s) original values are given on the data sheets, it is difficult to consider significant figures and analytical limitations in a completely consistant fashion. Therefore, the reader should be aware that in most cases the number of significant figures used for calculations was not that given by the original author(s). This was done to maintain coherence and consistency as data were given to varying significant figures. In many cases graphic data of sufficient size and clarity are reproduced. The data can be regarded of sufficient accuracy to serve as a starting point for more precise determinations. In many instances, the effect of additive concentration, pH, temperature, etc. can be depicted.

POLYMORPHISM:

Many sulfonamides exhibit several cyrstalline forms or polymorphs. There are several studies referenced in these volumes that specifically deal with the solubility difference between polymorphic modifications of the same compound. The solubility differences between polymorphs have been found to vary over a large range of values.

AMPHOLYTES:

Solubility of ampholytic sulfonamides as a function of pH varies enormously, sometimes by several orders of magnitude. Unless the pH is known experimentally, the solubility value may be suspect especially at low (1-3) and high pH (10-12) values. In these cases, the solubility is a rapidly changing value, frequently with small incremental changes of pH. The abrupt change of solubility with pH is usually associated with the formation of water soluble anionic and cationic species. Buffers, especially at higher concentrations may alter solubility by salting effect and the pH is also affected by ionic strength.

EQUILIBRATION TIME:

In general, it appears that many of these determined solubilities may not have been under equilibrium conditions. Unfortunately, in too many instances the equilibration time appears too low. Typically, solutes possess low aqueous solubilities and require long dissolution time to reach saturation. Saturation time should be experimentally determined in each case and for each compound. In many cases up to 24 hours may be required.

The editors consider the vast majority of the solubility values given in these volumes as tentative. It should be stressed, however, that they represent a useful starting point for more accurate determinations of a vast array of substituted 4-aminobenzene-sulfonamides, with many structurally and chemically related compound of various types. They amply illustrate the many factors and parameters affecting solubility and the direction and magnitude of these effects.

This compilation and evaluation is not only the result of the joint efforts of the compiler and evaluator, but also of all those who read the manuscripts, expressed their criticism, who procured copies of hard-to-get journals, who translated texts from Japanese as well as of those who in any other way assisted in the compilation and evaluation. We would like to express our gratitude in particular to the following colleagues: Prof. S. Kertes, Dr. M. Salomon, Prof. S. Yalkowsky, Prof. H. Akaiwa, Prof. C. Kalidas, Prof. W. Riess, Prof. A. Guerrero-Laverat, Prof. P. Rohdewald, Prof. J. Pütter, Dr.K. L. Loening, Dr. A. Brodin, Dr. D. Zimma, Mr. K. Hazelton, Dr. R. Fernandez-Prini, and Mr. E. MacMullan.

REFERENCES TO THE PREFACE:

1. Council on Pharmacy and Chemistry, J. Am. Med. Assoc. 1937, *108*, 1888.
2. Schroeter, G. Ber. Dtsch. Chem. Ges. 1906, *39*, 1559.
3. Gelmo, P. J. Prakt, Chem. *[2]*, 1908, *77*, 369.
4. Domagk, G. Dtsch. Med. Wschr. 1935, *61*, 250 and 829.
5. Tréfouël, T. J.; Nitti, F.; Bovet, D. Compt. Rend. Soc. Biol. 1935, *120*, 756.
6. Fuller, A. T. Lancet, 1937, 194.
7. Fildes, P. Lancet, 1940, 955.
8. Woods, D. D. J. Expt. Path. 1940, *21*, 74.
9. Langecker, H. Arch. Exptl. Pat. Pharmakol. 1948, *205*, 291.
10. Rolski, S. Chemia Srodkow Leczniczych (Chemistry of Medicinal Agents), 3rd ed., PZWL, Warsaw, 1968.
11. Negwer, M. Organisch-chemische Arzneimittel und ihre Synonyma (Organic-chemical Drugs and their Synonyms), Akademie-Verlag, Berlin, 1978.
12. Nomenclature of Organic Chemistry, Definitive Rules for Section C. Characteristic Groups Containing Carbon, Hydrogen, Oxygen, Nitrogen, Halogen, Sulfur, Selenium, and/or Tellurium IUPAC Commission on the Nomenclature of Organic Chemistry, London, Butterworths, 1971, rule 641.8.
13. J. Chem. Documentation 1974, *14*, 3.

INTRODUCTION TO THE SERIES ON SOLUBILITY OF SOLIDS IN LIQUIDS: SUBSERIES ON PHARMACEUTICALS

Nature of the Project

The Solubility Data Project (SDP) has as its aim a comprehensive search of the literature for solubilities of gases, liquids, and solids in liquids or solids. Data of suitable precision are compiled on data sheets in a uniform format. The data for each system are evaluated, and where data from different sources agree sufficiently, recommended values are proposed. The evaluation sheets, recommended values, and compiled data sheets are published on consecutive pages.

For *phamaceuticals*, the definitions, thermodynamics and methods of analysis are the same as those for the study of solubility of solids in liquids in general. For this subseries, special sections deal with matters of interest for pharmaceuticals, including discussions of polymorphism, factors influencing the rate of dissolution of drugs, and methods used to inhibit or enhance the rate of dissolution.

Definitions

A *mixture* (1, 2) describes a gaseous, liquid, or solid phase containing more than one substance, when the substances are all treated in the same way.

A *solution* (1, 2) describes a liquid or solid phase containing more than one substance, when for convenience one of the substances, which is called the *solvent*, and may itself be a mixture, is treated differently than the other substances, which are called *solutes*. If the sum of the mole fractions of the solutes is small compared to unity, the solution is called a *dilute solution*.

The *solubility* of a substance B is the relative proportion of B (or a substance related chemically to B) in a mixture which is saturated with respect to solid B at a specified temperature and pressure. *Saturated* implies the existence of equilibrium with respect to the processes of dissolution and precipitation; the equilibrium may be stable or metastable. The solubility of a substance in metastable equilibrium is usually greater than that of the corresponding substance in stable equilibrium. (Strictly speaking, it is the activity of the substance in metastable equilibrium that is greater.) Care must be taken to distinguish true metastability from supersaturation, where equilibrium does not exist.

Either point of view, mixture or solution, may be taken in describing solubility. The two points of view find their expression in the quantities used as measures of solubility and in the reference states used for definition of activities, activity coefficients and osmotic coefficients.

The qualifying phrase "substance related chemically to B" requires comment. The composition of the saturated mixture (or solution) can be described in terms of any suitable set of thermodynamic components. Thus, the solubility of a salt hydrate in water is usually given as the relative proportion of anhydrous salt in solution, rather than the relative proportions of hydrated salt and water.

For *pharmaceuticals*, the solubility of a drug substance in a given medium is of special importance in designing a suitable dosage form for a drug or in determination of a regimen for its administration. The solubility and rate of dissolution will determine the rate of appearance of the drug in various body fluids and at various sites of action. Therefore, the bioavailability of a drug is often determined by its solubility and rate of dissolution.

The solubility is a constant for a given substance in a given medium at constant temperature and pressure. Frequently it is possible to alter the solubility and rate of dissolution dramatically through changes in structure, degree of crystallinity or morphology, or by the addition of a solubilizing agent (cosolvent) to the dissolution medium. The appearance of a drug in adequate concentration at its site of action is a requirement for testing clinical efficiency; thus, enhancement of solubility may be required to render a substance clinically useful.

For reviews of recent literature on solubility and solubilization of drug substances, see (3, 4).

Quantities Used as Measures of Solubility

1. *Mole fraction of substance B*, x_B:

$$x_B = n_B / \sum_{s=1}^{c} n_s \qquad [1]$$

where n_s is the amount of substance of s, and c is the number of distinct substances present (often the number of thermodynamic components in the system). *Mole per cent* of *B* is 100 x_B.

2. *Mass fraction of substance B*, w_B:

$$w_B = m_B' / \sum_{s=1}^{c} m_s' \qquad [2]$$

where m_s' is the mass of substance s. *Mass per cent* is 100 w_B. The equivalent terms *weight fraction* and *weight per cent* are not used.

3. *Solute mole (mass) fraction of solute B* (5, 6):

$$x_{S,B} = n_B / \sum_{s=1}^{c'} n_s = x_B / \sum_{s=1}^{c'} x_s \qquad [3]$$

$$w_{S,B} = m_B' / \sum_{s=1}^{c'} m_s' = w_B / \sum_{s=1}^{c'} w_s \qquad [3a]$$

where the summation is over the solutes only. For the solvent *A*, $x_{S,A} = x_A/(1 - x_A)$, $w_{S,A} = w_A/(1 - w_A)$. These quantities are called *Jänecke mole (mass) fractions* in many papers.

4. *Molality of solute B* (1, 2) in a solvent *A*:

$$m_B = n_B / n_A M_A \qquad \text{SI base units: mol kg}^{-1} \qquad [4]$$

where M_A is the molar mass of the solvent.

5. *Concentration* of solute *B* (1, 2) in a solution of volume *V*:

$$c_B = [B] = n_B/V \qquad \text{SI base units: mol m}^{-3} \qquad [5]$$

The symbol c_B is preferred to [*B*], but both are used. The terms *molarity* and *molar* are not used.

Mole and mass fractions are appropriate to either the mixture or the solution point of view. The other quantities are appropriate to the solution point of view only. Conversions among these quantities can be carried out using the equations given in Table I-1 following this Introduction. Other useful quantities will be defined in the prefaces to individual volumes or on specific data sheets.

In addition to the quantities defined above, the following are useful in conversions between concentrations and other quantities.

6. *Density*: $\rho = m/V$ SI base units: kg m^{-3} [6]

7. *Relative density*: *d*; the ratio of the density of a mixture to the density of a reference substance under conditions which must be specified for both (1). The symbol $d_{t'}^{t}$ will be used for the density of a mixture at *t*°C, 1 bar divided by the density of water at *t'*°C, 1 bar. (In some cases, 1 atm = 101.325 kPa is used instead of 1 bar = 100 kPa.)

8. *A note on nomenclature*. The above definitions use the nomenclature of the IUPAC *Green Book* (1), in which a solute is called *B* and a solvent *A* In compilations and evaluations, the first-named component (component 1) is the solute, and the second (component 2 for a two-component system) is the solvent. The reader should bear these distinctions in nomenclature in mind when comparing nomenclature and theoretical equations given in this Introduction with equations and nomenclature used on the evaluation and compilation sheets.

Thermodynamics of Solubility

The principal aims of the Solubility Data Project are the tabulation and evaluation of: (a) solubilities as defined above; (b) the nature of the saturating phase. Thermodynamic analysis of solubility phenomena has two aims: (a) to provide a rational basis for the construction of functions to represent solubility data; (b) to enable thermodynamic quantities to be extracted from solubility data. Both these are difficult to achieve in many cases because of a lack of experimental or theoretical information concerning activity coefficients. Where thermodynamic quantities can be found, they are not evaluated critically, since this task would involve critical evaluation of a large body of data that is not directly relevant to solubility. The following is an outline of the principal thermodynamic relations encountered in discussions of solubility. For more extensive discussions and references, see books on thermodynamics, e.g., (7-14).

Activity Coefficients (1)

(a) *Mixtures*. The activity coefficient f_B of a substance B is given by

$$RT \ln (f_B x_B) = \mu_B - \mu_B^* \qquad [7]$$

where μ_B^* is the chemical potential of pure B at the same temperature and pressure. For any substance B in the mixture,

$$\lim_{x_B \to 1} f_B = 1 \qquad [8]$$

(b) *Solutions*.

(i) *Solute B*. The molal activity coefficient γ_B is given by

$$RT \ln(\gamma_B m_B) = \mu_B - (\mu_B - RT \ln m_B)^\infty \qquad [9]$$

where the superscript $^\infty$ indicates an infinitely dilute solution. For any solute B,

$$\gamma_B^\infty = 1 \qquad [10]$$

Activity coefficients y_B connected with concentrations c_B, and $f_{x,B}$ (called the *rational activity coefficient*) connected with mole fractions x_B, are defined in analogous ways. The relations among them (1, 9) are, where ρ^* is the density of the pure solvent:

$$f_B = (1 + M_A \sum_s m_s)\gamma_B = [\rho + \sum_s (M_A - M_s) c_s] y_B/\rho^* \qquad [11]$$

$$\gamma_B = (1 - \sum_s x_s) f_{x,B} = (\rho - \sum_s M_s c_s) y_B/\rho^* \qquad [12]$$

$$y_B = \rho^* f_{x,B}[1 + \sum_s (M_s/M_A - 1)x_B]/\rho = \rho^*(1 + \sum_s M_s m_s)\gamma_B/\rho \qquad [13]$$

For an electrolyte solute $B \equiv C_{\nu+}A_{\nu-}$, the activity on the molality scale is replaced by (11):

$$\gamma_B m_B = \gamma_\pm^\nu m_B^\nu Q^\nu \qquad [14]$$

where $\nu = \nu_+ + \nu_-$, $Q = (\nu_+^{\nu_+} \nu_-^{\nu_-})^{1/\nu}$, and $\gamma_\pm$ is the mean ionic activity coefficient on the molality scale. A similar relation holds for the concentration activity, $y_B c_B$. For the mole fractional activity,

$$f_{x,B} x_B = Q^\nu f_\pm^\nu x_\pm^\nu \qquad [15]$$

where $x_\pm = (x_+ x_-)^{1/\nu}$. The quantities x_+ and x_- are the ionic mole fractions (11), which are:

$$x_+ = \nu_+ x_B/[1 + \sum_s (\nu_s - 1)x_s]; \quad x_- = \nu_- x_B[1 + \sum_s (\nu_s - 1)x_s] \qquad [16]$$

where ν_s is the sum of the stoichiometric coefficients for the ions in a salt with mole fraction x_s. Note that the mole fraction of solvent is now

$$x_A' = (1 - \sum_s \nu_s x_s)/[1 + \sum_s (\nu_s - 1)x_s] \qquad [17]$$

so that

$$x_A' + \sum_s \nu_s x_s = 1 \qquad [18]$$

The relations among the various mean ionic activity coefficients are:

$$f_\pm = (1 + M_A \sum_s \nu_s m_s)\gamma_\pm = [\rho + \sum_s (\nu_s M_A - M_s)c_s] y_\pm/\rho^* \qquad [19]$$

$$\gamma_\pm = \frac{(1 - \sum_s x_s) f_\pm}{1 + \sum_s (\nu_s - 1)x_s} = (\rho - \sum_s M_s c_s) y_\pm/\rho^* \qquad [20]$$

$$y_\pm = \frac{\rho^*[1 + \sum_s (M_s/M_A - 1)x_s] f_\pm}{\rho[1 + \sum_s (\nu_s - 1)x_s]} = \rho^*(1 + \sum_s M_s m_s)\gamma_\pm/\rho \qquad [21]$$

(ii) *Solvent, A:*

The *osmotic coefficient*, ϕ , of a solvent A is defined as (1):

$$\phi = (\mu_A^* - \mu_A)/RT\, M_A \sum_s m_s \qquad [22]$$

where μ_A^* is the chemical potential of the pure solvent.

The *rational osmotic coefficient*, ϕ_x, is defined as (1):

$$\phi_x = (\mu_A - \mu_A^*)/RT\ln x_A = \phi M_A \sum_s m_s/\ln(1 + M_A \sum_s m_s) \qquad [23]$$

The activity, a_A, or the activity coefficient, f_A, is sometimes used for the solvent rather than the osmotic coefficient. The activity coefficient is defined relative to pure A, just as for a mixture.

For a mixed solvent, the molar mass in the above equations is replaced by the average molar mass; i.e., for a two-component solvent with components J, K, M_A becomes

$$M_A = M_J + (M_K - M_J)x_{v,K} \qquad [24]$$

where $x_{v,K}$ is the *solvent* mole fraction of component K.

The osmotic coefficient is related directly to the vapor pressure, p, of a solution in equilibrium with vapor containing A only by (14, p.306):

$$\phi M_A \sum_s \nu_s m_s = -\ln(p/p_A^*) + (V_{m,A}^* - B_{AA})(p - p_A^*)/RT \qquad [25]$$

where p_A^* is the vapor pressure of pure solvent A, $V_{m,A}^*$ is the molar volume of pure A in the liquid phase, and B_{AA} is the second virial coefficient of the vapor.

The Liquid Phase

A general thermodynamic differential equation which gives solubility as a function of temperature, pressure and composition can be derived. The approach is similar to that of Kirkwood and Oppenheim (9); see also (13, 14). Consider a solid mixture containing c thermodynamic components i. The Gibbs-Duhem equation for this mixture is:

$$\sum_{i=1}^{c} x_i'(S_i'dT - V_i'dp + d\mu_i') = 0 \qquad [26]$$

A liquid mixture in equilibrium with this solid phase contains c' thermodynamic components i, where $c' \geqslant c$. The Gibbs-Duhem equation for the liquid mixture is:

$$\sum_{i=1}^{c} x_i(S_i dT - V_i dp + d\mu_i') + \sum_{i=c+1}^{c'} x_i(S_i dT - V_i dp + d\mu_i) = 0 \qquad [27]$$

Subtract [26] from [27] and use the equation

$$d\mu_i = (d\mu_i)_{T,p} - S_i dT + V_i dp \qquad [28]$$

and the Gibbs-Duhem equation at constant temperature and pressure:

$$\sum_{i=1}^{c} x_i(d\mu_i')_{T,p} + \sum_{i=c+1}^{c'} x_i(d\mu_i)_{T,p} = 0 \qquad [29]$$

The resulting equation is:

$$RT\sum_{i=1}^{c} x_i'(d\ln a_i)_{T,p} = \sum_{i=1}^{c} x_i'(H_i - H_i')dT/T - \sum_{i=1}^{c} x_i'(V_i - V_i')dp \qquad [30]$$

where

$$H_i - H_i' = T(S_i - S_i') \qquad [31]$$

is the enthalpy of transfer of component i from the solid to the liquid phase at a given temperature, pressure and composition, with H_i and S_i the partial molar enthalpy and entropy of component i.

Use of the equations

$$H_i - H_i^o = -RT^2(\partial \ln a_i/\partial T)_{x,p} \qquad [32]$$

and

$$V_i - V_i^o = RT(\partial \ln a_i/\partial p)_{x,T} \qquad [33]$$

where superscript o indicates an arbitrary reference state gives:

$$RT\sum_{i=1}^{c} x_i' d\ln a_i = \sum_{i=1}^{c} x_i'(H_i^o - H_i')dT/T - \sum_{i=1}^{c} x_i'(V_i^o - V_i')dp \qquad [34]$$

where

$$d\ln a_i = (d\ln a_i)_{T,p} + (\partial \ln a_i/\partial T)_{x,p} + (\partial \ln a_i/\partial p)_{x,T} \qquad [35]$$

The terms involving enthalpies and volumes in the solid phase can be written as:

$$\sum_{i=1}^{c} x_i' H_i' = H_s^* \qquad \sum_{i=1}^{c} x_i' V_i' = V_s^* \qquad [36]$$

With eqn [36], the final general solubility equation may then be written:

$$R\sum_{i=1}^{c} x_i' d\ln a_i = (H_s^* - \sum_{i=1}^{c} x_i' H_i^0)d(1/T) - (V_s^* - \sum_{i=1}^{c} x_i' V_i^0)dp/T \qquad [37]$$

Note that those components which are not present in both phases do not appear in the solubility equation. However, they do affect the solubility through their effect on the activities of the solutes.

Several applications of eqn [37] (all with pressure held constant) will be discussed below. Other cases will be discussed in individual evaluations.

(a) *Solubility as a function of temperature.*

Consider a binary solid compound A_nB in a single solvent A. There is no fundamental thermodynamic distinction between a binary compound of A and B which dissociates completely or partially on melting and a solid mixture of A and B; the binary compound can be regarded as a solid mixture of constant composition. Thus, with $c = 2$, $x_A' = n/(n + 1)$, $x_B' = 1/(n + 1)$, eqn [37] becomes:

$$d\ln(a_A^n a_B) = -\Delta H_{AB}^0 d(1/RT) \qquad [38]$$

where

$$\Delta H_{AB}^0 = nH_A + H_B - (n + 1)H_s^* \qquad [39]$$

is the molar enthalpy of melting and dissociation of pure solid A_nB to form A and B in their reference states. Integration between T and T_0, the melting point of the pure binary compound A_nB, gives:

$$\ln(a_A^n a_B) = \ln(a_A^n a_B)_{T=T_0} - \int_{T_0}^{T} \Delta H_{AB}^0 d(1/RT) \qquad [40]$$

(i) Non-electrolytes

In eqn [32], introduce the pure liquids as reference states. Then, using a simple first-order dependence of ΔH_{AB}^* on temperature, and assuming that the activitity coefficients conform to those for a simple mixture (8):

$$RT\ \ln f_A = wx_B^2 \qquad RT\ \ln f_B = wx_A^2 \qquad [41]$$

then, if w is independent of temperature, eqn [32] and [33] give:

$$\ln\{x_B(1-x_B)^n\} + \ln\left\{\frac{n^n}{(1 + n)^{n+1}}\right\} = G(T) \qquad [42]$$

where

$$G(T) = -\left\{\frac{\Delta H_{AB}^* - T^*\Delta C_p^*}{R}\right\}\left\{\frac{1}{T} - \frac{1}{T^*}\right\} + \frac{\Delta C_p^*}{R}\ln(T/T^*) - \frac{w}{R}\left\{\frac{x_A^2 + nx_B^2}{T} - \frac{n}{(n + 1)T^*}\right\} \qquad [43]$$

where ΔC_p^* is the change in molar heat capacity accompanying fusion plus decomposition of the pure compound to pure liquid A and B at temperature T^*, (assumed here to be independent of temperature and composition), and ΔH_{AB}^* is the corresponding change in enthalpy at $T = T^*$. Equation [42] has the general form:

$$\ln\{x_B(1-x_B)^n\} = A_1 + A_2/(T/K) + A_3\ln(T/K) + A_4(x_A^2 + nx_B^2)/(T/K) \qquad [44]$$

If the solid contains only component B, then $n = 0$ in eqn [42] to [44].

If the infinite dilution reference state is used, then:

$$RT \ln f_{x,B} = w(x_A^2 - 1) \qquad [45]$$

and [39] becomes

$$\Delta H_{AB}^{\infty} = nH_A^* + H_B^{\infty} - (n + 1)H_s^* \qquad [46]$$

where ΔH_{AB}^{∞} is the enthalpy of melting and dissociation of solid compound A_nB to the infinitely dilute reference state of solute B in solvent A; H_A^* and H_B^{∞} are the partial molar enthalpies of the solute and solvent at infinite dilution. Clearly, the integral of eqn [32] will have the same form as eqn [35], with ΔH_{AB}^{∞} replacing ΔH_{AB}^*, ΔC_p^{∞} replacing ΔCp^*, and $x_A^2 - 1$ replacing x_A^2 in the last term.

See (7) and (13) for applications of these equations to experimental data.

(ii) Electrolytes

(a) Mole fraction scale

If the liquid phase is an aqueous electrolyte solution, and the solid is a salt hydrate, the above treatment needs slight modification. Using rational mean activity coefficients, eqn [34] becomes:

$$\ln\left\{\frac{x_B^{\nu}(1 - x_B)^n}{[1 + (\nu - 1)x_B]^{n+\nu}}\right\} - \ln\left\{\frac{n^n}{(n + \nu)^{n+\nu}}\right\} + \ln\left\{\left[\frac{f_B}{f_B^*}\right]^{\nu}\left[\frac{f_A}{f_A}\right]^n\right\}$$
$$= -\left\{\frac{\Delta H_{AB}^* - T^*\Delta C_p^*}{R}\right\}\left\{\frac{1}{T} - \frac{1}{T^*}\right\} + \frac{\Delta Cp^*}{R}\ln(T/T^*) \qquad [47]$$

where superscript * indicates the pure salt hydrate. If it is assumed that the activity coefficients follow the same temperature dependence as the right-hand side of eqn [47] (15-17), the thermochemical quantities on the right-hand side of eqn [47] are not rigorous thermodynamic enthalpies and heat capacities, but are apparent quantities only. Data on activity coefficients (11) in concentrated solutions indicate that the terms involving these quantities are not negligible, and their dependence on temperature and composition along the solubility-temperature curve is a subject of current research.

A similar equation (with $\nu = 2$ and without the heat capacity terms) or activity coefficients) has been used to fit solubility data for some $MOH-H_2O$ systems, where M is an alkali metal (15); enthalpy values obtained agreed well with known values. The full equation has been deduced by another method in (16) and applied to MCl_2-H_2O systems in (16) and (17). For a summary of the use of equation [47] and similar equations, see (18).

(2) Molality scale

Substitution of the mean activities on the molality scale in eqn [40] gives:

$$\nu\ln\left\{\frac{\gamma_{\pm}m_B}{\gamma_{\pm}^* m_B^*}\right\} - \nu(m_B/m_B^* - 1) - \nu\{m_B(\phi - 1)/m_B^* - \phi^* + 1\}$$
$$= G(T) \qquad [48]$$

where $G(T)$ is the same as in eqn [47], $m_B^* = 1/nM_A$ is the molality of the anhydrous salt in the pure salt hydrate and $\gamma_{\pm}$ and ϕ are the mean activity coefficient and the osmotic coefficient, respectively. Use of the osmotic coefficient for the activity of the solvent leads, therefore, to an equation that has a different appearance to [47]; the content is identical. However, while eqn [47] can be used over the whole range of composition ($0 \leqslant x_B \leqslant 1$), the molality in eqn [48] becomes infinite at $x_B = 1$; use of eqn [48] is therefore confined to solutions sufficiently dilute that the molality is a useful measure of composition. The essentials of eqn [48] were deduced by Williamson (19); however, the form used here appears first in the *Solubility Data Series*. For typical applications (where activity and osmotic coefficients are not considered explicitly, so that the enthalpies and heat capacities are apparent values, as explained above), see (20).

The above analysis shows clearly that a rational thermodynamic basis exists for functional representation of solubility-temperature curves in two-component systems, but may be difficult to apply because of lack of experimental or theoretical knowledge of activity coefficients and partial molar enthalpies. Other phenomena which are related ultimately to the stoichiometric activity coefficients and which complicate interpretation include ion pairing, formation of complex ions, and hydrolysis. Similar considerations hold for the variation of solubility with pressure, except that the effects are relatively smaller at the pressures used in many investigations of solubility (7).

(b) *Solubility as a function of composition.*

At constant temperature and pressure, the chemical potential of a saturating solid phase is constant:

$$\mu_{A_nB}^* = \mu_{A_nB}(\text{sln}) = n\mu_A + \mu_B \qquad [49]$$

$$= (n\mu_A^* + \nu_+\mu_+^\infty + \nu_-\mu_-^\infty) + nRT \ln f_A x_A$$

$$+ \nu RT \ln(\gamma_\pm m_\pm Q)$$

for a salt hydrate A_nB which dissociates to water (A), and a salt (B), one mole of which ionizes to give ν_+ cations and ν_- anions in a solution in which other substances (ionized or not) may be present. If the saturated solution is sufficiently dilute, $f_A = x_A = 1$, and the quantity K_s in

$$\Delta G^\infty = (\nu_+\mu_+^\infty + \nu_-\mu_-^\infty + n\mu_A^* - \mu_{AB}^*)$$

$$= -RT \ln K_s$$

$$= -\nu RT \ln(Q\gamma_\pm m_B) \qquad [50]$$

is called the *solubility product* of the salt. (It should be noted that it is not customary to extend this definition to hydrated salts, but there is no reason why they should be excluded.) Values of the solubility product are often given on mole fraction or concentration scales. In dilute solutions, the theoretical behavior of the activity coefficients as a function of ionic strength is often sufficiently well known that reliable extrapolations to infinite dilution can be made, and values of K_s can be determined. In more concentrated solutions, the same problems with activity coefficients that were outlined in the section on variation of solubility with temperature still occur. If these complications do not arise, the solubility of a hydrate salt $C_\nu A_\nu \cdot nH_2O$ in the presence of other solutes is given by eqn [50] as

$$\nu \ln\{m_B/m_B(0)\} = -\nu \ln\{\gamma_\pm/\gamma_\pm(0)\} - n \ln\{a_A/a_A(0)\} \qquad [51]$$

where a_A is the activity of water in the saturated solution, m_B is the molality of the salt in the saturated solution, and (0) indicates absence of other solutes. Similar considerations hold for non-electrolytes.

Consideration of *complex mixed ligand equilibria* in the solution phase is also frequently of importance in the interpretation of solubility equilibria. For nomenclature connected with these equilibria (and solubility equilibria as well) see (21, 22).

(c) *Alteration of the dissolution medium for pharmaceuticals*

Many substances which are only slightly soluble in water may be made more soluble by the addition of a cosolvent, surface-active agents, or complexing agents.

(i) Addition of a cosolvent. It is frequently necessary to dissolve a quantity of drug in a small volume of liquid so that it may be administered parenterally by injection. If the drug is not sufficiently soluble in water because of its hydrophobicity, the addition of a quantity of water-miscible, but less polar solvent may render the drug soluble in a small quantity of the mixed solvent. Solvents used for this purpose have included propylene glycol, glycerol, ethanol, polyethylene glycol and glycofural. Solubilities of many drug substances in water-organic solvent mixtures have been tabulated by Yalkowsky and Roseman (23).

(ii) Surface-active agents. Another approach to increasing the solubility and rate of dissolution of drug substances is to add a surface-active agent. There is an extensive literature on the application of surfactants and micellar dissolution, which has been summarized recently by Florence (24). Cationic, anionic or neutral surfactants are available. In choosing a surfactant, the possibility of charge-charge interactions between the drug and the surfactant must be considered, as well as the degree of ionization of each species as a function of pH. Micellar dissolution of drugs or additives may protect the dissolved species from hydrolytic degradation by the aqueous solvent. The stability of drugs may therefore be enhanced considerably by the addition of a surfactant. Surfactants may also facilitate the transport of drugs across biological membranes. Examples of substantially improved bioavailability of drugs under the influence of micellar dissolution have been reported (24).

(iii) Other modifications of the dissolution medium. The solubility of weak acid and weak base drugs will usually depend on the pH of the medium. Within reasonable limits for pharmaceutical preparations, pH may be adjusted to obtain the drug in the charged (and usually more soluble) form. The addition of complexing agents such as chelating agents, organic salts, cyclodextrins, or ion-pairing agents may be used to enhance solubility and rate of dissolution. Examples are given in the chapter by A.J. Repta in (3).

The Solid Phase

The definition of solubility permits the occurrence of a single solid phase which may be a pure anhydrous compound, a salt hydrate, a non-stoichiometric compound, or a solid mixture (or solid solution, or "mixed crystals"), and may be stable or metastable. As well, any number of solid phases consistent with the requirements of the phase rule may be present. Metastable solid phases are of widespread occurrence, and may appear as polymorphic (or allotropic) forms or crystal solvates whose rate of transition to more stable forms is very slow. Surface heterogeneity may also give rise to metastability, either when one solid precipitates on the surface of another, or if the size of the solid particles is sufficiently small that surface effects become important. In either case, the solid is not in stable equilibrium with the solution. See (25) for the modern formulation of the effect of particle size on solubility. The stability of a solid may also be affected by the atmosphere in which the system is equilibrated.

Many of these phenomena require very careful, and often prolonged, equilibration for their investigation and elimination. A very general analytical method, the "wet residues" method of Schreinemakers (26), is often used to investigate the composition of solid phases in equilibrium with salt solutions. This method has been reviewed in (27), where [see also (28)] least-squares methods for evaluating the composition of the solid phase from wet residue data (or initial composition data) and solubilities are described. In principle, the same method can be used with systems of other types. Many other techniques for examination of solids, in particular X-ray, optical, and thermal analysis methods, are used in conjunction with chemical analyses (including the wet residues method).

Solid State Manipulation in Pharmaceuticals

(i) Polymorphism. Many drug substances may crystallize in more than one form, a phenomenon called *polymorphism*. The different modifications (polymorphs) arise because of the relative positions and bonding of the molecules in their crystal lattices; true polymorphs do not differ in chemical composition. Polymorphs of the same substance frequently have different physical properties such as solubility and rate of dissolution. Ultimately, the solubility of all forms will revert to that of the form with the lowest Gibbs energy; the solubility of a less-stable form will thus be an initial solubility. The rate of reversion to the most stable form is often very slow, and a form with higher Gibbs energy may exhibit its higher solubility for hours. This phenomenon may be used to advantage by choosing the polymorph with the desired solubility or rate of dissolution. Examples of polymorphism and methods of characterization have been reviewed by Haleblian (29) and Burger (30).

(ii) Crystallinity. In many cases, drug substances may occur in the solid state as amorphous or partly crystalline forms. This is a special case of polymorphism, and may result from rapid precipitation or from freeze-drying. These amorphous or partly crystalline materials are unstable relative to the crystalline form. However, reversion to the crystalline form may be slow, and the less stable forms may be used to enhance solubility and rate of dissolution (31).

(iii) Choice of salt form. Many drug substances are organic salts. In most cases the drug moiety is the organic cation or anion, such as a quaternary ammonium cation or a carboxylate or sulfonate anion. The counterion is frequently an inorganic ion such as sodium or chloride. It is possible to obtain large variations in initial solubility depending on the choice of the salt form of the drug.

COMPILATIONS AND EVALUATIONS

The formats for the compilations and critical evaluations have been standardized for all volumes. A brief description of the data sheets has been given in the FOREWORD; additional explanation is given below.

Guide to the Compilations

The format used for the compilations is, for the most part, self-explanatory. The details presented below are those which are not found in the FOREWORD or which are not self-evident.

Components. Each component is listed according to IUPAC or Chemical Abstracts (CA) name and CA Registry Number. The formula is given either in terms of the IUPAC or Hill (32) system and the choice of formula is governed by what is usual for most current users: i.e., IUPAC for inorganic compounds, and Hill system for organic compounds. Components are ordered according to:

(a) saturating components;
(b) non-saturating components in alphanumerical order;
(c) solvents in alphanumerical order.

The saturating components are arranged in order according to a 18-column periodic table with two additional rows:

Columns 1 and 2:	H, alkali elements, ammonium, alkaline earth elements
3 to 12:	transition elements
13 to 17:	boron, carbon, nitrogen groups; chalcogenides, halogens
18:	noble gases
Row 1:	Ce to Lu
Row 2:	Th to the end of the known elements, in order of atomic number.

Salt hydrates are generally not considered to be saturating components since most solubilities are expressed in terms of the anhydrous salt. The existence of hydrates or solvates is carefully noted in the text, and CA Registry Numbers are given where available, usually in the critical evaluation. Mineralogical names are also quoted, along with their CA Registry Numbers, again usually in the critical evaluation.

Original Measurements. References are abbreviated in the forms given by *Chemical Abstracts Service Source Index* (CASSI). Names originally in other than Roman alphabets are given as transliterated by *Chemical Abstracts*.

Experimental Values. Data are reported in the units used in the original publication, with the exception that modern *names* for units and quantities are used; e.g., mass per cent for weight per cent; mol dm^{-3} for molar; etc. Both mass and molar values are given. Usually, only one type of value (e.g., mass per cent) is found in the original paper, and the compiler has added the other type of value (e.g., mole per cent) from computer calculations based on 1983 atomic weights (33).

Errors in calculations and fitting equations in original papers have been noted and corrected, by computer calculations where necessary.

Method. Source and Purity of Materials. Abbreviations used in *Chemical Abstracts* are often used here to save space.

Estimated Error. If these data were omitted by the original authors, and if relevant information is available, the compilers have attempted to estimate errors from the internal consistency of data and type of apparatus used. Methods used by the compilers for estimating and and reporting errors are based on the papers by Ku and Eisenhart (34).

Comments and/or Additional Data. Many compilations include this section which provides short comments relevant to the general nature of the work or additional experimental and thermodynamic data which are judged by the compiler to be of value to the reader.

References. See the above description for Original Measurements.

Guide to the Evaluations

The evaluator's task is to check whether the compiled data are correct, to assess the reliability and quality of the data, to estimate errors where necessary, and to recommend "best" values. The evaluation takes the form of a summary in which all the data supplied by the compiler have been critically reviewed. A brief description of the evaluation sheets is given below.

Components. See the description for the Compilations.

Evaluator. Name and date up to which the literature was checked.

Critical Evaluation

(a) Critical text. The evaluator produces text evaluating *all* the published data for each given system. Thus, in this section the evaluator reviews the merits or shortcomings of the various data. Only published data are considered; even published data can be considered only if the experimental data permit an assessment of reliability.

(b) Fitting equations. If the use of a smoothing equation is justifiable the evaluator may provide an equation representing the solubility as a function of the variables reported on all the compilation sheets.

(c) Graphical summary. In addition to (b) above, graphical summaries are often given.

(d) Recommended values. Data are *recommended* if the results of at least two independent groups are available and they are in good agreement, and if the evaluator has no doubt as to the adequacy and reliability of the applied experimental and computational procedures. Data are considered as *tentative* if only one set of measurements is available, or if the evaluator considers some aspect of the computational or experimental method as mildly undesirable but estimates that it should cause only minor errors. Data are considered as *doubtful* if the evaluator considers some aspect of the computational or experimental method as undesirable but still considers the data to have some value in those instances where the order of magnitude of the solubility is needed. Data determined by an inadequate method or under ill-defined conditions are *rejected*. However references to these data are included in the evaluation together with a comment by the evaluator as to the reason for their rejection.

(e) References. All pertinent references are given here. References to those data which, by virtue of their poor precision, have been rejected and not compiled are also listed in this section.

(f) Units. While the original data may be reported in the units used by the investigators, the final recommended values are reported in S.I. units (1, 35) when the data can be accurately converted.

References

1. Whiffen, D.H., ed., *Manual of Symbols and Terminology for Physicochemical Quantities and Units. Pure Applied Chem.* 1979, 51, No. 1.
2. McGlashan, M.L. *Physicochemical Quantities and Units.* 2nd ed. Royal Institute of Chemistry. London. 1971.
3. Yalkowsky, S.H., ed. *Techniques of Solubilization of Drugs.* Marcel Dekker. New York. 1981.
4. Byron, S.R. *Solid State Chemistry of Drugs.* Academic Press. New York. 1982.
5. Jänecke, E. *Z. Anorg. Chem.* 1906, *51*, *132*.
6. Friedman, H.L. *J. Chem. Phys.* 1960, 32, 1351.
7. Prigogine, I.; Defay, R. *Chemical Thermodynamics.* D.H. Everett, transl. Longmans, Green. London, New York, Toronto. 1954.
8. Guggenheim, E.A. *Thermodynamics.* North-Holland. Amsterdam. 1959. 4th ed.
9. Kirkwood, J.G.; Oppenheim, I. *Chemical Thermodynamics.* McGraw-Hill. New York, Toronto, London. 1961.
10. Lewis, G.N.; Randall, M. (rev. Pitzer, K.S.; Brewer, L.). *Thermodynamics.* McGraw Hill. New York, Toronto, London. 1961. 2nd. ed.
11. Robinson, R.A.; Stokes, R.H. *Electrolyte Solutions.* Butterworths. London. 1959. 2nd ed.
12. Harned, H.S.; Owen, B.B. *The Physical Chemistry of Electrolytic Solutions.* Reinhold. New York. 1958. 3rd ed.
13. Haase, R.; Schönert, H. *Solid-Liquid Equilibrium.* E.S. Halberstadt, trans. Pergamon Press, London, 1969.
14. McGlashan, M.L. *Chemical Thermodynamics.* Academic Press. London. 1979.
15. Cohen-Adad, R.; Saugier, M.T.; Said, J. *Rev. Chim. Miner.* 1973, *10*, 631.
16. Counioux, J.-J.; Tenu, R. *J. Chim. Phys.* 1981, *78*, 815.
17. Tenu, R.; Counioux, J.-J. *J. Chim. Phys.* 1981, *78*, 823.
18. Cohen-Adad, R. *Pure Appl. Chem.* 1985, *57*, 255.
19. Williamson, A.T. *Faraday Soc. Trans.* 1944, *40*, 421.
20. Siekierski, S.; Mioduski, T.; Salomon, M. *Solubility Data Series.* Vol. 13. *Scandium, Yttrium, Lanthanum and Lanthanide Nitrates.* Pergamon Press. 1983.
21. Marcus, Y., ed. *Pure Appl. Chem.* 1969, *18*, 459.
22. IUPAC Analytical Division. *Proposed Symbols for Metal Complex Mixed Ligand Equilibria (Provisional). IUPAC Inf. Bull.* 1978, No. *3*, 229.

23. Yalkowsky, S.H.; Roseman, T.J. *in* S.H. Yalkowsky, ed. *Techniques of Solubilization of Drugs*. Marcel Dekker. New York. 1981.
24. Florence, A.T. *in* S.H. Yalkowsky, ed. *Techniques of Solubilization of Drugs*. Marcel Dekker. New York. 1981.
25. Enüstün, B.V.; Turkevich, J. *J. Am. Chem. Soc.* 1960, *82*, 4502.
26. Schreinemakers. F.A.H. *Z. Phys. Chem., Stoechiom. Verwandschaftsl.* 1893, 11, 75.
27. Lorimer, J.W. *Can. J. Chem.* 1981, 59, 3076.
28. Lorimer, J.W. *Can. J. Chem.* 1982, *60*, 1978.
29. Haleblian, J.K. *J. Pharm. Sci.* 1975, 64, 1269.
30. Burger, A. *Pharm. Int.* 1982, *3*, 158.
31. Shefter, E. *in* S.H. Yalkowsky, ed. *Techniques of Solubilization of Drugs*. Marcel Dekker. New York. 1981.
32. Hill, E.A. *J. Am. Chem. Soc.* 1900, *22*, *478*.
33. IUPAC Commission on Atomic Weights. *Pure Appl. Chem.* 1984, 56, 653.
34. Ku, H.H., p. 73; Eisenhart, C., p. 69; in Ku, H.H., ed. *Precision Measurement and Calibration*. NBS Special Publication 300. Vol. 1. Washington. 1969.
35. *The International System of Units*. Engl. transl. approved by the BIPM of *Le Système International d'Unités*. H.M.S.O. London. 1970.

September, 1986

R. Cohen-Adad,
Villeurbanne, France

S. Lindenbaum,
Lawrence, Kansas, U.S.A.

J.W. Lorimer,
London, Ontario, Canada

A.N. Paruta,
Kingston, R.I., U.S.A.

R. Piekos,
Gdansk, Poland

M. Salomon,
Fair Haven, New Jersey, U.S.A.

Table I-1

Quantities Used as Measures of Solubility
Conversion Table for 2-Component Systems
Containing Solvent A and Solute B

	mole fraction $x_B =$	mass fraction $w_B =$	molality $m_B =$	concentration $c_B =$
x_B	x_B	$\frac{1}{1 - M_A(1 - 1/x_B)/M_B}$	$\frac{1}{M_A(1/x_B - 1)}$	$\frac{\rho}{M_B + M_A(1/x_B - 1)}$
w_B	$\frac{1}{1 + M_B(1/w_B - 1)}$	w_B	$\frac{1}{M_B(1/w_B - 1)}$	$\rho w_B/M_B$
m_B	$\frac{1}{1 + 1/m_B M_A}$	$\frac{1}{1 + 1/M_B m_B}$	m_B	$\frac{\rho}{M_B + 1/m_B}$
c_B	$\frac{1}{1 + (\rho/c_B - M_B)/M_A}$	$M_B c_B/\rho$	$\frac{1}{\rho/c_B - M_B}$	c_B

ρ = density of solution
M_A, M_B = molar masses of solvent, solute
Formulas are given in forms suitable for rapid computation; all calculations should be made using SI base units.

STRUCTURES

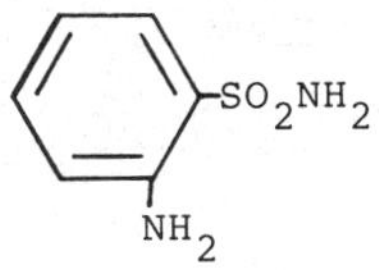 [3306-62-5] M.W.=172.20	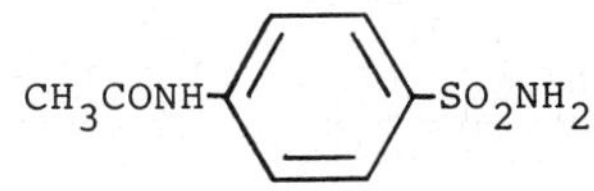 [121-61-9] M.W.=214.24
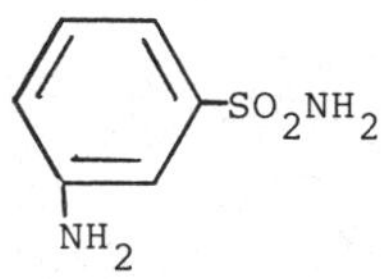 [98-18-0] M.W.=172.20	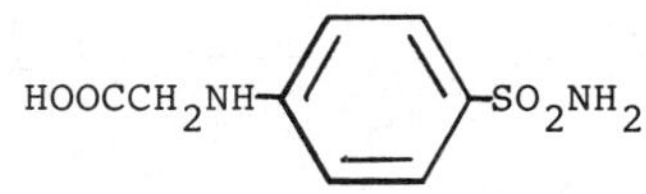 [6138-11-0] M.W.=230.24
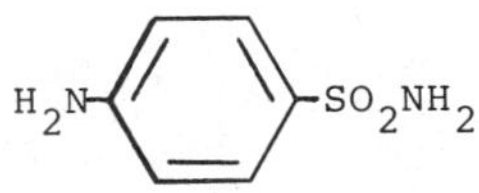 [63-74-1] M.W.=172.20	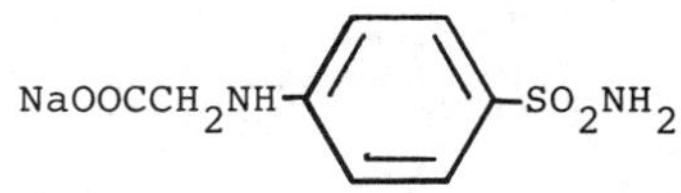 [60364-23-0] M.W.=252.22
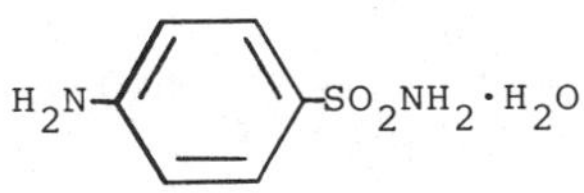 [20203-81-0] M.W.=190.22	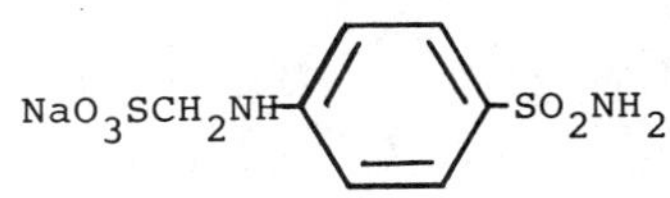 [138-43-2] M.W.=288.27
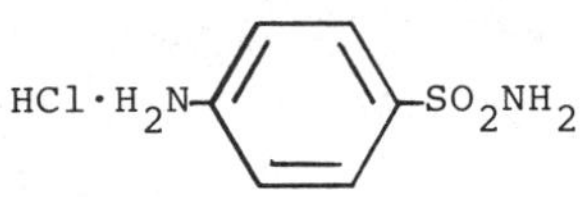 [6101-31-1] M.W.=208.66	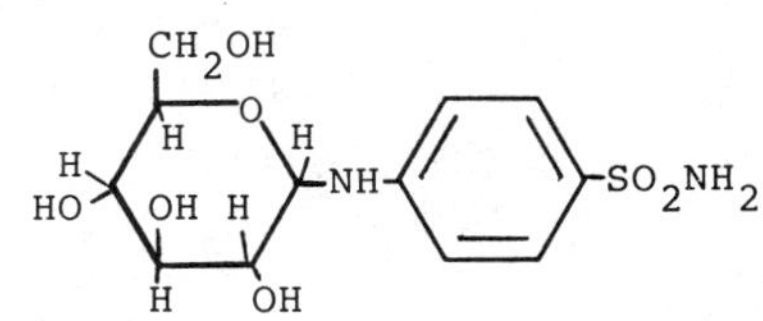 [77400-75-0] M.W.=334.34
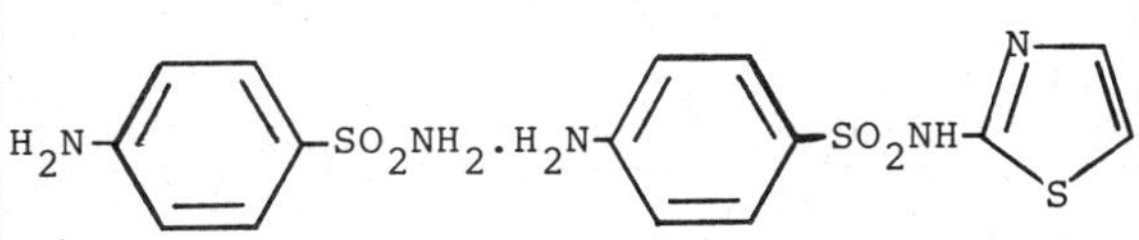 [1704-78-3] M.W.=427.51	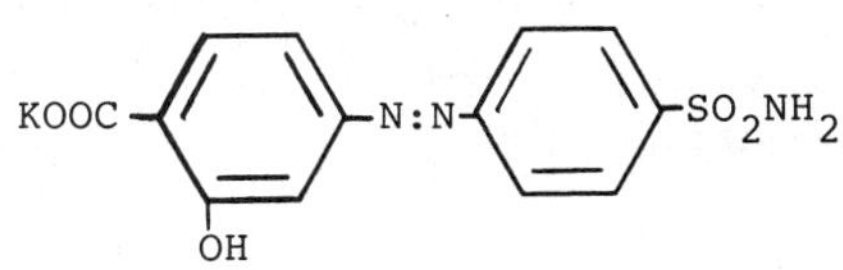 [77400-72-7] M.W.=359.40
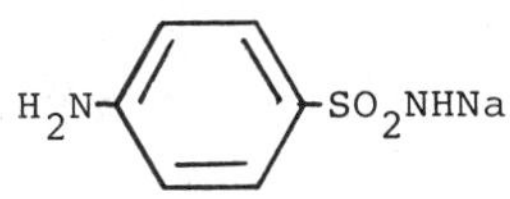 [10103-15-8] M.W.=194.18	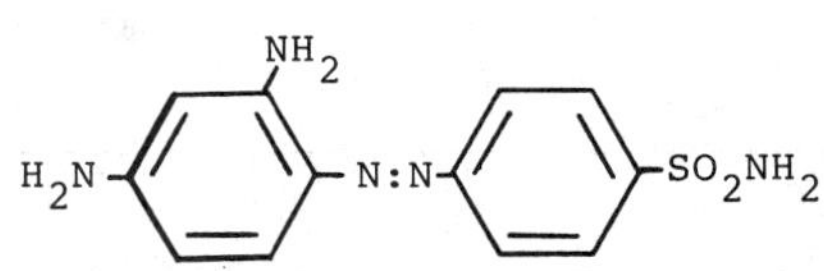 [103-12-8] M.W.=291.33
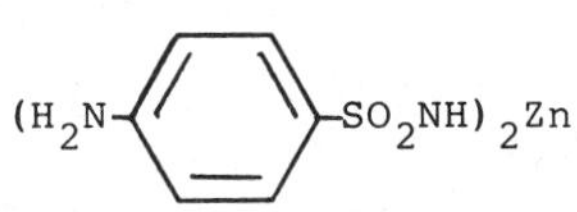 [78739-60-3] M.W.=407.77	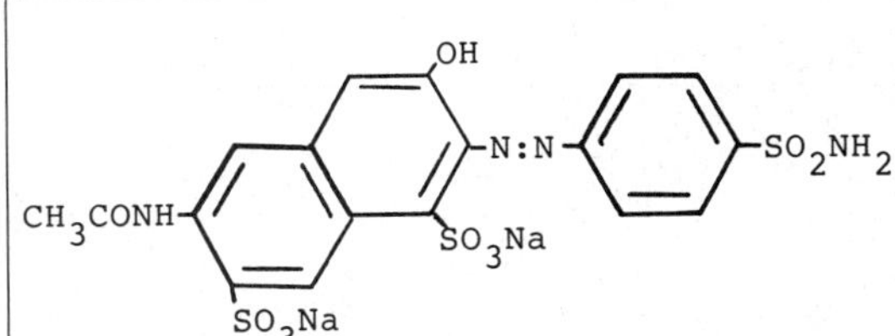 [133-60-8] M.W.=588.49

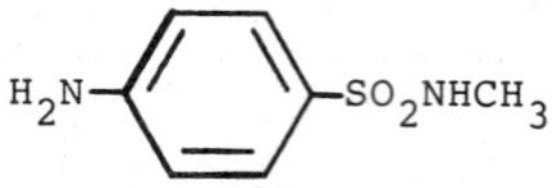

[1709-52-0] M.W.=186.23

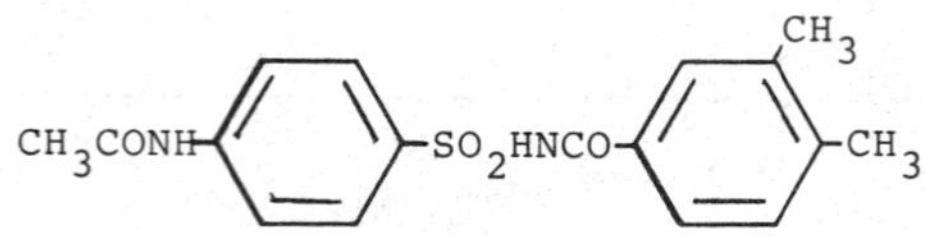

[71119-40-9] M.W.=346.40

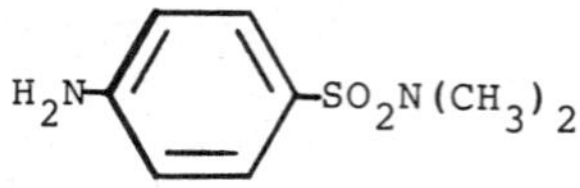

[1709-39-7] M.W.=200.25

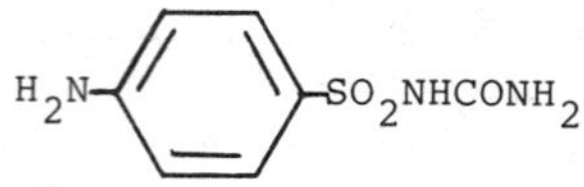

[547-44-4] M.W.=215.23

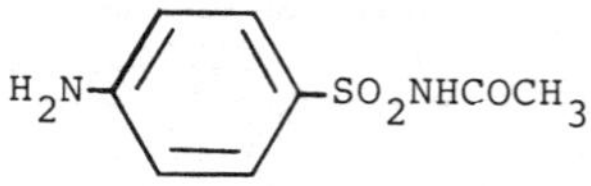

[144-80-9] M.W.=214.24

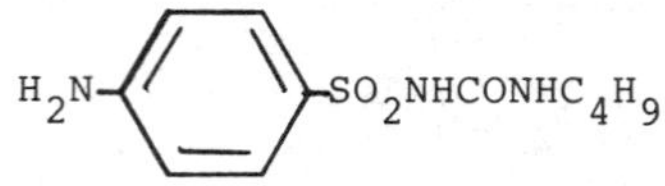

[339-43-5] M.W.=271.34

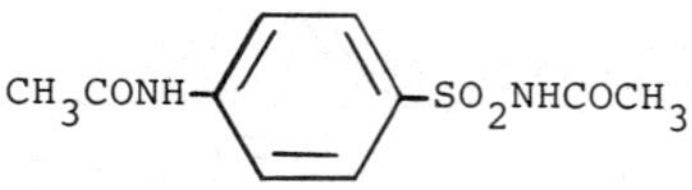

[5626-90-4] M.W.=256.28

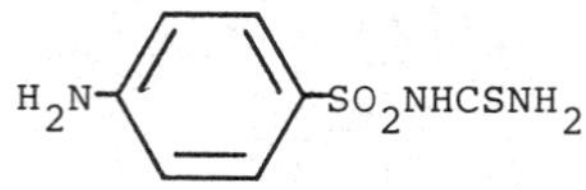

[515-49-1] M.W.=231.29

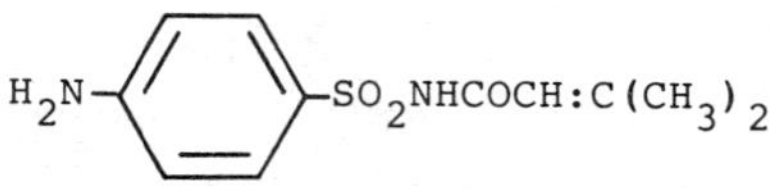

[115-68-4] M.W.=254.30

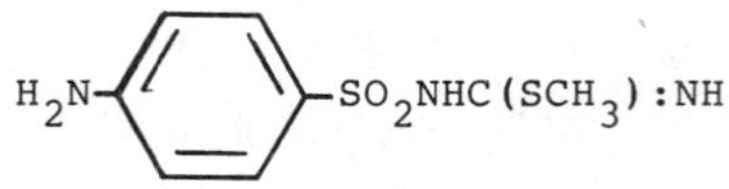

[2651-18-5] M.W.=245.31

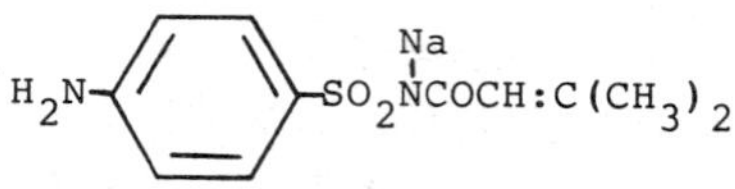

[78739-59-0] M.W.=276.44

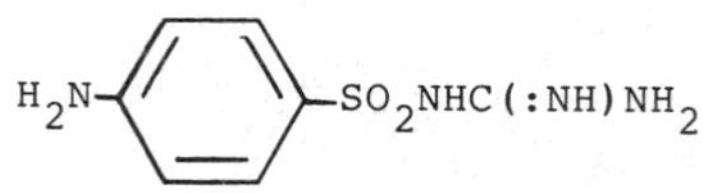

[57-67-0] M.W.=214.24

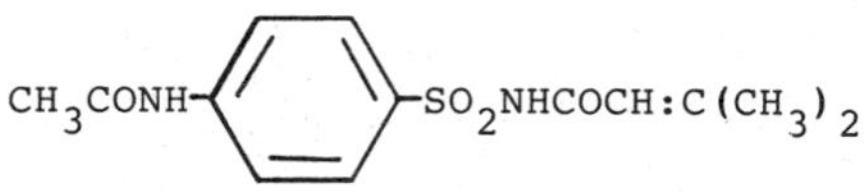

[71119-41-0] M.W.=296.48

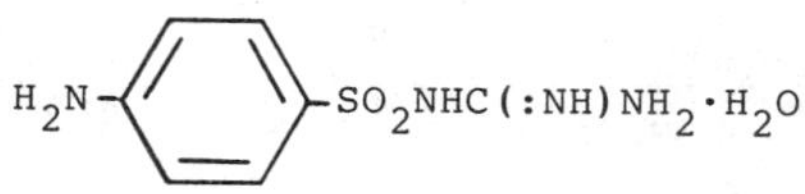

[6190-55-2] M.W.=232.26

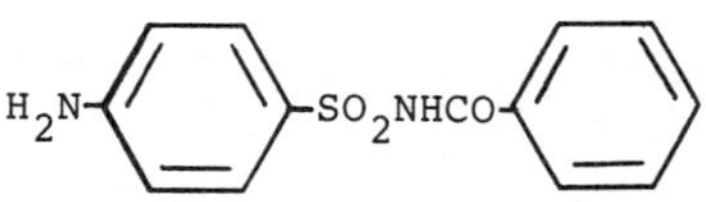

[127-71-9] M.W.=276.45

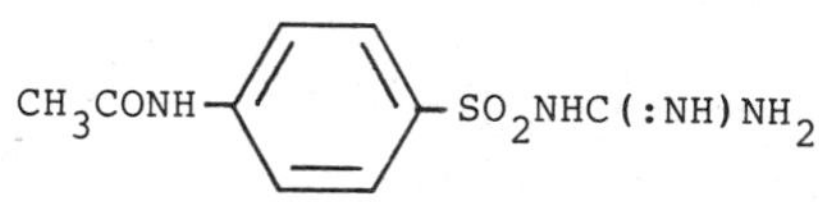

[19077-97-5] M.W.=256.28

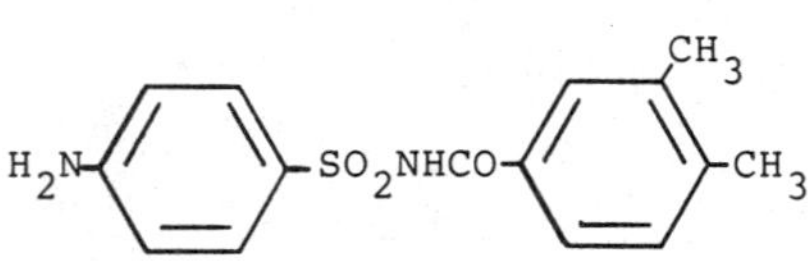

[120-34-3] M.W.=304.50

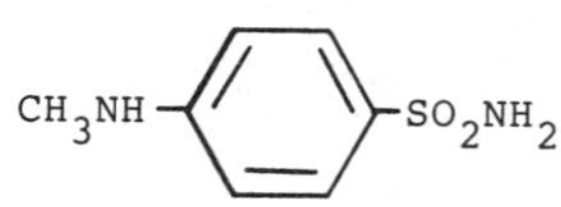

[138-39-6] M.W.=186.08

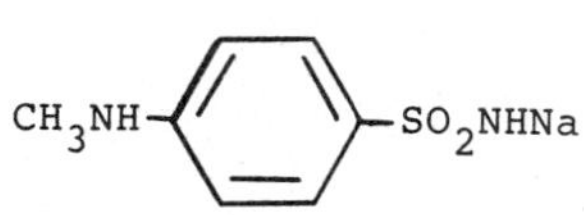

[60758-21-6] M.W.=208.01

COMPONENTS:	ORIGINAL MEASUREMENTS:
(1) Benzenesulfonamide, 2-amino-, (orthanilamide); $C_6H_8N_2O_2S$; [3306-62-5] (2) Water; H_2O; [7732-18-5]	Kienle, R. H.; Sayward, J. M., *J. Am. Chem. Soc.* 1942, *64*, 2464-8
VARIABLES:	PREPARED BY:
Temperature	R. Piekos

EXPERIMENTAL VALUES:

t/°C	g/100 g soln		mol kg^{-1} (compiler)	
23.0	0.65		0.038	
24.0	0.67		0.039	
26.0	0.75		0.044	
28.0	0.82; 0.82		0.048	
30.5	0.91		0.053	
33.0	1.01		0.0586	
34.0	1.05		0.0610	
35.5	1.11		0.0645	
37.0	1.20; 1.20		0.0697	
37.0	1.20[a]	± 0.024	0.0697	±0.0014
37.0	1.18[b]		0.0685	
37.05	1.19		0.0691	
42.0	1.46		0.0848	
46.0	1.70		0.0987	
50.0	2.00[b]		0.116	

[a] Equilibrium approached from below.
[b] Duration less than 12 hours.

AUXILIARY INFORMATION

METHOD/APPARATUS/PROCEDURE:

An excess of solid was rotated with water usually overnight. Equilibrium was approached from above. Sampling was accomplished by forcing the soln through a filter into a pycnometer. From the pycnometer the contents were flushed into a volumetric flask. Duplicate aliquots were acidified, iced below 15°C and titrated with 0.04 mol dm^{-3} $NaNO_2$ to first blue on starch - iodide paper.

SOURCE AND PURITY OF MATERIALS:

Orthanilamide, m.p. 155.2°C, was prepd by the authors. Titrn. with nitrite indicated that the compd was 100.0 ±0.3% pure. Elemental analysis and mixed m.p. confirmed this value. Purity of the water was not specified.

ESTIMATED ERROR: Temp: ±0.02°C (authors). Soly: ±0.01 g/100 g soln (authors) or ±0.012 x 10^{-3} in mole fraction. The 2σ value for 37°C indicated in results.

REFERENCES:

COMPONENTS:	ORIGINAL MEASUREMENTS:
(1) Benzenesulfonamide, 2-amino- (orthanilamide); $C_6H_8N_2O_2S$; [3306-62-5] (2) Phosphoric acid, disodium salt; Na_2HPO_4; [7558-94-4] (3) Phosphoric acid, monopotassium salt; KH_2PO_4; [7778-77-0] (4) Water; H_2O; [7732-18-5]	Kienle, R. H.; Sayward, J. M. *J. Am. Chem. Soc.* 1942, *64*, 2464-8.
VARIABLES: One temperature: 37.0°C; one pH: 6.9	PREPARED BY: R. Piekos

EXPERIMENTAL VALUES:

Solubility of orthanilamide in a buffer solution prepared by mixing together 55.2 cm^3 of 1/15 M Na_2HPO_4 with 44.8 cm^3 of 1/15 M KH_2PO_4 (pH 6.9 , ionic strength calculated from dissociation constants 0.03[a]) at 37.0°C is 1.19 g/100 cm^3 solution (6.91 x 10^{-2} mol dm^{-3}, compiler).

[a] Not specified for which reactions were the dissociation constants calculated - compiler.

AUXILIARY INFORMATION

METHOD/APPARATUS/PROCEDURE:

An excess of orthanilamide was rotatd with the buffer soln usually overnight. Equilibrium was approached from above. Sampling was accomplished by forcing the soln through a filter into a pycnometer. From the pycnometer the contents were flushed into a volumetric flask. Duplicate aliquots were acidified, iced below 15°C and titrated with a 0.04 mol dm^{-3} $NaNO_2$ soln to first blue on a starch - iodide paper.

SOURCE AND PURITY OF MATERIALS:

Orthanilamide, mp 155.2°C, was prepd by the authors. Titrn with nitrite indicated that the compd was 100.0±0.3% pure. Elemental analysis and mixed mp confirmed this value.
Source and purity of the remaining materials was not specified.

ESTIMATED ERROR: Soly: ±0.01 g/100 g soln or ±0.012 x 10^{-3} in mole fraction (authors). Temp: ±0.02°C (authors).

REFERENCES:

COMPONENTS:

(1) Benzenesulfonamide, 2-amino- (orthanilamide); $C_6H_8N_2O_2S$; [3306-62-5]

(2) Hydrochloric acid; HCl; [7647-01-0]

(3) Potassium chloride; KCl; [7447-40-7]

(4) Water; H_2O; [7732-18-5]

ORIGINAL MEASUREMENTS:

Kienle, R. H.; Sayward, J. M.

J. Am. Chem. Soc. 1942, *64*, 2464-8.

VARIABLES:

One temperature: 37.0°C; one pH: 1.2

PREPARED BY:

R. Piekos

EXPERIMENTAL VALUES:

Solubility of orthanilamide in a solution prepared by mixing together 25 cm^3 of 0.2 M KCl with 42.5 cm^3 of 0.2 M HCl and diluting up to 100 cm^3 with water (pH 1.2, ionic strength calculated from dissociation constants 0.12[a]) at 37.0°C is 1.92 g/100 cm^3 solution (0.111 mol dm^{-3}, compiler).

[a]Not specified for which reactions were the dissociation constants calculated - compiler.

AUXILIARY INFORMATION

METHOD/APPARATUS/PROCEDURE:

An excess of orthanilamide was rotated with the buffer soln usually overnight. Equilibrium was approached from above. Sampling was accomplished by forcing the soln through a filter into a pycnometer. From the pycnometer the contents were flushed into a volumetric flask. Duplicate aliquots were acidified, iced below 15°C and titrated with a 0.04 mol dm^{-3} $NaNO_2$ soln to first blue on starch - iodide paper.

SOURCE AND PURITY OF MATERIALS:

Orthanilamide, mp 155.2°C, was prepd by the authors. Titrn with nitrite indicated that the compd was 100.0±0.3% pure. Elemental analysis and mixed mp confirmed this value. Purity of the water was not specified.
The source and purity of the remaining materials was not specified.

ESTIMATED ERROR:

Soly: ±0.01 g/100 g soln or ±0.012 x 10^{-3} in mole fraction (authors). Temp: ±0.02°C (authors).

REFERENCES:

COMPONENTS:	ORIGINAL MEASUREMENTS:
(1) Benzenesulfonamide, 2-amino- (orthanilamide); $C_6H_8N_2O_2S$; [3306-62-5] (2) Boric acid; H_3BO_3; [10043-35-3] (3) Potassium chloride; KCl; [7447-40-7] (4) Sodium hydroxide; NaOH; [1310-73-2] (5) Water; H_2O; [7732-18-5]	Kienle, R. H.; Sayward, J. M. *J. Am. Chem. Soc.* 1942, *64*, 2464-8
VARIABLES:	PREPARED BY:
pH; ionic strength	R. Piekos

EXPERIMENTAL VALUES:

pH of borate buffer	Ionic strength[a]	Solubility at 37.0°C: g/100 cm^3 solution	Solubility at 37.0°C: 10^2 mol dm^{-3} [b]
9.4[c]	0.08	1.34	7.78
9.7[d]	0.09	1.39	8.07

[a]Calculated from dissociation constants (reactions not specified).

[b]Calculated by compiler.

[c]Obtained by mixing together 50 cm^3 of a 0.1 M solution in both H_3BO_3 and KCl with 32.1 cm^3 of 0.1 M NaOH and diluting with water up to 100 cm^3.

[d]Obtained by mixing together 50 cm^3 of a 0.1 M solution in both H_3BO_3 and KCl with 38.75 cm^3 of 0.1 M NaOH and diluting with water up to 100 cm^3.

AUXILIARY INFORMATION

METHOD/APPARATUS/PROCEDURE:

An excess of orthanilamide was rotated with the buffer soln usually overnight. Equilibrium was approached from above. Sampling was accomplished by forcing the soln through a filter into a pycnometer. From the pycnometer the contents were flushed into a volumetric flask. Duplicate aliquots were acidified, iced below 15°C and titrated with a 0.04 mol dm^{-3} $NaNO_2$ soln to first blue on a starch - iodine paper.

SOURCE AND PURITY OF MATERIALS:

Orthanilamide, mp 155.2°C, was prepd by the authors. Titrn with nitrite indicated that the compd was 100.0$\pm$ 0.3% pure. Elemental analysis and mixed mp confirmed this value. Source and purity of the remaining materials was not specified.

ESTIMATED ERROR: Soly: $\pm$ 0.01 g/100 g soln or $\pm$0.012 x 10^{-3} in mole fraction (authors). Temp: $\pm$0.02°C (authors).

REFERENCES:

COMPONENTS:

(1) Benzenesulfonamide, 2-amino- (orthanilamide); $C_6H_8N_2O_2S$; [3306-62-5]

(2) Phosphoric acid, disodium salt; Na_2HPO_4; [7558-94-4]

(3) 1, 2, 3-Propanetricarboxylic acid, 2-hydroxy- (citric acid); $C_6H_8O_7$; [77-92-9]

(4) Water; H_2O; [7732-18-5]

ORIGINAL MEASUREMENTS:

Kienle, R. H.; Sayward, J. M.

J. Am. Chem. Soc. 1942, *64*, 2464-8.

VARIABLES:

One temperature: 37.0°C; one pH: 4.2

PREPARED BY:

R. Piekos

EXPERIMENTAL VALUES:

Solubility of orthanilamide in a solution prepared by mixing together 41.4 cm^3 of 0.2 M Na_2HPO_4 with 58.6 cm^3 of 0.1 M citric acid (pH 4.2 ionic strength calculated from dissociation constants 0.84[a]) at 37.0°C is 1.08 g/100 cm^3 solution (6.27×10^{-2} mol dm^{-3}, compiler).

[a]Not specified for which reactions were the dissociation constants calculated - compiler.

AUXILIARY INFORMATION

METHOD/APPARATUS/PROCEDURE:

An excess of orthanilamide was rotated with the buffer soln usually overnight. Equilibrium was approached from above. Sampling was accomplished by forcing the soln through a filter into a pycnometer. From the pycnometer the contents were flushed into a volumetric flask. Duplicate aliquots were acidified, iced below 15°C and titrated with a 0.04 mol dm^{-3} $NaNO_2$ soln to first blue on a starch - iodide paper.

SOURCE AND PURITY OF MATERIALS:

Orthanilamide, mp 155.2°C, was prepd by the authors. Titrn with nitrite indicated that the compd was 100.0±0.3% pure. Elemental analysis and mixed mp confirmed this value.

The source and purity of the remaining materials was not specified.

ESTIMATED ERROR:

Soly: ±0.01 g/100 g soln or $\pm 0.012 \times 10^{-3}$ in mole fraction (authors). Temp: ±0.02°C (authors).

REFERENCES:

COMPONENTS:

(1) Benzenesulfonamide, 2-amino- (orthanilamide); $C_6H_8N_2O_2S$; [3306-62-5]

(2) 1,2-Benzenedicarboxylic acid, monopotassium salt; $C_8H_5KO_4$; [877-24-7]

(3) Hydrochloric acid; HCl; [7647-01-0]

(4) Water; H_2O; [7732-18-5]

ORIGINAL MEASUREMENTS:

Kienle, R. H.; Sayward, J. M.

J. Am. Chem. Soc. 1942, *64*, 2464-8.

VARIABLES:

One temperature: 37.0°C; one pH: 2.2

PREPARED BY:

R. Piekos

EXPERIMENTAL VALUES:

Solubility of orthanilamide in a buffer soln prepared by mixing together 50 cm^3 of 0.1 M monopotassium 1,2-benzenedicarboxylate with 49.5 cm^3 of 0.1 M HCl and diluting up to 100 cm^3 with water (pH 2.2, ionic strength calculated from dissociation constants 0.06[a]) at 37.0°C is 1.31 g/100 cm^3 solution (7.61×10^{-2} mol dm^{-3}, compiler).

[a]Not specified for which reactions were the dissociation constants calculated - compiler.

AUXILIARY INFORMATION

METHOD/APPARATUS/PROCEDURE:

An excess of orthanilamide was rotated with the buffer soln usually overnight. Equilibrium was approached from above. Sampling was accomplished by forcing the soln through a filter into a pycnometer. From the pycnometer the contents were flushed into a volumetric flask. Duplicate aliquots were acidified, iced below 15°C and titrated with a 0.04 mol dm^{-3} $NaNO_2$ soln to first blue on a starch - iodide paper.

SOURCE AND PURITY OF MATERIALS:

Orthanliamide, mp 155.2°C, was prepd by the authors. Titrn with nitrite indicated that the compd was 100.0±0.3% pure. Elemental analysis and mixed mp confirmed this value. The source and purity of the remaining materials was not specified.

ESTIMATED ERROR:

Soly: ± 0.01 g/100 g soln or $\pm 0.012 \times 10^{-3}$ in mole fraction (authors). Temp: ±0.02°C (authors).

REFERENCES:

COMPONENTS:	ORIGINAL MEASUREMENTS:
(1) Benzenesulfonamide, 3-amino- (metanilamide); $C_6H_8N_2O_2S$; [98-18-0] (2) Water; H_2O [7732-18-5]	Kienle, R. H.; Sayward, J. M., *J. Am. Chem. Soc.* 1942, *64*, 2464-8
VARIABLES:	PREPARED BY:
Temperature	R. Piekos

EXPERIMENTAL VALUES:

t/°C	g/100 g soln		mol kg^{-1} (compiler)	
23.0	1.14		0.0662	
24.0	1.21		0.0703	
26.0	1.34		0.0778	
28.0	1.48	$\pm$0.018	0.0859	$\pm$0.001
28.0	1.49		0.0865	
33.0	1.89		0.110	
35.5	2.19		0.127	
37.0	2.37	$\pm$0.031	0.138	$\pm$0.002
37.0	2.36[a]		0.137	
37.0	2.35[a]		0.136	
37.0	2.34[b]		0.136	
39.0	2.58		0.150	
42.0	3.01		0.175	
46.0	3.70		0.215	
50.0	4.58[b]		0.266	

[a] Equilibrium approached from below.

[b] Duration less than 12 hours.

AUXILIARY INFORMATION

METHOD/APPARATUS/PROCEDURE:

An excess of solid was rotated with water usually overnight. Equilibrium was approached from above. Sampling was accomplished by forcing the soln through a filter into a pycnometer. From the pycnometer the contents were flushed into a volumetric flask. Duplicate aliquots were acidified, iced below 15°C and titrated with 0.04 mol dm^{-3} $NaNO_2$ to first blue on starch - iodide paper.

SOURCE AND PURITY OF MATERIALS:

Metanilamide, m.p. 142.1°C, was prepd. by the authors. Titrn with nitrite indicated the compd to be 100.0$\pm$0.3% pure. Elemental analysis and mixed m.p. detns confirmed this value. Purity of the water was not specified.

ESTIMATED ERROR:

Temp: $\pm$0.02°C (authors)
Soly: $\pm$0.01 g/100g soln or $\pm$0.012 x 10^{-3} in mole fraction. The values of 2 varied from $\pm$0.018 to $\pm$0.031/100 g soln

REFERENCES:

COMPONENTS:	ORIGINAL MEASUREMENTS:
(1) Benzenesulfonamide, 3-amino- (metanilamide); $C_6H_8N_2O_2S$; [98-18-0] (2) Phosphoric acid, disodium salt; Na_2HPO_4; [7558-94-4] (3) Phosphoric acid, monopotassium salt; KH_2PO_4; [7778-77-0] (4) Water; H_2O; [7732-18-5]	Kienle, R. H.; Sayward, J. M. *J. Am. Chem. Soc.* 1942, *64*, 2464-8.
VARIABLES:	**PREPARED BY:**
One temperature: 37.0°C; one pH: 6.9	R. Piekos

EXPERIMENTAL VALUES:

Solubility of metanilamide in a buffer solution prepared by mixing together 55.2 cm^3 of 1/15 M Na_2HPO_4 with 44.8 cm^3 of 1/15 M KH_2PO_4 (pH 6.9, ionic strength calculated from dissociation constants 0.03[a]) at 37.0°C is 2.30 g/100 cm^3 solution (0.134 mol dm^{-3}, compiler).

[a]Not specified for which reactions were the dissociation constants calculated - compiler.

AUXILIARY INFORMATION

METHOD/APPARATUS/PROCEDURE:

An excess of metanilamide was rotated with the buffer soln usually overnight. Equilibrium was approached from above. Sampling was accomplished by forcing the soln through a filter into a pycnometer. From the pycnometer the contents were flushed into a volumetric flask. Duplicate aliquots were acidified, iced below 15°C and titrated with a 0.04 mol dm^{-3} $NaNO_2$ soln to first blue on a starch - iodide paper.

SOURCE AND PURITY OF MATERIALS:

Metanilamide, mp 142.1°C, was prepd by the authors. Titrn with nitrite indicated that the compd was 100.0±3% pure. Elemental analysis and mixed mp detns confirmed this value. Source and purity of the remaining materials was not specified.

ESTIMATED ERROR:

Soly: ±0.01 g/100 g soln or ±0.012 x 10^{-3} in mole fraction (authors). Temp: ±0.02°C (authors).

REFERENCES:

COMPONENTS:

(1) Benzenesulfonamide, 3-amino- (metanilamide); $C_6H_8N_2O_2S$; [98-18-0]

(2) Hydrochloric acid; HCl; [7647-40-7]

(3) Potassium chloride; KCl; [7447-40-7]

(4) Water; H_2O; [7732-18-5]

ORIGINAL MEASUREMENTS:

Kienle, R. H.; Sayward, J. M.,

J. Am. Chem. Soc. 1942, *64*, 2464-8.

VARIABLES:

One temperature, one pH

PREPARED BY:

R. Piekos

EXPERIMENTAL VALUES:

Solubility of metanilamide in a solution prepared by mixing together 25 cm^3 of 0.2 M KCl with 42.5 cm^3 0.2 M HCl and diluting up to 100 cm^3 with water (pH 1.2, ionic strength calculated from dissociation constants 0.12[a]) at 37.0°C is 4.48 g/100 cm^3 solution (0.260 mol dm^{-3} - compiler).

[a]Not specified for which reactions were the dissociation constants calculated - compiler.

AUXILIARY INFORMATION

METHOD/APPARATUS/PROCEDURE:

An excess of solid was rotated with water usually overnight. Equilibrium was approached from above. Sampling was accomplished by forcing the soln through a filter into a pycnometer. From the pycnometer the contents were flushed into a volumetric flask. Duplicate aliquots were acidified, iced below 15°C and titrated with 0.04 mol dm^{-3} $NaNO_2$ to first blue on starch - iodide paper.

SOURCE AND PURITY OF MATERIALS:

Metanilamide, m.p. 142.1°C, was prepd by the authors. Titrn with nitrite indicated that the compd was 100.0±0.3% pure. Elemental analysis and mixed m.p. detns confirmed this value. Source and purity of the remaining materials was not specified.

ESTIMATED ERROR:

Temp: ±0.02°C (authors)

Soly: accuracy ±0.01 g/100 g soln (authors)

REFERENCES:

COMPONENTS:	ORIGINAL MEASUREMENTS:
(1) Benzenesulfonamide, 3-amino- (metanil - amide); $C_6H_8N_2O_2S$; [98-18-0] (2) Boric acid; H_3BO_3; [10043-35-3] (3) Potassium chloride; KCl; [7447-40-7] (4) Sodium hydroxide; NaOH; [1310-73-2] (5) Water; H_2O; [7732-18-5]	Kienle, R. H.; Sayward, J. M. *J. Am. Chem. Soc.* 1942, *64*, 2464-8.
VARIABLES:	**PREPARED BY:**
pH; ionic strength	R. Piekos

EXPERIMENTAL VALUES:

pH of borate buffer	Ionic strength[a]	Solubility at 37.0°C: g/100 cm^3 solution	Solubility at 37.0°C: 10^2 mol dm^{-3} [b]
9.4[c]	0.08	2.61	15.2
9.7[d]	0.09	2.60	15.1

[a] Calculated from dissociation constants (reactions not specified).

[b] Calculated by compiler.

[c] Obtained by mixing together 50 cm^3 of a 0.1 M solution in both H_3BO_3 and KCl with 32.1 cm^3 of 0.1 M NaOH and diluting with water up to 100 cm^3.

[d] Obtained by mixing together 50 cm^3 of a 0.1 M solution in both H_3BO_3 and KCl with 38.75 cm^3 of 0.1 M NaOH and diluting with water up to 100 cm^3.

AUXILIARY INFORMATION

METHOD/APPARATUS/PROCEDURE:

An excess of metanilamide was rotated with the buffer soln usually overnight. Equilibrium was approached from above. Sampling was accomplished by forcing the soln through a filter into a pycnometer. From the pycnometer the contents were flushed into a volumetric flask. Duplicate aliquots were acidified, iced below 15°C and titrated with a 0.04 mol dm^{-3} $NaNO_2$ soln to first blue on a starch - iodide paper.

SOURCE AND PURITY OF MATERIALS:

Metanilamide, mp 142.1°C, was prepd by the authors. Titrn with nitrite indicated that the compd was 100.01±0.3% pure. Elemental analysis and mixed mp detns confirmed this value. Source and purity of the remaining materials was not specified.

ESTIMATED ERROR:

Soly: ±0.01 g/100 g soln or ±0.012 x 10^{-3} in mole fraction (authors). Temp: ±0.02°C (authors).

REFERENCES:

COMPONENTS:

(1) Benzenesulfonamide, 3-amino- (metanilamide); $C_6H_8N_2O_2S$; [98-18-0]

(2) Phosphoric acid, disodium salt; Na_2HPO_4; [7558-94-4]

(3) 1,2,3-Propanetricarboxylic acid, 2-hydroxy- (citric acid); $C_6H_8O_7$; [77-92-9]

(4) Water; H_2O; [7732-18-5]

ORIGINAL MEASUREMENTS:

Kienle, R. H.; Sayward, J. M.

J. Am. Chem. Soc. 1942, *64*, 2464-8.

VARIABLES:

One temperature: 37.0°C; one pH: 4.2

PREPARED BY:

R. Piekos

EXPERIMENTAL VALUES:

Solubility of metanilamide in a solution prepared by mixing together 41.4 cm^3 of 0.2 M Na_2HPO_4 with 58.6 cm^3 of 0.1 M citric acid (pH 4.2, ionic strength calculated from dissociation constants 0.84[a]) at 37.0°C is 2.26 g/100 cm^3 solution (0.131 mol dm^{-3}, compiler).

[a]Not specified for which reactions were the dissociation constants calculated - compiler.

AUXILIARY INFORMATION

METHOD/APPARATUS/PROCEDURE:

An excess of metanilamide was rotated with the buffer soln usually overnight. Equilibrium was approached from above. Sampling was accomplished by forcing the soln through a filter into a pycnometer. From the pycnometer the contents were flushed into a volumetric flask. Duplicate aliquots were acidified, iced below 15°C and titrated with a 0.04 mol dm^{-3} $NaNO_2$ soln to first blue on a starch - iodide paper.

SOURCE AND PURITY OF MATERIALS:

Metanilamide, mp 142.1°C was prepd by the authors. Titrn with nitrite indicated that the compd was 100.0±0.3% pure. Elemental analysis and mixed mp detns confirmed this value. Source and purity of the remaining materials was not specified.

ESTIMATED ERROR:

Soly: ±0.01 g/100 g soln or ±0.012 x 10^{-3} in mole fraction (authors). Temp: ±0.02°C (authors).

REFERENCES:

COMPONENTS:	ORIGINAL MEASUREMENTS:
(1) Benzenesulfonamide, 3-amino- (metanilamide); $C_6H_8N_2O_2S$; [98-18-0] (2) 1,2-Benzenedicarboxylic acid, monopotassium salt; $C_8H_5KO_4$; [877-24-7] (3) Hydrochloric acid; HCl; [7647-01-0] (4) Water; H_2O; [7732-18-5]	Kienle, R. H.; Sayward, J. M. *J. Am. Chem. Soc.* 1942, *64*, 2464-8.
VARIABLES:	PREPARED BY:
One temperature: 37.0°C; one pH: 2.2	R. Piekos

EXPERIMENTAL VALUES:

Solubility of metanilamide in a buffer solution prepared by mixing together 50 cm^3 of 0.1 M monopotassium 1,2-benzenedicarboxylate with 49.5 cm^3 of 0.1 M HCl and diluting up to 100 cm^3 with water (pH 2.2, ionic strength calculated from dissociation constants 0.06[a]) at 37.0°C is 3.07 g/100 cm^3 aolution (0.178 mol dm^{-3}, compiler).

[a]Not specified for which reactions were the dissociation constants calculated - compiler.

AUXILIARY INFORMATION

METHOD/APPARATUS/PROCEDURE:	SOURCE AND PURITY OF MATERIALS:
An excess of metanilamide was rotated with the buffer soln usually overnight. Equilibrium was approached from above. Sampling was accomplished by forcing the soln through a filter into a pycnometer. From the pycnometer the contents were flushed into a volumetric flask. Duplicate aliquots were acidified, iced below 15°C and titrated with a 0.04 mol dm^{-3} $NaNO_2$ soln to first blue on a starch - iodide paper.	Metanilamide, mp 142.1°C was prepd by the authors. Titrn with nitrite indicated that the compd was 100.0±0.3% pure. Elemental analysis and mixed mp detns confirmed this value. Source and purity of the remaining materials was not specified. ESTIMATED ERROR: Soly: ±0.01 g/100 g soln or ±0.012 x 10^{-3} in mole fraction (authors). Temp: ±0.02°C (authors). REFERENCES:

COMPONENTS:	EVALUATOR:
(1) Benzenesulfonamide, 4-amino- (sulfanilamide); $C_6H_8N_2O_2S$; [63-74-1] (2) Water (3) Aqueous sodium salicylate	Anthony N. Paruta Department of Pharmaceutics University of Rhode Island Kingston, Rhode Island, USA and Ryszard Piekos Faculty of Pharmacy, University of Gdansk Gdansk, Poland 1986

CRITICAL EVALUATION:

The solubilities of sulfanilamide, the parent compound of the derivatives found in these volumes, in water at various temperatures are given in Tables I-III on the mol dm^{-3} concentration scale. The originating author's values are always stated in the units reported. All recommended values are given as the simple average of the referenced magnitudes on the mol dm^{-3} scale.

Table I: Solubility of Sulfanilamide in water, 293K

Reference	10^2 mol dm^{-3} (*indicates mol kg^{-1})
5	3.1*
6	2.3*
7	3.5 (291-292K)
10	3.1
11	3.07*
12	3.1
14	3.07*
17	3.07
18	3.34
20	3.662 (polymorph)
21	3.1
22	3.07 (294-298K)
23	3.06*
24	3.447
27	3.539* (beta polymorph)

The solubility given by Weinstein and McDonald (6) is about two-thirds of the other values and was therefore rejected. The results of Becher and Leya (7), and of Gusyakov et al. (22) were also rejected since they refer to a temperature range of two to four degrees. Further, Becher and Leya (7), give a value about 15% higher than the apparent average. In 1944, Sapozhnikova and Postovskii (5), and Knazko (18) in 1966, used equilibrium times of one hour and three to four hours, respectively. These equilibrium times are considered inadequate for a saturation to be reached, and even though the values appear to be valid, they were not considered further. In 1971, Rohdewald (24) used a two hour equilibrium time which is considered as a pre-saturation time, yet Rohdewald (24) reported a value that was about 15% too high with respect to the apparent average. Ito and Sekiguchi (20), and Sekiguchi et al. (27) in 1967 and 1975, respectively, used polymorphic forms of sulfanilamide which may not be identifiable with the solute under consideration. Since the reports (20,27) do not give melting points for the polymorphs, they cannot be used in this evaluation. It is interesting to note that the beta form used by Sekiguchi et al. (27) is the stable form and should be quite close to the sulfanilamide value. However, the value provided is calculated rather than experimental, and would not be appropriate to mix with experimental results. The remaining values are sufficiently close in magnitude to evaluate further. The values of Gusyakov and Sukmans'ka in 1960 and 1961 (10-12), and Shkadova (23) in 1969, indicate that equilibrium had been reached, and may be assumed reasonably accurate. However, in two cases (10,12), the solubility technique of adding solute to a solution of the solute until saturation has been reached, is considered invalid, and were not used although the values given were "good" (concurring with the apparent average) numbers. The final pool of values (11,14,17,21), using at least an 8 hour equilibration time and appropriate analytical methods, allow for an average to be calculated and the recommended value for sulfanilamide in water at 293K is given as 3.08×10^{-2} mol dm^{-3}. It is interesting to note that even if the values from other workers (10,12,23) were used in the calculation, there would be no difference in the final result in the recommended value due obviously to their concurrence with the average value.

Aqueous solubility of sulfanilamide was studied at room temperature, 298K by eight workers (3,8,9,13,15,16,20,27), and the reported numerical data are summarized in Table II. There are three reports by Sekiguchi et al. (16,20,27) dealing with the polymorphic forms of sulfanilamide. While the value of 4.711×10^{-2} mol dm^{-3} in reference 20 is quite close to other values, as was the value in reference 27, none of the polymorphic forms were considered further. The value given by Khawan et al. (13) for 297K, appears to be too low by about 15% and was not placed in the final pool of acceptable values. The remaining four values by Clark et al. (3), Dolique and Foucault (8), Matsuura and Sekiguchi (9) and Paruta et al. (15) using equilibrium times longer than 9 hours, up to 72 hours, were determined by appropriate analytical methods. The initial average value of the four acceptable values allow for a tentative recommeded value of 4.72×10^{-2} mol dm^{-3} at 298K.

Table II: Solubility of Sulfanilamide in water, 298K

Reference	10^2 mol dm^{-3} (*indicates mol kg^{-1})
3	4.86*
8	4.7* (299K)
9	4.55
13	4.16 (297K)
15	4.8 (296.7K)
16	4.1925 (beta polymorph)
20	4.711 (unspecified polymorph)
27	4.677* (beta polymorph)
27	5.451* (alpha polymorph)

The tentative designation results from the slight differences in temperature reported in these papers (8,15), which were 299K and 297.6K, as given by Dolique and Foucault (8) and Paruta et al. (15). The authors give values within 1% of the average value using Clark et al. (3) and Matsuura and Sekiguchi (9) values alone leading to an average value of 4.70×10^{-2} mol dm^{-3}.

The solubility of sulfanilamide at body temperature, 310K in water as reported by various workers are given in Table III.

Table III: Solubility of Sulfanilamide in water, 310K

Reference	10^2 mol dm^{-3} (*indicates mol kg^{-1})
1	8.59
2	11*
3	8.48*
4	8.64*
5	8.4*
26	8.19
28	8.65
29	9.93

The value given by Tréfouël (2) is obviously too high. That of Goto et al. (29) is also higher than the apparent average by about 15% and neither is considered in the final pool of acceptable values. Sapozhnikova and Postovskii (5), and Kitao et al. (26) provide reasonable values, though the former used an equilibrium time of one hour only and the latter give no specification relative to saturation. Clark et al. (3), Kienle and Sayward (4) and Kaneniwa et al. (28) used adequate equilibration times and accurate analytical methods. Durel and Allinne (1) used 24 hours for saturation, however, no analytical method was specified. Thus, the saturation solubility for sulfanilamide at 310K can be given as 8.59×10^{-2} mol dm^{-3}. There were two additional solubility determinations in water. Yamazaki et al. (19) reported a value of 6.1×10^{-2} mol dm^{-3} at 303K, which is about the correct magnitude. Burger (25) gave solubilities of polymorphic Form II with a melting point of 429K over a temperature range of 313K-358K.

Solubility was also determined in an aqueous hydrotropic 1 mol dm^{-3} sodium salicylate (30-32) at 293K and a tentative value of 8.10×10^{-2} mol dm^{-3} can be suggested. The 1 mol dm^{-3} solution of sodium salicylate causes a 2.63 fold increase in the solubility of sulfanilamide.

REFERENCES:

(1) Durel, M.P.; Allinne, M. *Bull. Soc. Med. Hop. Paris III* 1941, 251-9.
(2) Tréfouël, M. *Bull. Acad. Med. Paris* 1941, *124*, 546-54.
(3) Clark, W.G.; Strakosch, E.A.; Levitan, N.I. *J. Lab. Clin. Med.* 1942, *28*, 188-9.
(4) Kienle, R.H.; Sayward, J.M. *J. Am. Chem. Soc.* 1942, *64*, 2464-8.
(5) Sapozhnikova, N.V.; Postovskii, I.Ya. *Zh. Prikl. Khim.* 1944, *17*, 427-34.
(6) Weinstein, L.; McDonald, A. *Science* 1945, *101*, 44-5.
(7) Becher, R.; Leya, S. *Experientia* 1946, *2*, 459-60.
(8) Dolique, R.; Foucault, J. *Trav. soc. pharm. Montpellier* 1952, *12*, 145-53.
(9) Matsuura, H.; Sekiguchi, K. *Yakuzaigaku* 1960, *20*, 213-18.
(10) Gusyakov, V.P.; Sukmans'ka, I.V. *Farm. Zh. (Kiev)* 1960, *15(1)*, 20-23.
(11) Gusyakov, V.P.; Likhol'ot, N.M. *Farm. Zh. (Kiev)* 1960, *15(3)*, 21-4.
(12) Gusyakov, V.P.; Sukmans'ka, I.V. *Farm. Zhur. (Kiev)* 1961, *16*, 25-8.
(13) Khawan, M.N.; Tawashi, R.; Czetsch-Lindenwald, H.V. *Sci. Pharm.* 1964, *32*, 271-9.
(14) Likhol'ot, N.M.; Gusyakov, V.P. *Farm. Zh. (Kiev)* 1964, *19(1)*, 52-5.

(14) Likhol'ot, N.M.; Gusyakov, V.P. *Farm. Zh. (Kiev)* 1964, *19(1)*, 52-5.
(15) Paruta, A.N. *J. Pharm. Sci.* 1964, *53(10)*, 1252-4.
(16) Sekiguchi, K.; Ito, K. *Chem. Pharm. Bull.* 1965, *13(4)*, 405-13.
(17) Likhol'ot, N.M. *Farm. Zh. (Kiev)* 1965, *20(5)*, 44-6.
(18) Knazko, L. *Farm. Obzor* 1966, *35*, 298-311.
(19) Yamazaki, M.; Aoki, M.; Kamada, A.; Yata, N. *Yakuzaigaku* 1967, *27(1)*, 37-40.
(20) Ito, K.; Sekiguchi, K. *Chem. Pharm. Bull.* 1967, *15(4)*, 420-6.
(21) Gusyakov, V.P.; Likhol'ot, N.M.; Kutna, I.M. *Farm. Zh. (Kiev)* 1967, *22(3)*, 34-9.
(22) Gusyakov, V.P.; Likhol'ot, N.M.; Kutna, I.M. *Farm. Zh. (Kiev)* 1968, *23(6)*, 56-61.
(23) Shkadova, A.I. *Farm. Zh. (Kiev)* 1969, *24(3)*, 39-41.
(24) Rohdewald, P. *Pharm. Ztg.* 1971, *No. 38*, 1342-4.
(25) Burger, A. *Pharm. Ind.* 1973, *35*, 626-33.
(26) Kitao, K.; Kubo, K.; Morishita, T.; Yata, N.; Kamada, A. *Chem. Pharm. Bull.* 1973, *21*, 2417-26.
(27) Sekiguchi, T.; Tsuda, Y.; Kanke, M. *Chem. Pharm. Bull.* 1975, *23*, 1353-62.
(28) Kaneniwa, N.; Watari, N.; Iijima, H. *Chem. Pharm. Bull.* 1978, *26(9)*, 2603-14.
(29) Goto, S.; Komatsu, M.; Tagawa, K.; Kawata, M. *Chem. Pharm. Bull.* 1983, *31(1)*, 256-61.
(30) Knazko, L. *Farm. Obzor* 1966, *35*, 298-311.
(31) Gusyakov, V.P.; Sukmans'ka, I.V. *Farm. Zh. (Kiev)* 1960, *15(1)*, 20-23.
(32) Gusyakov, V.P.; Sukmans'ka, I.V. *Farm. Zhur. (Kiev)* 1961, *16*, 25-8.

COMPONENTS:	ORIGINAL MEASUREMENTS:
(1) Benzenesulfonamide, 4-amino- (sulfanilamide); $C_6H_8N_2O_2S$; [63-74-1] (2) Water; H_2O; [7732-18-5]	Allport, N. L. *Quart. J. Pharm. Pharmacol.* 1936, *9*, 360-6.
VARIABLES:	PREPARED BY:
Temperature	R. Piekos

EXPERIMENTAL VALUES:

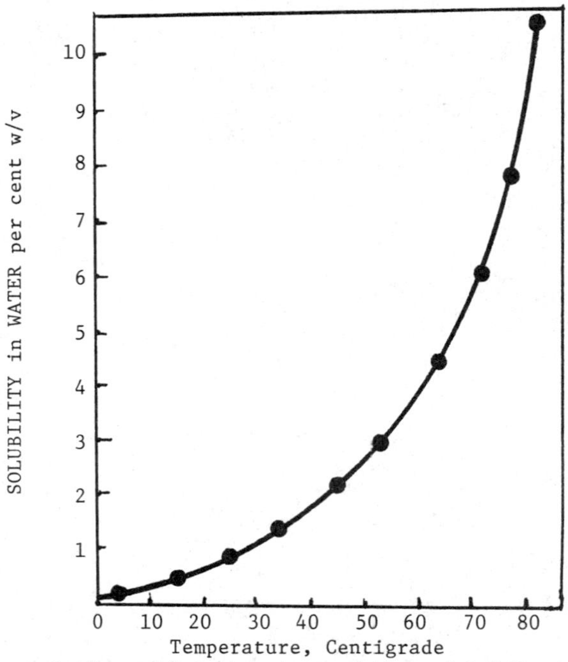

Fig. 1. - The variation with temperature of the solubility in water of *p*-aminobenzenesulphonamide.

AUXILIARY INFORMATION

METHOD/APPARATUS/PROCEDURE:

Nothing specified.

SOURCE AND PURITY OF MATERIALS:

The sulfanilamide was prepd in the author's laboratory on a semitechnical scale and melted quite sharply at 166.0°C. Specimens recrystd from dil alcohol melted at 166.5°C. No other details were given. Purity of the water was not specified.

ESTIMATED ERROR:

Temp: not specified.

Soly: not specified.

REFERENCES:

COMPONENTS:	ORIGINAL MEASUREMENTS:
(1) Benzenesulfonamide, 4-amino- (sulfanilamide); $C_6H_8N_2O_2S$; [63-74-1] (2) Water; H_2O; [7732-18-5]	Durel, M.P.; Allinne. M. *Bull. Soc. Med. Hop. Paris III* 1941, 251-9.
VARIABLES:	PREPARED BY:
One temperature: 37°C	R. Piekos

EXPERIMENTAL VALUES:

Solubility of sulfanilamide in water at 37°C is 14.8 g/liter (8.59 x 10^{-2} mol dm^{-3}, compiler)

AUXILIARY INFORMATION

METHOD/APPARATUS/PROCEDURE:	SOURCE AND PURITY OF MATERIALS:
A mixt of sulfanilamide and water was agitated for 24 hours at 37°C.	Source and purity of sulfanilamide was not specified. Distilled water was used.
	ESTIMATED ERROR:
	Nothing specified.
	REFERENCES:

COMPONENTS:	ORIGINAL MEASUREMENTS:
(1) Benzenesulfonamide, 4-amino- (sulfanilamide); $C_6H_8N_2O_2S$; [63-74-1] (2) Water; H_2O; [7732-18-5]	Tréfouël, M. *Bull. Acad. Med. Paris* 1941, *124*, 546-54.
VARIABLES:	PREPARED BY:
One temperature: 37°C	R. Piekos

EXPERIMENTAL VALUES:

Solubility of sulfanilamide in water at 37°C is 1.8 parts per 100 parts water (0.11 mol kg^{-1} water, compiler).

AUXILIARY INFORMATION

METHOD/APPARATUS/PROCEDURE:	SOURCE AND PURITY OF MATERIALS:
Sulfanilamide was diazotized, coupled with N-naphthyl-1-N,N-diethyl-3-propylenediamine and assayed colorimetrically.	Nothing specified. ESTIMATED ERROR: Nothing specified. REFERENCES:

COMPONENTS:	ORIGINAL MEASUREMENTS:
(1) Benzenesulfonamide, 4-amino- (sulfanilamide); $C_6H_8N_2O_2S$; [63-74-1] (2) Water; H_2O; [7732-18-5]	Clark. W.G.; Strakosch, E.A.; Levitan, N.I. *J. Lab. Clin. Med.* 1942, *28*, 188-9.
VARIABLES:	PREPARED BY:
Temperature	R. Piekos

EXPERIMENTAL VALUES:

t/°C	Solubility	
	g/100 g water	10^2 mol kg^{-1} water[a]
25	0.836	4.86
37	1.460	8.48

[a]Calculated by compiler.

AUXILIARY INFORMATION

METHOD/APPARATUS/PROCEDURE:

A small tinted glass container contg excess sulfanilamide in water was shaken in a water bath thermostat for 24 h. The satd soln was then filtered by aspiration through a washed and dried asbestos filter stick into a weighed weighing bottle. The entire app was kept at the temp at which the compd was dissolved. The amt dissolved was then detd by the method of Bratton and Marshall (1), using a photoelectric colorimeter.

SOURCE AND PURITY OF MATERIALS:

Neither source nor purity of sulfanilamide was specified.

CO_2-free distd water was used.

ESTIMATED ERROR:

Soly: not specified.

Temp: ± 0.1°C (authors).

REFERENCES:

1. Bratton, A.C.; Marshall, E.K., Jr. *J. Biol. Chem.* 1939, *128*, 537.

COMPONENTS:

(1) Benzenesulfonamide, 4-amino- (sulfanilamide); $C_6H_8N_2O_2S$; [63-74-1]

(2) Water; H_2O; [7732-18-5]

ORIGINAL MEASUREMENTS:

Kienle, R. H.; Sayward, J. M.

J. Am. Chem. Soc. 1942, *64*, 2464-8.

VARIABLES:

Temperature

PREPARED BY:

R. Piekos

EXPERIMENTAL VALUES:

$t/^\circ C$	Solubility g/100 g soln (2σ)		Solubility mol kg^{-1} [a]	
23.0	0.64; 0.64		0.0374	
24.0	0.69		0.0403	
26.0	0.77		0.0451	
27.0	0.82; 0.82[c,b]	±0.012	0.0480	±0.00070
27.0	0.82; 0.83		0.0480; 0.0486	
28.0	0.87; 0.87		0.0510	
30.5	1.01		0.0592	
31.7	1.08; 1.08[b]		0.0634	
33.0	1.19		0.0699	
34.0	1.26[b]	±0.015	0.0741	±0.000871
34.0	1.27[b]; 1.27		0.0747	
35.5	1.37		0.0807	
37.0	1.47; 1.47	±0.010	0.0866	±0.000581
37.0	1.47[b] 1.47[b]		0.0866	
37.0	1.46[c]		0.0860	
39.0	1.61; 1.61		0.0950	
42.0	1.84		0.1088	
46.0	2.21		0.1312	
50.0	2.68[c]		0.1599	

[a] calculated by compiler; [b] equilibrium approached from below;

[c] duration less than twelve hours.

AUXILIARY INFORMATION

METHOD/APPARATUS/PROCEDURE:

An excess of sulfanilamide was rotated with water usually overnight. Equilibrium was approached usually from above. Sampling was accomplished by forcing the soln through a filter into a pycnometer. From the pycnometer the contents were flushed into a volumetric flask. Duplicate aliquots were acidified, iced below 15°C and titrated with a 0.04 mol dm^{-3} $NaNO_2$ soln to first blue on starch - iodide paper.

SOURCE AND PURITY OF MATERIALS:

Sulfanilamide (U.S.P.) from plant production was recrystd from alcohol and from hot water; mp 165.9°C. Titrn with nitrite indicated that the compd was 100$\pm$0.3% pure. Elemental analysis and mixed mp detns confirmed this value. Purity of the water was not specified.

ESTIMATED ERROR:

Soly: $\pm$0.01 g/100 g soln or $\pm0.012 \times 10^{-3}$ in mole fraction (authors). Temp: $\pm0.02^\circ$C (authors).

REFERENCES:

COMPONENTS:

(1) Benzenesulfonamide, 4-amino- (sulfanilamide); $C_6H_8N_2O_2S$; [63-74-1]

(2) Water; H_2O; [7732-18-5]

ORIGINAL MEASUREMENTS:

Sapozhnikova, N.V.; Postovskii, I. Ya.
Zh. Prikl. Khim. 1944, *17*, 427-34.

VARIABLES:

Temperature

PREPARED BY:

R. Piekos

EXPERIMENTAL VALUES:

t/°C	Solubility	
	Weight %	mol kg^{-1}water[a]
20	0.53	0.031
37	1.42	0.084
50	2.63	0.157
75	7.58	0.476
99	19.2	1.380

[a]calculated by compiler.

AUXILIARY INFORMATION

METHOD/APPARATUS/PROCEDURE:

Sulfanilamide was dissolved in water to form a satd soln which was occasionally agitated in a glass vessel immersed in a thermostat. The equilibrium was usually attained after 1 h. Five to 100-cm^3 samples of the satd soln were placed in Pt crucibles or dishes and evapd to dryness at temps lower than 110-115°C. The residue was dried to const wt at 105-110°C and weighed.

SOURCE AND PURITY OF MATERIALS:

Pure, recrystd sulfanilamide was used. Its mp conformed to that reported in the literature.

Purity of the water was not specified.

ESTIMATED ERROR:

Soly: quite reliable results were obtained over the temp range 20-75°C. At higher temps the accuracy was poor due to evapn of water during sampling (authors). Temp: ±0.05°C (authors).

REFERENCES:

COMPONENTS:	ORIGINAL MEASUREMENTS:
(1) Benzenesulfonamide, 4-amino- (sulfanilamide); $C_6H_8N_2O_2S$; [63-74-1] (2) Water; H_2O; [7732-18-5]	Weinstein, L.; McDonald, A. *Science* 1945, *101*, 44-5.
VARIABLES:	PREPARED BY:
One temperature: 20°C	R. Piekos

EXPERIMENTAL VALUES:

Solubility of sulfanilamide in water at 20°C is 400 mg/100 cm^3 water (2.3 x 10^{-2} mol kg^{-1}, compiler).

AUXILIARY INFORMATION

METHOD/APPARATUS/PROCEDURE:	SOURCE AND PURITY OF MATERIALS:
Nothing specified.	Nothing specified.
	ESTIMATED ERROR:
	Nothing specified.
	REFERENCES:

COMPONENTS:	ORIGINAL MEASUREMENTS:
(1) Benzenesulfonamide, 4-amino- (sulfanilamide); $C_6H_8N_2O_2S$; [63-74-1] (2) Water: H_2O [7732-18-5]	Becher, R.; Leya, S. *Experientia* 1946 *2*, 459-60
VARIABLES: One temperature: 18-19°C	PREPARED BY: R. Piekos

EXPERIMENTAL VALUES:

Solubility of sulfanilamide in water at room temperature (18-19°C) is 610 mg% (3.5×10^{-2} mol dm^{-3}, compiler).

AUXILIARY INFORMATION

METHOD/APPARATUS/PROCEDURE:

After standing for more than two days the soln of sulfanilamide in water was filtered and sulfanilamide was assayed in the filtrate colorimetrically by the method of Druey and Oesterheld (1).

SOURCE AND PURITY OF MATERIALS:

Nothing specified.

ESTIMATED ERROR:

Nothing specified.

REFERENCES:

1. Druey, J.; Oesterheld, G. *Helv. Chim. Acta* 1942, *25*, 753.

COMPONENTS:	ORIGINAL MEASUREMENTS:
(1) Benzenesulfonamide, 4-amino- (sulfanilamide); $C_6H_8N_2O_2S$; [63-74-1] (2) Water; H_2O; [7732-18-5]	Dolique, R; Foucault, J. *Trav. soc. pharm. Montpellier* 1952, 12, 145-53.
VARIABLES:	PREPARED BY:
One temperature: 26°C	R. Piekos

EXPERIMENTAL VALUES:

Solubility of sulfanilamide in water at 26°C is 0.8 g/100 g water (4.7×10^{-2} mol kg^{-1} water - compiler).

AUXILIARY INFORMATION

METHOD/APPARATUS/PROCEDURE:	SOURCE AND PURITY OF MATERIALS:
The satd soln of sulfanilamide was shaken in a test tube for 12 h at 26°C and filtered. The filtrate was evapd at 100-110°C and the residue was weighed.	Nothing specified.
	ESTIMATED ERROR:
	Nothing specified.
	REFERENCES:

COMPONENTS:

(1) Benzenesulfonamide, 4-amino- (sulfanilamide); $C_6H_8N_2O_2S$; [63-74-1]

(2) Water; H_2O; [7732-18-5]

ORIGINAL MEASUREMENTS:

Matsuura, H.; Sekiguchi, K.

Yakuzaigaku 1960, *20*, 213-18.

VARIABLES:

One temperature: 25°C

PREPARED BY:

R. Piekos

EXPERIMENTAL VALUES:

Solubility of sulfanilamide in water at 25°C is 0.782 g/100 ml (4.55×10^{-2} mol L^{-1}).

AUXILIARY INFORMATION

METHOD/APPARATUS/PROCEDURE:

An excess of sulfanilamide was agitated in water at 25°C for more than 9 h. Aliquots of the satd soln were withdrawn with a pipet fitted with a filter and sulfanilamide was detd spectrophotometrically using the Tsuda reagent for producing color.

SOURCE AND PURITY OF MATERIALS:

Sulfanilamide was of the Japanese Pharmacopeia grade and was recrystd. Purity of the water was not specified.

ESTIMATED ERROR:

Soly: the error was 1.5% (authors).

Temp: ± 0.05°C (authors).

REFERENCES:

COMPONENTS:	ORIGINAL MEASUREMENTS:
(1) Benzenesulfonamide, 4-amino- (sulfanilamide); $C_6H_8N_2O_2S$; [63-74-1] (2) Water; H_2O; [7732-18-5]	Gusyakov, V. P.; Sukmans'ka, I.V. *Farm. Zh. (Kiev)* 1960, *15(1)*, 20-23.
VARIABLES:	PREPARED BY:
One temperature: 20°C	R. Piekos

EXPERIMENTAL VALUES:

Solubility of sulfanilamide in water at 20°C is 0.53 g/100 ml water (3.1×10^{-2} mol dm^{-3}, compiler).

AUXILIARY INFORMATION

METHOD/APPARATUS/PROCEDURE:	SOURCE AND PURITY OF MATERIALS:
To 50-ml tightly stoppered test tubes contg 25 ml water, placed in a thermostat, accurately weighed 0.02-0.002-g quantities of sulfanilamide were added under agitation until satn was attained.	Sulfanilamide conformed to the requirements of the State Pharmacopeia VIII. Distilled water was used.
	ESTIMATED ERROR: Temp: ±0.1°C (authors). Soly: the accuracy of the detn of the concn was similar to that attained by volumetric method (authors).
	REFERENCES:

COMPONENTS:	ORIGINAL MEASUREMENTS:
(1) Benzenesulfonamide, 4-amino- (sulfanilamide); $C_6H_8N_2O_2S$; [63-74-1] (2) Water; H_2O; [7732-18-5]	Gusyakov, V. P.; Likhol'ot, N. M. *Farm. Zh. (Kiev)* 1960, *15(3)*, 21-4.
VARIABLES:	**PREPARED BY:**
One temperature: 20°C	R. Piekos

EXPERIMENTAL VALUES:

Solubility of sulfanilamide in water at 20°C is 0.528 g/100 g water (3.07×10^{-2} mol kg^{-1}, compiler).

AUXILIARY INFORMATION

METHOD/APPARATUS/PROCEDURE:

A small excess of sulfanilamide was equilibrated for 8 h in a 50-ml test tube with 20 ml of water. Aliquots were taken with a pipet fitted with a filter. Sulfanilamide was detd at 285 nm using a SF-4 spectrophotometer.

SOURCE AND PURITY OF MATERIALS:

Sulfanilamide was crystd three times from hot water. Its purity conformed to the requirements of the State Pharmacopeia VIII.

Purity of the water was not specified.

ESTIMATED ERROR:

Soly: the accuracy corresponded to that of colorimetric detns (authors).

Temp: not specified.

REFERENCES:

COMPONENTS:	ORIGINAL MEASUREMENTS:
(1) Benzenesulfonamide, 4-amino- (sulfanilamide); $C_6H_8N_2O_2S$; [63-74-1] (2) Water; H_2O; [7732-18-5]	Gusyakov, V. P.; Sukmans'ka, I. V. *Farm. Zh. (Kiev)* 1961, *16*, 25-8.
VARIABLES:	PREPARED BY:
One temperature: 20°C	R. Piekos

EXPERIMENTAL VALUES:

Solubility of sulfanilamide in water at 20°C is 0.53 g/100 ml water (3.1×10^{-2} mol dm^{-3}, compiler).

AUXILIARY INFORMATION

METHOD/APPARATUS/PROCEDURE:	SOURCE AND PURITY OF MATERIALS:
Sulfanilamide was added in small portions (0.02 - 0.002 g) to a known volume of water, held on a water bath, until satn was attained. Moreover, the concn of sulfanilamide was detd by means of a FEK-M photoelectrocolorimeter.	Sulfanilamide was a pharmacopeial product. Purity of the water was not specified. ESTIMATED ERROR: Soly: results of the colorimetric and gravimetric runs differed by 1-3% (authors). Temp: ±0.1°C (authors). REFERENCES:

COMPONENTS:

(1) Benzenesulfonamide, 4-amino- (sulfanilamide); $C_6H_8N_2O_2S$; [63-74-1]

(2) Water; H_2O; [7732-18-5]

ORIGINAL MEASUREMENTS:

Khawan, M. N.; Tawashi, R.; Czetsch-Lindenwald, H. V.

Sci. Pharm. 1964, *32*, 271-9.

VARIABLES:

Temperature

PREPARED BY:

R. Piekos

EXPERIMENTAL VALUES:

$t/^{\circ}C$	Solubility	
	g/l	10^2 mol dm^{-3} [a]
13	3.58	2.08
15	3.79	2.20
16	3.85	2.24
24	7.17	4.16
34	12.71	7.38
44	20.36	11.82

[a] calculated by compiler.

AUXILIARY INFORMATION

METHOD/APPARATUS/PROCEDURE:

Sulfanilamide was added in excess to water and the mixture was stirred for 30 min. The soln was then kept in a drying cabinet for 24 h and occasionally shaken. After filtration the solution was assayed in the filtrate by the USP XVI method based on diazotization. The end point was detected by means of a starch paste as an indicator.

SOURCE AND PURITY OF MATERIALS:

Sulfanilamide conformed to the requirements of USP XVI.

Purity of the water was not specified.

ESTIMATED ERROR:

Soly: Measurements were made in duplicate (authors).

Temp: $\pm 1^{\circ}C$ (authors).

REFERENCES:

COMPONENTS:

(1) Benzenesulfonamide, 4-amino- (sulfanilamide); $C_6H_8N_2O_2S$; [63-74-1]

(2) Water; H_2O; [7732-18-5]

ORIGINAL MEASUREMENTS:

Likhol'ot, N. M.; Gusyakov, V. P.
Farm. Zh. (Kiev) 1964, *19(1)*, 52-5.

VARIABLES:

One temperature: 20°C

PREPARED BY:

R. Piekos

EXPERIMENTAL VALUES:

Solubility of sulfanilamide in water at 20°C is 0.528 g/100 g water (3.07×10^{-2} mol kg^{-1} water, compiler).

AUXILIARY INFORMATION

METHOD/APPARATUS/PROCEDURE:

The previously described method (1) was employed whereby a small excess of sulfanilamide was equilibrated with 20 ml of water in a 50-ml test tube for 8 h. Aliquots of the satd soln were removed through a filter and sulfanilamide was assayed bromatometrically (2).

SOURCE AND PURITY OF MATERIALS:

Sulfanilamide (source not specified) was recrystd from water.
Purity of the water was not specified.

ESTIMATED ERROR:

Nothing specified.

REFERENCES:

1. Gusyakov, V. P.; Likhol'ot, N. M. *Farm. Zh. (Kiev)* 1960, *15(3)*, 21.
2. *The Extra Pharmacopeia (Martindale)* 1955, *2(23)*, 353, 389.

COMPONENTS:	ORIGINAL MEASUREMENTS:
(1) Benzenesulfonamide, 4-amino-(sulfanilamide); $C_6H_8N_2O_2S$; [63-74-1] (2) Water; H_2O; [7732-18-5]	Paruta, A. N. *J. Pharm. Sci.* 1964, *53(10)* 1252-4.
VARIABLES: One temperature: 24.6°C	PREPARED BY: R. Piekos

EXPERIMENTAL VALUES:

Solubility of sulfanilamide in water at 24.6°C is 8.3 mg/ml (4.8 x 10^{-2} mol dm^{-3} - compiler).

AUXILIARY INFORMATION

METHOD/APPARATUS/PROCEDURE:

The soly was detd in 15-ml vials attached to a submerged rotating plastic disk. The plastic disk unit was attached to a dispersator motor fitted with an Al shaft and submerged with attached samples in an 8-gal water bath. The temp of the bath was maintained at 24.6±0.2°C by a Sargent thermomonitor unit. After equilibration for about 72 h, samples were withdrawn through a fine glass-wool plug fitted to a pipet, placed in a volumetric flask and dild for subsequent analysis. The concn of the solute was detd by a Beckman DK-2 spectrophotometer from sample absorbance and a previously detd Beer-Lambert law plot. The absorption max for sulfanilamide was 259 mμ.

SOURCE AND PURITY OF MATERIALS:

Sulfanilamide was recrystd from an EtOH - water mixt and dried to const wt at 40°C. Its source was not specified.

Dist water was used.

ESTIMATED ERROR:

Soly: not specified.

Tem: ±0.2°C (author).

REFERENCES:

COMPONENTS:

(1) Benzenesulfonamide, 4-amino- (sulfanilamide); $C_6H_8N_2O_2S$; [63-74-1]

(2) Water; H_2O; [7732-18-5]

ORIGINAL MEASUREMENTS:

Sekiguchi, K.: Ito, K.

Chem. Pharm. Bull. 1965, *13(4)*, 405-13.

VARIABLES:

Temperature

PREPARED BY:

R. Piekos

EXPERIMENTAL VALUES:

t/°C	Solubility 10^3 mol dm^{-3} solution	g dm^{-3} [a]
15	23.111	3.9797
25	41.925	7.2195
35	77.928	13.419

[a] Calculated by compiler.

AUXILIARY INFORMATION

METHOD/APPARATUS/PROCEDURE:

In a 200-ml egg-plant type flask, immersed in a thermostat, an excess of sulfanilamide was placed with 100 ml of redistd water (pH 5.7∿5.9) which was previously kept at appropriate temp. Immediately after addn of water, the mixt was vigorously agitated with an elec stirrer. Aliquots were withdrawn at certain time intervals with a pipet equipped with a filter, and the concn of solute was detd spectrophotometrically at 258 mμ.

SOURCE AND PURITY OF MATERIALS:

Sulfanilamide was a comm product of J.P. grade. The most stable polymorphic modification was used.

Redistd water was used.

ESTIMATED ERROR:

Soly: not specified.

Temp: $\pm 0.05^\circ C$ (authors.

REFERENCES:

COMPONENTS:	ORIGINAL MEASUREMENTS:
(1) Benzenesulfonamide, 4-amino- (sulfanilamide); $C_6H_8N_2O_2S$; [63-74-1] (2) Water; H_2O; [7732-18-5]	Likhol'ot, N. M. *Farm. Zh. (Kiev)* 1965, *20(5)*, 44-6.
VARIABLES:	**PREPARED BY:**
One temperature: 20°C	R. Piekos

EXPERIMENTAL VALUES:

Solubility of sulfanilamide in water at 20°C is 0.528 g/100 ml (3.07×10^{-2} mol dm^{-3}, compiler).

AUXILIARY INFORMATION

METHOD/APPARATUS/PROCEDURE:

An earlier described method was employed (1) whereby a small excess of sulfanilamide was equilibrated with 20 ml of water for 8 h is a 50-ml test tube. Aliquots were removed through a filter and sulfanilamide was assayed bromatometrically.

SOURCE AND PURITY OF MATERIALS:

Nothing specified.

ESTIMATED ERROR:

Soly: not specified.

Temp: ±0.1°C (authors).

REFERENCES:

1. Gusyakov, V. P.; Likhol'ot, N. M. *Farm. Zh. (Kiev)* 1960, *15(8)*, 21.

COMPONENTS:

(1) Benzenesulfonamide, 4-amino- (sulfanilamide); $C_6H_8N_2O_2S$; [63-74-1]

(2) Water; H_2O; [7732-18-5]

ORIGINAL MEASUREMENTS:

Kňažko, L. *Farm. Obzor* 1966, *35*, 298-311.

VARIABLES:

One temperature: 20°C

PREPARED BY:

R. Piekos

EXPERIMENTAL VALUES:

Solubility of sulfanilamide in water at 20°C is 0.575 g/100 ml water (3.34×10^{-2} mol dm^{-3}, compiler).

AUXILIARY INFORMATION

METHOD/APPARATUS/PROCEDURE:

Sulfanilamide was added in small portions to a known volume of water until reaching satn. The equilibration time was 3-4 h under stirring. The temp was held const by means of the Höppler ultrathermostat.

SOURCE AND PURITY OF MATERIALS:

Sulfanilamide conformed to the requirements of the Czechoslovak Pharmacopeia 2 (1954), Suppl. 1959. Purity of the water was not specified.

ESTIMATED ERROR:

Nothing specified.

REFERENCES:

COMPONENTS:	ORIGINAL MEASUREMENTS:
(1) Benzenesulfonamide, 4-amino- (sulfanilamide); $C_6H_8N_2O_2S$; [63-74-1] (2) Water; H_2O; [7732-18-5]	Yamazaki, M.; Aoki, M.; Kamada, A.; Yata, N. *Yakuzaigaku* 1967, *27(1)*, 37-40.
VARIABLES:	**PREPARED BY:**
One temperature: 30°C	R. Piekos

EXPERIMENTAL VALUES:

Solubility of sulfanilamide in water at 30°C is 61.0 mmol/L (10.50 g dm^{-3}, compiler).

AUXILIARY INFORMATION

METHOD/APPARATUS/PROCEDURE:

Sulfanilamide (0.5 g) was placed in an L-shaped tube together with 20 ml of water. The mixt was shaken in a thermostat until equilibrium was attained. The sulfanilamide was then assayed in the supernatant spectrophotometrically at 545 nm on a Beckman DU spectrophotometer. The results were taken from a calibration graph.

SOURCE AND PURITY OF MATERIALS:

Nothing specified.

ESTIMATED ERROR:

Soly: not specified.

Temp: $\pm$1°C (authors).

REFERENCES:

COMPONENTS:

(1) Benzensulfonamide, 4-amino- (sulfanilamide); $C_6H_8N_2O_2S$; [63-74-1]

(2) Water; H_2O; [7732-18-5]

ORIGINAL MEASUREMENTS:

Ito, K.; Sekiguchi, K.

Chem. Pharm. Bull. 1967, *15(4)*, 420-6.

VARIABLES:

Temperature

PREPARED BY:

R. Piekos

EXPERIMENTAL VALUES:

$t/^{\circ}C$	Solubility	
	10^3 mol dm^{-3} solution	g dm^{-3} [a]
20	3.622	0.6237
25	4.711	0.8112
30	5.988	1.0311
35	7.703	1.3265

[a] Calculated by compiler.

AUXILIARY INFORMATION

METHOD/APPARATUS/PROCEDURE:

The earlier described method (1) was used; in a 200-ml egg-plant type flask, immersed in a thermostat, an excess of sulfanilamide was placed with 100 ml of distd water which was previously kept at appropriate temp. Immediately after addn of water the mixt was vigorously agitated by an elec stirrer. Aliquots were withdrawn at certain time intervals with a pipet equipped with a filter, and the concn of solute was detd spectrophotometrically at 258 mμ.

SOURCE AND PURITY OF MATERIALS:

Polymorphic modifications of sulfanilamide (source not specified) were prepd by the method of Watanabe (2). Distd water was used.

ESTIMATED ERROR:

Nothing specified.

REFERENCES:

1. Sekiguchi, K.; Ito, K. *Chem. Pharm. Bull.* 1965, *13(4)*, 405.

2. Watanabe, A.; Kamio, H. *Yakugaku Zasshi* 1942, *62* 501.

COMPONENTS:	ORIGINAL MEASUREMENTS:
(1) Benzenesulfonamide, 4-amino- (sulfanilamide); $C_6H_8N_2O_2S$; [63-74-1] (2) Water; H_2O; [7732-18-5]	Gusyakov, V.P.; Likhol'ot, N.M.; Kutna, I.M. *Farm. Zh. (Kiev)* 1967, *22(3)* 34-9.
VARIABLES:	PREPARED BY:
One temperature: 20°C	R. Piekos

EXPERIMENTAL VALUES:

Solubility of sulfanilamide in water at 20°C is 0.53 g/100 ml (3.1×10^{-2} mol dm^{-3}, compiler).

AUXILIARY INFORMATION

METHOD/APPARATUS/PROCEDURE:	SOURCE AND PURITY OF MATERIALS:
An excess of sulfanilamide in water was equilibrated for 24 h in an ampul immersed in a water thermostat. Aliquots of the satd soln were withdrawn through a filter and the sulfanilamide content was assayed in the filtrate photometrically.	Sulfanilamide conformed to the requirements of the State Pharmacopeia IX. Purity of the water was not specified.
	ESTIMATED ERROR:
	Soly: not specified. Temp: $\pm$0.1°C (authors).
	REFERENCES:

COMPONENTS:	ORIGINAL MEASUREMENTS:
(1) Benzenesulfonamide, 4-amino- (sulfanilamide); $C_6H_8N_2O_2S$; [63-74-1] (2) Water; H_2O; [7732-18-5]	Gusyakov, V.P.; Likhol'ot, N.M.; Kutna, I.M. *Farm. Zh. (Kiev)* 1968, *23(6)* 56-61.

VARIABLES:	PREPARED BY:
One temperature: 21-25°C	R. Piekos

EXPERIMENTAL VALUES:

Solubility of sulfanilamide in water at room temperature (21-25°C) is 0.528 g/100 ml (3.07×10^{-2} mol dm^{-3}, compiler).

AUXILIARY INFORMATION

METHOD/APPARATUS/PROCEDURE:	SOURCE AND PURITY OF MATERIALS:
Small quantities (2-4 mg) of sulfanilamide were added to a known quantity of water under stirring until satn was achieved.	Sulfanilamide: neither source nor purity was specified. Purity of the water was not specified.
	ESTIMATED ERROR:
	Nothing specified.
	REFERENCES:

COMPONENTS:	ORIGINAL MEASUREMENTS:
(1) Benzenesulfonamide, 4-amino- (sulfanilamide); $C_6H_8N_2O_2S$; [63-74-1] (2) Water; H_2O; [7732-18-5]	Shkadova, A.I. *Farm. Zh. (Kiev)* 1969, *24(3)*, 39-41.
VARIABLES:	PREPARED BY:
One temperature: 20°C	R. Piekos

EXPERIMENTAL VALUES:

Solubility of sulfanilamide in water at 20°C is 3.06 x 10^{-2} mol/kg (0.53 g/100 g, compiler).

AUXILIARY INFORMATION

METHOD/APPARATUS/PROCEDURE:	SOURCE AND PURITY OF MATERIALS:
A satd aqueous solution of sulfanilamide was equilibrated in a water thermostat at 20±0.1°C. The concn of sulfanilamide was detd bromatometrically.	Purity of sulfanilamide conformed to the requirements of the State Pharmacopeia IX. Distd water was used.
	ESTIMATED ERROR: Soly: not specified. Temp: ±0.1°C (author).
	REFERENCES:

COMPONENTS:	ORIGINAL MEASUREMENTS:
(1) Benzenesulfonamide, 4-amino- (sulfanilamide); $C_6H_8N_2O_2S$; [63-74-1] (2) Water; H_2O; [7732-18-5]	Rohdewald, P. *Pharm. Ztg.* 1971, *No. 38* 1342-4.
VARIABLES:	PREPARED BY:
One temperature: 20°C	R. Piekos

EXPERIMENTAL VALUES:

Solubility of sulfanilamide in water at 20°C is 0.296_7 g/50 ml (3.447×10^{-2} mol dm^{-3}, compiler).

AUXILIARY INFORMATION

METHOD/APPARATUS/PROCEDURE:	SOURCE AND PURITY OF MATERIALS:
The soln was equilibrated by agitation for 2 h at 20°C and the sulfanilamide was assayed by differential gravimetric analysis. No details were given.	The source and purity of the materials was not specified.
	ESTIMATED ERROR: Soly: mean std deviation 68.3% of results deviating by S g), S = 0.028; relative std deviation 9.37%; no of detns 131 (author). Temp: ±0.05°C (author).
	REFERENCES:

COMPONENTS:

(1) Benzenesulfonamide, 4-amino- (sulfanilamide); $C_6H_8N_2O_2S$; [63-74-1]

(2) Water; H_2O; [7732-18-5]

ORIGINAL MEASUREMENTS:

Burger, A.

Pharm. Ind. 1973, *35*, 626-33.

VARIABLES:

Temperature

PREPARED BY:

R. Piekos

EXPERIMENTAL VALUES:

t/°C	Saturation solubility, C_s[a], of crystalline form II	
	mg/100 ml solution	mol dm^{-3}[b]
40.0	1440	0.084
45.0	1870	0.109
50.0	2390	0.140
55.0	3080	0.180
60.0	3950	0.230
65.0	5100	0.296
70.0	6600	0.383
75.0	8900	0.517

[a] $C_s = [HA] + [A^-]$, where [HA] is the molar concentration of the dissolved, undissociated molecules of sulfanilamide and $[A^-]$ is the concentration of the dissolved anion of sulfanilamide.

[b] Calculated by compiler.

AUXILIARY INFORMATION

METHOD/APPARATUS/PROCEDURE:

Sulfanilamide was assayed spectrophotometrically at 258.5 nm using a Zeiss PMQ II spectrophotometer and a 1/15 M phosphate buffer of pH 7.00 ($E^{1\%}_{1\ cm} = 945$).

SOURCE AND PURITY OF MATERIALS:

Form II of sulfanilamide, mp 156°C, was obtained by the known procedure (1).

Purity of the water was not specified.

ESTIMATED ERROR:

Soly: not specified.

Temp: ±0.1°C (author).

REFERENCES:

1. Burger, A. *Sci. Pharm.* 1973, *41*, 290 and 303.

COMPONENTS:	ORIGINAL MEASUREMENTS:
(1) Benzenesulfonamide, 4-amino- (sulfanilamide); $C_6H_8N_2O_2S$; [63-74-1] (2) Water; H_2O; [7732-18-5]	Kitao, K.; Kubo, K.; Morishita, T.; Yata, N.; Kamada, A. *Chem. Pharm. Bull.* 1973, *21*, 2417-26.
VARIABLES:	**PREPARED BY:**
One temperature: 37°C	R. Piekos

EXPERIMENTAL VALUES:

Solubility of sulfanilamide in water at 37°C is 1.41 g/100 cm^{-3} solution.

AUXILIARY INFORMATION

METHOD/APPARATUS/PROCEDURE:	SOURCE AND PURITY OF MATERIALS:
Sulfanilamide was assayed by diazotization. No details were given.	Comm available sulfanilamide was used as supplied. Deionized water was used.
	ESTIMATED ERROR: Soly: not specified. Temp: ± 1°C (authors).
	REFERENCES:

COMPONENTS:	ORIGINAL MEASUREMENTS:
(1) Benzenesulfonamide, 4-amino- (sulfanilamide); $C_6H_8N_2O_2S$; [63-74-1] (2) Water; H_2O; [7732-18-5]	Sekiguchi, K.; Tsuda, Y.; Kanke, M. *Chem. Pharm. Bull.* 1975, *23*, 1353-62.
VARIABLES:	PREPARED BY:
Temperature	R. Piekos

EXPERIMENTAL VALUES:

t/°C	Solubility			
	α - form		β - form	
	g/100 g soln	10^3 mol kg^{-1} water[a]	g/100 g soln	10^3 mol kg^{-1} water[a]
20	0.68	0.3976	0.615[b]	0.3593
25	0.93	0.5451	0.799[b]	0.4677
30	1.15	0.6756	1.024[b]	0.6008
35	1.49	0.8783	1.299	0.7643
40	1.90	1.1247	1.639	0.9676
45	2.17	1.2881	2.069	1.2269
50	3.33	2.0004	2.562	1.5269

[a] Calculated by compiler.

[b] Calculated from the equation $C'_s = C_s \times \frac{dE'/dt}{dE/dt}$, where C_s and dE/dt, and C'_s and dE'/dt are the solubility and dissolution rate of the stable (β) and metastable (α) crystalline forms of sulfanilamide, respectively.

AUXILIARY INFORMATION

METHOD/APPARATUS/PROCEDURE:

A sufficient amt of sample powder was placed in a 50 g of distd water in a dissoln measurement cell and stirred at 600 rpm. At appropriate time intervals samples were taken by glass syringes until the concn attained equilibrium. The sample solns were immediately filtered through a 0.45-μ membrane filter. The filtrate was weighed and dild for spectrophotometric assay at 225 nm on a Hitachi-139 UV spectrophotometer.

SOURCE AND PURITY OF MATERIALS:

The α-form of sulfanilamide was obtained by crystn of a comm product of the JP VIII grade from isoamyl, n-amyl or n-butyl alcohol, by holding the warm soln at room temp or cooling it immediately. The β-form was crystd from EtOH by gradual cooling of its warm soln to room temp and maintaining for 2-3 days. The forms were characterized by instrumental methods.

Distilled water was used.

ESTIMATED ERROR:

Nothing specified.

REFERENCES:

COMPONENTS:	ORIGINAL MEASUREMENTS:
(1) Benzenesulfonamide, 4-amino- (sulfanilamide); $C_6H_8N_2O_2S$; [63-74-1] (2) Water; H_2O; [7732-18-5]	Kaneniwa, N.; Watari, N.; Iijima, H. *Chem. Pharm. Bull.* 1978, *26(9)*, 2603-14.
VARIABLES:	PREPARED BY:
One temperature: 37°C	R. Piekos

EXPERIMENTAL VALUES:

Solubility of sulfanilamide in water at 37°C is 14.9 mg/ml solution (8.65×10^{-2} mol dm^{-3}, compiler).

AUXILIARY INFORMATION

METHOD/APPARATUS/PROCEDURE:

An excess of sulfanilamide was placed in a flask contg 25 ml of water. The flask was shaken (2 strokes/s at the amplitude of 3 cm) in a thermostatically controlled water bath at 37°C. One-ml sample was withdrawn every 6 h (total equilibration period was 3-5 days) using a warmed Millipore filter syringe with a filter pore size of 0.45 μ (Millipore HAWP 01300) and the filtrate was dild with water and assayed spectrophotometrically (1).

SOURCE AND PURITY OF MATERIALS:

Commercial sulfanilamide of the Japanese Pharmacopeia grade and distd water were used.

ESTIMATED ERROR:

Soly: not specified.
Temp: ±0.05°C (authors).

REFERENCES:

1. Kaneniwa, N.; Watari, N. *Chem. Pharm. Bull.* 1974, *22*, 1699.

COMPONENTS:

(1) Benzenesulfonamide, 4-amino- (sulfanilamide); $C_6H_8N_2O_2S$; [63-74-1]

(2) Water; H_2O; [7732-18-5]

ORIGINAL MEASUREMENTS:

Goto, S.; Komatsu, M.; Tagawa, K.; Kawata, M.

Chem. Pharm. Bull. 1983, *31(1)*, 256-61.

VARIABLES:

Temperature

PREPARED BY:

R. Piekos

EXPERIMENTAL VALUES:

t/°C	Solubility	
	g/l	10^2 mol dm^{-3} [a]
37	17.10	9.930
55	39.86	23.150

[a] Calculated by compiler

AUXILIARY INFORMATION

METHOD/APPARATUS/PROCEDURE:

A 3 g sample of sulfanilamide powder was accurately weighed into a 20-ml ampule and 10 ml of water was added. the ampule was sealed, placed in a const temp (37° or 55°C) bath and allowed to stand for several days. The equilibrium concn of the solute was measured spectrophotometrically at 540 nm after diazotization with the 0.1% Tsuda reagent (1).

SOURCE AND PURITY OF MATERIALS:

Sulfanilamide had mp 163-6°C.

The purity of water was not specified.

ESTIMATED ERROR:

Nothing specified.

REFERENCES:

1. Tsuda, K.; Matsunaga, S. *Yakugaku Zasshi* 1942, *62*, 362.

COMPONENTS:

(1) Benzenesulfonamide, 4-amino- (sulfanilimide); $C_6H_8N_2O_2S$; [63-74-1]

(2) Hydrochloric acid; HCl; [7647-01-0]

(3) Water; H_2O; [7732-18-5]

ORIGINAL MEASUREMENTS:

Paál, T.; Regősz, P.

Gyógyszerészet 1973, *17*, 59-63.

VARIABLES:

Concentration of HCl

PREPARED BY:

R. Piekos

EXPERIMENTAL VALUES:

Concentration of HCl N	Concentration of the most concentrated real solution of sulfanilamide at 26°C mol dm^{-3} solvent
5	0.3
1	1.15
0.1	0.13

AUXILIARY INFORMATION

METHOD/APPARATUS/PROCEDURE:

Satd solns were prepd by addn of increasing amts of aq HCl to weighed quantities of sulfanilamide. After the dissoln was completed, the soln was stirred with a magnetic stirrer and allowed to stand for 24 h. The soln was considered stable, if it remained clear during a 24-h period.

SOURCE AND PURITY OF MATERIALS:

Sulfanilamide was a com product (source not specified), doubly crystd from 98% EtOH. Its purity was > 99.5% as detd by diazotization. The source and purity of the remaining materials was not specified.

ESTIMATED ERROR:

Soly: accuracy ±10% (authors).

Temp: ±3°C (authors).

REFERENCES:

COMPONENTS:

(1) Benzenesulfonamide, 4-amino- (sulfanilamide); $C_6H_8N_2O_2S$; [63-74-1]

(2) Perchloric acid; $HClO_4$; [7601-90-3]

(3) Water; H_2O; [7732-18-5]

ORIGINAL MEASUREMENTS:

Paál, T.; Regősz, P.

Gyógyszerészet 1973, *17*, 59-63.

VARIABLES:

Concentration of $HClO_4$

PREPARED BY:

R. Piekos

EXPERIMENTAL VALUES:

Concentration of $HClO_4$ N	Concentration of the most concentrated real solution of sulfanilamide at 26°C mol dm^{-3} solvent
5	0.4 (1.75)[a]
1	1.15
0.1	0.13

[a]Concentration of the most concentrated metastable solution that could be prepared without precipitation of the solute.

AUXILIARY INFORMATION

METHOD/APPARATUS/PROCEDURE:

Satd solns were prepd by addn of increasing amts of aq $HClO_4$ to weighed quantities of sulfanilamide. After the dissoln was completed, the soln was stirred with a magnetic stirrer and allowed to stand for 24 h. The soln was considered stable, if it remained clear during a 24-h period. If the solute fell out from the clear soln, the soln was considered metastable.

SOURCE AND PURITY OF MATERIALS:

Sulfanilamide was a comm product (source not specified), double crystd from 98% EtOH. Its purity was > 99.5% as detd by diazotization. The source and purity of the remaining materials was not specified.

ESTIMATED ERROR:

Soly: accuracy ±10% (authors).

Temp: ±3°C (authors).

REFERENCES:

COMPONENTS:

(1) Benzenesulfonamide, 4-amino- (sulfanilamide); $C_6H_8N_2O_2S$; [63-74-1]

(2) Nitric acid HNO_3; [7697-37-2]

(3) Water; H_2O; [7732-18-5]

ORIGINAL MEASUREMENTS:

Paál, T.; Regősz, P.

Gyógyszerészet 1973, *17*, 59-63

VARIABLES:

Concentration of HNO_3

PREPARED BY:

R. Piekos

EXPERIMENTAL VALUES:

Concentration of HNO_3 N	Concentration of the most concentrated real solution of sulfanilamide at 26°C mol dm^{-3} solvent
5	0.15
1	1.15
0.1	9×10^{-2}

AUXILIARY INFORMATION

METHOD/APPARATUS/PROCEDURE:

Satd solns were prepd by addn of increasing amts of aq HNO_3 to weighed quantities of sulfanilamide. After the dissolution was completed, the soln was stirred with a magnetic stirrer and allowed to stand for 24 h. The soln was considered stable, if it remained clear during a 24-h period.

SOURCE AND PURITY OF MATERIALS:

Sulfanilamide was a comm product (source not specified), doubly crystd from 98% EtOH. Its purity was > 99.5% as detd by diazotization. The source and purity of the remaining materials was not specified.

ESTIMATED ERROR:

Soly: accuracy ±10% (authors).

Temp: ±3°C (authors).

REFERENCES:

COMPONENTS:

(1) Benzenesulfonamide, 4-amino-(sulfanilamide); $C_6H_8N_2O_2S$; [63-74-1]

(2) Lithium chloride; LiCl; [7447-41-8]

(3) Water; H_2O; [7732-18-5]

ORIGINAL MEASUREMENTS:

Likholet, M. A.; Gusyakov, V. P.

Med. Prom. SSSR 1963, *17(5)*, 28-31.

VARIABLES:

Concentration of LiCl

T/K = 293

PREPARED BY:

R. Piekos

EXPERIMENTAL VALUES:

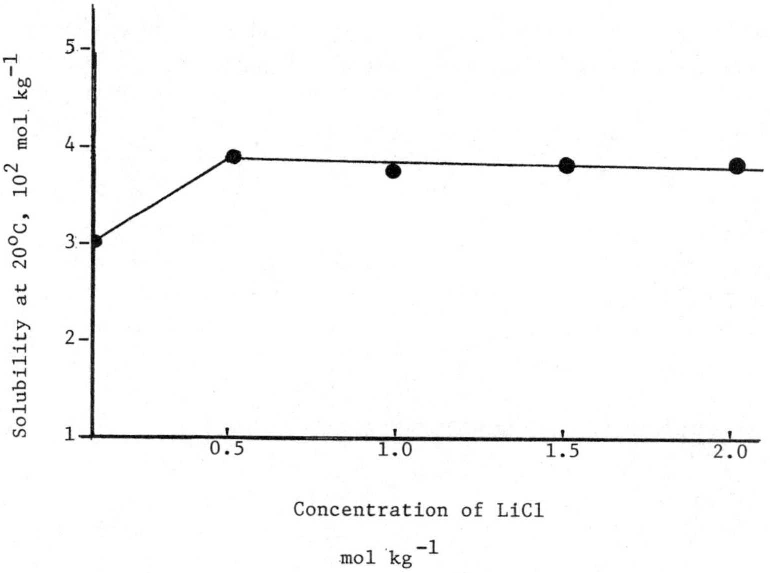

AUXILIARY INFORMATION

METHOD/APPARATUS/PROCEDURE:

Satd solns of sulfanilamide were prepd in 50-ml tightly closed ampuls in which 20 ml of a LiCl soln was placed and a small excess of sulfanilamide. the mixts were equilibrated for 18 h at 20°C. Aliquots were pipetted out through a filter, diluted, and assayed spectrophotometrically at 285 nm on a SF-IV spectrophotometer.

SOURCE AND PURITY OF MATERIALS:

Sulfanilamide was recrystd. Its purity was 99.22%. LiCl was purified by a recomeded procedure (1).

Purity of the water was not specified.

ESTIMATED ERROR:

Soly: measurements were repeated several times (authors).

Temp: ±0.1°C (authors).

REFERENCES:

1. Karysakin, Ya. V.; Angelov, I. I. *Chistye khimicheskye reaktivy*, Moscow, 1955.

COMPONENTS:	ORIGINAL MEASUREMENTS:
(1) Benzenesulfonamide, 4-amino- (sulfanilamide); $C_6H_8N_2O_2S$; [63-74-1] (2) Sodium chloride; NaCl; [7647-14-5] (3) Water; H_2O; [7732-18-5]	Becher, R.; Leya, S. *Experientia* 1946, *2*, 459-60.
VARIABLES:	PREPARED BY:
One temperature: 18-19 $^{\circ}$C	R. Piekos

EXPERIMENTAL VALUES:

Solubility of sulfanilamide in a 5% NaCl solution at room temperature (18-19 $^{\circ}$C) is 610 mg% (3.5 x 10^{-2} mol dm^{-3}, compiler).

AUXILIARY INFORMATION

METHOD/APPARATUS/PROCEDURE:

After standing for more than two days the soln of sulfanilamide was filtered and sulfanilamide was assayed in the filtrate colorimetrically by the method of Druey and Oesterheld (1).

SOURCE AND PURITY OF MATERIALS:

Nothing specified.

ESTIMATED ERROR:

Nothing specified.

REFERENCES:

1. Druey, J.; Oesterheld, G. *Helv. Chim. Acta* 1942, *25*, 753.

COMPONENTS:

(1) Benzenesulfonamide, 4-amino- (sulfanilamide); $C_6H_8N_2O_2S$; [63-74-1]

(2) Sodium chloride; NaCl; [7647-14-5]

(3) Water; H_2O; [7732-18-5]

ORIGINAL MEASUREMENTS:

Langecker, H.

Arch. Exptl. Fath. Pharmakol. 1948, *205*, 291-301.

VARIABLES:

One temperature: 37°C

PREPARED BY:

R. Piekos

EXPERIMENTAL VALUES:

Solubility of sulfanilamide in physiological saline (0.9% w/w NaCl solution) at 37°C is 1098 mg% (6.376×10^{-2} mol dm^{-3}, compiler).

AUXILIARY INFORMATION

METHOD/APPARATUS/PROCEDURE:

An excess of sulfanilamide was added to physiological saline and boiled for 1 h in a sealed ampul followed by keeping the ampul at 37°C. Sulfanilamide was assayed colorimetrically by the method of Bratton and Marshall (1) using a Havemann colorimeter (2), and by microanal detn of the solid residue.

SOURCE AND PURITY OF MATERIALS:

Source and purity of sulfanilamide was not specified.

The water was free of oxidants.

ESTIMATED ERROR:

Nothing specified.

REFERENCES:

1. Bratton, A.G.; Marshall, E.K. *J. Biol. Chem.* 1939, *128*, 537.

2. Havemann, R. *Klin. Wochenschr.* 1940, *p. 503*.

COMPONENTS:	ORIGINAL MEASUREMENTS:
(1) Benzenesulfonamide, 4-amino- (sulfanilamide); $C_6H_8N_2O_2S$; [63-74-1] (2) Sodium chloride; NaCl; [7647-14-5] (3) Water; H_2O; [7732-18-5]	Matsuura, H.; Sekiguchi, K. *Yakuzaigaku* 1960, *20*, 213-18
VARIABLES:	PREPARED BY:
Concentration of NaCl	R. Piekos

EXPERIMENTAL VALUES:

Concentration of NaCl Formula weight/L	Solubility of sulfanilamide at 25°C g/100 ml	10^2 mol/L
0	0.782	4.55
0.5	0.720	4.18
1.0	0.623	3.77
1.5	0.580	3.37
2.0	0.470	2.94
2.5	0.387	2.25
3.0	0.244	1.58

AUXILIARY INFORMATION

METHOD/APPARATUS/PROCEDURE:

Sulfanilamide was added to NaCl solns in such amts as to obtain satn. The mixts were agitated for more than 9 h. Aliquots were withdrawn with a pipet fitted with a filter and sulfanilamide was detd spectrophotometrically using the Tsuda reagent for producing color.

SOURCE AND PURITY OF MATERIALS:

Sulfanilamide was of the Japanese Pharmacopeia grade and was recrystd. NaCl was an extra pure reagent conforming to the first degree of the Japanese Industrial Standard.
Purity of the water was not specified.

ESTIMATED ERROR:

Soly: the error was 1.5% (authors).
Temp: $\pm 0.05^\circ C$ (authors).

REFERENCES:

COMPONENTS:

(1) Benzenesulfonamide, 4-amino- (sulfanilamide); $C_6H_8N_2O_2S$; [63-74-1]

(2) Sodium chloride; NaCl; [7647-14-5]

(3) Water; H_2O; [7732-18-5]

ORIGINAL MEASUREMENTS:

Likholet, M. N.; Gusyakov, V. P.

Med. Prom. SSSR 1963, *17(5)*, 28-31.

VARIABLES:

Concentration of NaCl

PREPARED BY:

R. Piekos

EXPERIMENTAL VALUES:

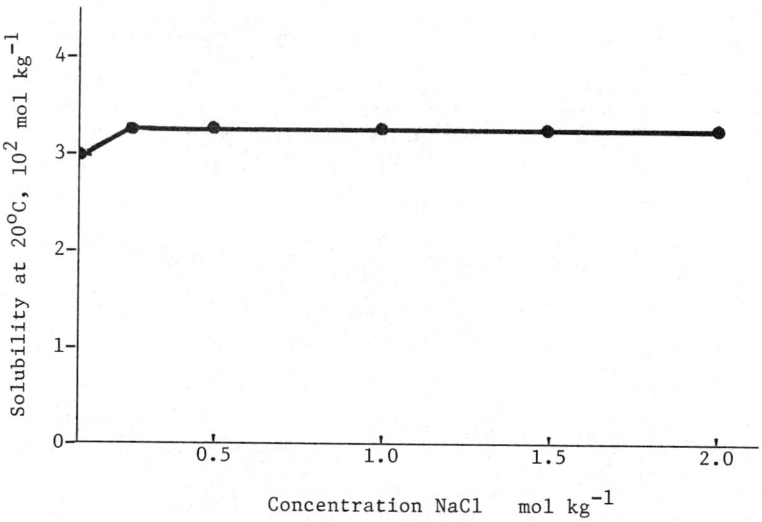

AUXILIARY INFORMATION

METHOD/APPARATUS/PROCEDURE:

Satd solns of sulfanilamide were prepd in 50-ml tightly closed ampuls in which 20 ml of a NaCl soln was placed and as small excess of sulfanilamide. The mixts were equilibrated for 18 h at 20°C. Aliquots were pipetted out through a filter, diluted and assayed spectrophotometrically at 285 nm on SF-IV spectrophotometer.

SOURCE AND PURITY OF MATERIALS:

Sulfanilamide was recrystd. Its purity was 99.22%. NaCl was purified by a recommended procedure (1).

Purity of the water was not specified.

ESTIMATED ERROR:

Soly: measurements were repeated several times (authors).

Temp: ±0.1°C (authors).

REFERENCES:

1. Karyakin, Ya. V.; Angelov, I. I. *Chistye khimicheskye reaktivy*, Moscow, 1955.

COMPONENTS:	ORIGINAL MEASUREMENTS:
(1) Benzenesulfonamide, 4-amino- (sulfanilamide); $C_6H_8N_2O_2S$; [63-74-1] (2) Sodium bromide; NaBr; [7647-15-6] (3) Water; H_2O; [7732-18-5]	Matsuura, H.; Sekiguchi, K. *Yakuzaigaku* 1960, *20*, 213-18.
VARIABLES:	PREPARED BY:
Concentration of NaBr	R. Piekos

EXPERIMENTAL VALUES:

Concentration of NaBr Formula weight/L	Solubility of sulfanilamide at 25°C g/100 ml	10^2 mol/L
0	0.782	4.55
0.448	0.786	4.57
0.673	0.778	4.52
0.345	0.784	4.58
2.018	0.786	4.56
2.690	0.788	4.58
3.363	0.767	4.45
3.766	0.750	4.36
5.380	0.692	4.02

AUXILIARY INFORMATION

METHOD/APPARATUS/PROCEDURE:

Sulfanilamide was added to NaBr solns in such amts as to obtain satn. The mixts were agitated for more than 9 h. Aliquots were withdrawn with a pipet fitted with a filter and sulfanilamide was detd spectrophotometrically using the Tsuda reagent for producing color.

SOURCE AND PURITY OF MATERIALS:

Sulfanilamide was of the Japanese Pharmacopeia grade and was recrystd. NaBr was an extra pure reagent conforming to the first degree of the Japanese Industrial Standard.

Purity of the water was not specified.

ESTIMATED ERROR:

Soly: the error was 1.5% (authors).

Temp: $\pm 0.05^\circ C$ (authors).

REFERENCES:

COMPONENTS:	ORIGINAL MEASUREMENTS:
(1) Benzenesulfonamide, 4-amino- (sulfanilamide); $C_6H_8N_2O_2S$; [63-74-1] (2) Sodium iodide; NaI; [7681-82-5] (3) Water; H_2O; [7732-18-5]	Matsuura, H.; Sekiguchi, K. *Yakuzaigaku* 1960, *20*, 213-18.
VARIABLES: Concentration of NaI	PREPARED BY: R. Piekos

EXPERIMENTAL VALUES:

Concentration of NaI		Solubility of sulfanilamide at 25°C	
g/100 ml	Formula wt/L	g/100 ml	10^2 mol/L
0	0	0.782	4.55
0.683	0.456	0.858	4.98
4.10	0.941	0.895	5.20
18.07	1.205	0.963	5.59
23.13	1.543	1.044	6.06
33.35	2.225	1.219	7.08
43.25	2.885	1.343	7.97

AUXILIARY INFORMATION

METHOD/APPARATUS/PROCEDURE:

Sulfanilamide was added to NaI solns in such amts as to obtain satn. The mixts were agitated for more than 9 h. Aliquots were withdrawn with a pipet fitted with a filter and sulfanilamide was detd spectrophotometrically using the Tsuda reagent for producing color.

SOURCE AND PURITY OF MATERIALS:

Sulfanilamide was of the Japanese Pharmacopeia grade and was recrystd. NaI was an extra pure reagent conforming to the first degree of the Japanese Industrial Standard.
Purity of the water was not specified.

ESTIMATED ERROR:

Soly: the error was 1.5% (authors).
Temp: $\pm 0.05^\circ C$ (authors).

REFERENCES:

COMPONENTS:	ORIGINAL MEASUREMENTS:
(1) Benzenesulfonamide, 4-amino- (sulfanilamide); $C_6H_8N_2O_2S$; [63-74-1] (2) Carbonic acid, monosodium salt; $NaHCO_3$; [144-55-8] (3) Water; H_2O; [7732-18-5]	Takubo, T.; Matsumaru, H.; Tsuchiya, S.; Hiura, M. *Chem. Pharm. Bull.* 1973, *21(7)*, 1440-5.
VARIABLES:	PREPARED BY:
One temperature: 37°C; one pH: 8.4	R. Piekos

EXPERIMENTAL VALUES:

Solubility of sulfanilamide in a $NaHCO_3$ solution (1.680 g $NaHCO_3$/100 ml water) of pH 8.4 at 37°C is 13.82 mg/ml solution[a] (8.026 x 10^{-2} mol dm^{-3} solution, compiler).

[a]Numerical value to the graphical one was given by one of the authors (S.T.) in personal communication.

AUXILIARY INFORMATION

METHOD/APPARATUS/PROCEDURE:	SOURCE AND PURITY OF MATERIALS:
Aliquots of the $NaHCO_3$ soln were placed in glass-stoppered flasks with excess of sulfanilamide. The flasks were allowed to stand at 37±1°C and shaken vigorously for 4 h until equilibrium was established. One ml of the supernatant was removed by means of a filter pipet and the sulfanilamide was assayed by the previously reported method (1).	The sulfanilamide was of pharmaceutical grade. Source and purity of $NaHCO_3$ was not specified. Distd water was used. ESTIMATED ERROR: Soly and pH: not specified. Temp: ±1°C (authors). REFERENCES: 1. Takubo, T.; Tsuchiya, S.; Hiura, M. *Yakuzaigaku* 1971, *31*, 298.

COMPONENTS:	ORIGINAL MEASUREMENTS:
(1) Benzenesulfonamide, 4-amino- (sulfanilamide); $C_6H_8N_2O_2S$; [63-74-1] (2) Carbonic acid, disodium salt; Na_2CO_3; [497-19-8] (3) Water; H_2O; [7732-18-5]	Takubo, T.; Matsumaru, H.; Tsuchiya, S.; Hiura, M. *Chem. Pharm. Bull.* 1973, *21(7)*, 1440-5.
VARIABLES:	**PREPARED BY:**
One temperature: 37°C; one pH: 11.3	R. Piekos

EXPERIMENTAL VALUES:

Solubility of sulfanilamide in a Na_2CO_3 solution (2.120 g Na_2CO_3/100 ml water) of pH 11.3 at 37°C is 15.39 mg/ml solution[a] (8.937 x 10^{-2} mol dm^{-3} solution, compiler).

[a]Numerical value to the graphical one was given by one of the authors (S.T.) in personal communication.

AUXILIARY INFORMATION

METHOD/APPARATUS/PROCEDURE:

Aliquots of the Na_2CO_3 soln were placed in glass-stoppered flasks with excess of sulfanilamide. The flasks were allowed to stand at 37±1°C and shaken vigorously for 4 h until equilibrium was established. One ml of the supernatant was removed by means of a filter and the sulfanilamide was assayed by the previously reported method (1).

SOURCE AND PURITY OF MATERIALS:

The sulfanilamide was of pharmaceutical grade. Source and purity of Na_2CO_3 was not specified.

Distd water was used.

ESTIMATED ERROR:

Soly and pH: not specified.

Temp: ±1°C (authors).

REFERENCES:

1. Takubo, T.; Tsuchiya, S.; Hiura, M. *Yakuzaigaku* 1971, *31*, 298.

COMPONENTS:	ORIGINAL MEASUREMENTS:
(1) Benzenesulfonamide, 4-amino- (sulfanilamide); $C_6H_8N_2O_2S$; [63-74-1] (2) Phosphoric acid, disodium salt; Na_2HPO_4; [7558-94-4] (3) Water; H_2O; [7732-18-5]	Krüger-Thiemer, E. *Arch. Dermatol. Syphilis* 1942, *183*, 90-116.
VARIABLES:	PREPARED BY:
One temperature: ca 20°C; one pH: 8.74	R. Piekos

EXPERIMENTAL VALUES:

Solubility of sulfanilamide in a 0.705 M (10%) Na_2HPO_4 solution of pH 8.74, at room temperature (about 20°C), is 0.57 g% (3.3×10^{-2} mol dm^{-3} solution, compiler).

AUXILIARY INFORMATION

METHOD/APPARATUS/PROCEDURE:

Sulfanilamide (0.5 g) was dissolved in 10 cm^3 of a 0.705M (10%) Na_2HPO_4 soln, shaken for 2 h, and filtered. A 1-cm^3 aliquot of the filtrate was then withdrawn, cooled, acidified with 2N HCl, and the sulfanilamide content was detd colorimetrically by the method of Marshall modified by Kimmig (1) using an Autenrieth colorimeter. The pH was detd on an ultraionograph using a glass electrode.

SOURCE AND PURITY OF MATERIALS:

Sulfanilamide was the product manufd by Schering AG. The source and purity of the remaining materials was not specified.

ESTIMATED ERROR:

Soly: precision ±5% (author)

Temp: not specified

pH: ±0.05 pH unit (author)

REFERENCES:

1. Kimmig, J. *Arch. Dermatol.* 1938, *176*, 722; *Erg. Hyg.* 1941, *24*, 398.

COMPONENTS:

(1) Benzenesulfonamide, 4-amino- (sulfanilamide); $C_6H_8N_2O_2S$; [63-74-1]

(2) Potassium chloride; KCl; [7747-40-7]

(3) Water; H_2O; [7732-18-5]

ORIGINAL MEASUREMENTS:

Gusyakov, V.P.; Likhol'ot, N.M.

Farm. Zh. (Kiev) 1960, *15(3)*, 21-4.

VARIABLES:

Concentration of KCl

PREPARED BY:

R. Piekos

EXPERIMENTAL VALUES:

Concentration of KCl	Solubility at 20°C	
Weight %	g/100 g water	10^2 mol kg^{-1} [a]
0.74	0.656	3.81
1.82	0.662	3.84
3.59	0.700	4.07
6.93	0.763	4.43
12.97	0.771	4.48
15.70	0.818	4.75

[a] calculated by compiler

AUXILIARY INFORMATION

METHOD/APPARATUS/PROCEDURE:

A small excess of sulfanilamide was equilibrated for 8 h in a 50-ml test tube with 20 ml of aqueous KCl soln. Aliquots were taken with a pipet fitted with a filter. Sulfanilamide was detd in the filtrate at 285 nm using a SF-4 spectrophotometer.

SOURCE AND PURITY OF MATERIALS:

Sulfanilamide was crystd three times from hot water. Its purity conformed to the requirements of the State Pharmacopeia VIII. KCl was doubly crystd. Purity of the water was not specified.

ESTIMATED ERROR:

Soly: the accuracy corresponded to that of colorimetric detns (authors).

Temp: not specified.

REFERENCES:

COMPONENTS:

(1) Benzenesulfonamide, 4-amino- (sulfanilamide); $C_6H_8N_2O_2S$; [63-74-1]

(2) Potassium chloride; KCl; [7447-40-7]

(3) Water; H_2O; [7732-18-5]

ORIGINAL MEASUREMENTS:

Likholet, N. M.; Gusyakov, V. P.

Med. Prom. SSSR 1963, *17(5)*, 28-31.

VARIABLES:

Concentration of KCl

PREPARED BY:

R. Piekos

EXPERIMENTAL VALUES:

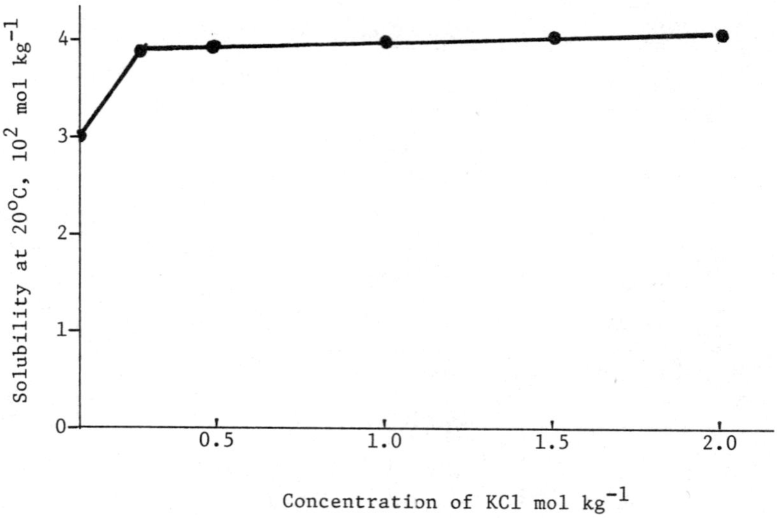

AUXILIARY INFORMATION

METHOD/APPARATUS/PROCEDURE:

Satd solns of sulfanilamide were prepd in 50-ml tightly closed ampuls in which 20 ml of a KCl was placed and a small excess of sulfanilamide. The mixts were equilibrated for 18 h at 20°C. Aliquots were pipetted out through a filter, dild, and assayed spectrophotometrically at 285 nm on a SF-IV spectrophotometer.

SOURCE AND PURITY OF MATERIALS:

Sulfanilamide was recrystd. Its purity was 99.22%. KCl was purified by a recommended procedure (1).

Purity of the water was not specified.

ESTIMATED ERROR:

Soly: measurements were repeated several times (authors).

Temp: ±0.1°C (authors).

REFERENCES:

1. Karyakin, Ya. V.; Angelov, I. I. *Chistye khimicheskye reaktivy*, Moscow, 1955.

COMPONENTS:	ORIGINAL MEASUREMENTS:
(1) Benzenesulfonamide, 4-amino- (sulfanilamide); $C_6H_8N_2O_2S$; [63-74-1] (2) Potassium bromide; KBr; [7759-02-3] (3) Water; H_2O [7732-18-5]	Gusyakov, V.P.; Likhol'ot, N.M. *Farm. Zh. (Kiev)* 1960, *15(3)* 21-4
VARIABLES: Concentration of KBr	PREPARED BY: R. Piekos

EXPERIMENTAL VALUES:

Concentration of KBr	Solubility at 20°C	
Weight %	g/100 g water	10^2 mol kg^{-1} [a]
1.17	0.738	4.29
2.88	0.739	4.29
5.61	0.798	4.63
10.63	0.844	4.90
19.22	0.993	5.77

[a] calculated by compiler

AUXILIARY INFORMATION

METHOD/APPARATUS/PROCEDURE:

A small excess of sulfanilamide was equilibrated for 8 h in a 50-ml test tube with 20 ml of aqueous KBr soln. Aliquots were taken with a pipet fitted with a filter. Sulfanilamide was detd in the filtrate at 285 nm using a SF-4 spectrophotometer.

SOURCE AND PURITY OF MATERIALS:

Sulfanilamide was crystd three times from hot water. Its purity conformed to the requirements of the State Pharmacopeia VIII. KBr was doubly crystd. Purity of the water was not specified.

ESTIMATED ERROR:

Soly: the accuracy corresponded to that of colorimetric detns (authors).

Temp: not specified.

REFERENCES:

COMPONENTS:

(1) Benzenesulfonamide, 4-amino- (sulfanilamide); $C_6H_8N_2O_2S$; [63-74-1]

(2) Potassium bromide; KBr; [7758-02-3]

(3) Water; H_2O; [7732-18-5]

ORIGINAL MEASUREMENTS:

Likholet, N. M.; Gusyakov, V. P.

Med. Prom. SSSR 1963, *17(5)*, 28-31.

VARIABLES:

Concentration of KBr

PREPARED BY:

R. Piekos

EXPERIMENTAL VALUES:

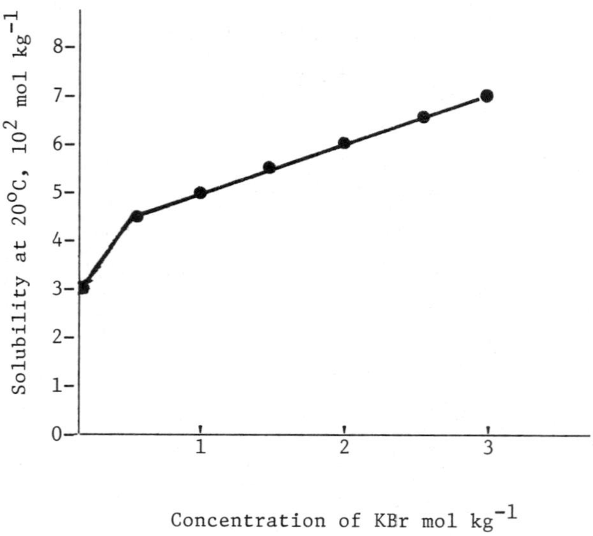

AUXILIARY INFORMATION

METHOD/APPARATUS/PROCEDURE:

Satd solns of sulfanilamide were prepd in 50-ml tightly closed ampuls in which 20 ml of a KBr soln was placed and a small excess of sulfanilamide. The mixts were equilibrated for 18 h at 20°C. Aliquots were pipetted out through a filter, dild, and assayed spectrophotometrically at 285 nm on a SF-IV spectrophotometer.

SOURCE AND PURITY OF MATERIALS:

Sulfanilamide was recrystd. Its purity was 99.22%. KBr was purified by a recommended procedure (1).

Purity of the water was not specified.

ESTIMATED ERROR:

Soly: measurements were repeated several times (authors).

Temp: ±0.1°C (authors).

REFERENCES:

1. Karyakin, Ya. V.; Angelov, I. I. *Chistye khimicheskye reaktivy*, Moscow, 1955.

COMPONENTS:

(1) Benzenesulfonamide, 4-amino- (sulfanilamide); $C_6H_8N_2O_2S$; [63-74-1]

(2) Potassium iodide; KI; [7681-11-0]

(3) Water; H_2O; [7732-18-5]

ORIGINAL MEASUREMENTS:

Gusyakov, V.P.; Likhol'ot, N.M.

Farm. Zh. (Kiev) 1960, *15(3)* 21-4

VARIABLES:

Concentration of KI

PREPARED BY:

R. Piekos

EXPERIMENTAL VALUES:

Concentration of KI	Solubility at 20°C	
Weight %	g/100 g water	10^2 mol kg^{-1} [a]
1.63	0.724	4.20
3.98	0.762	4.43
7.66	0.833	4.84
14.23	1.023	5.94

[a] calculated by compiler

AUXILIARY INFORMATION

METHOD/APPARATUS/PROCEDURE:

A small excess of sulfanilamide was equilibrated for 8 h in a 50-ml test tube with 20 ml of aqueous KI soln. Aliquots were taken with a pipet fitted with a filter. Sulfanilamide was detd in the filtrate at 285 nm using a SF-4 spectrophotometer.

SOURCE AND PURITY OF MATERIALS:

Sulfanilamide was crystd three times from hot water. Its purity conformed to the requirements of the State Pharmacopeia VIII. KI was doubly crystd. Purity of the water was not specified.

ESTIMATED ERROR:

Soly: the accuracy corresponded to that of colorimetric detns (authors).

Temp: not specified.

REFERENCES:

COMPONENTS:	ORIGINAL MEASUREMENTS:
(1) Benzenesulfonamide, 4-amino- (sulfanilamide); $C_6H_8N_2O_2S$; [63-74-1] (2) Potassium iodide; KI; [7681-11-0] (3) Water; H_2O; [7732-18-5]	Likholet, M. N.; Gusyakov, V. P. *Med. Prom. SSSR* 1963, *17(5)*, 28-31.
VARIABLES: Concentration of KI	PREPARED BY: R. Piekos

EXPERIMENTAL VALUES:

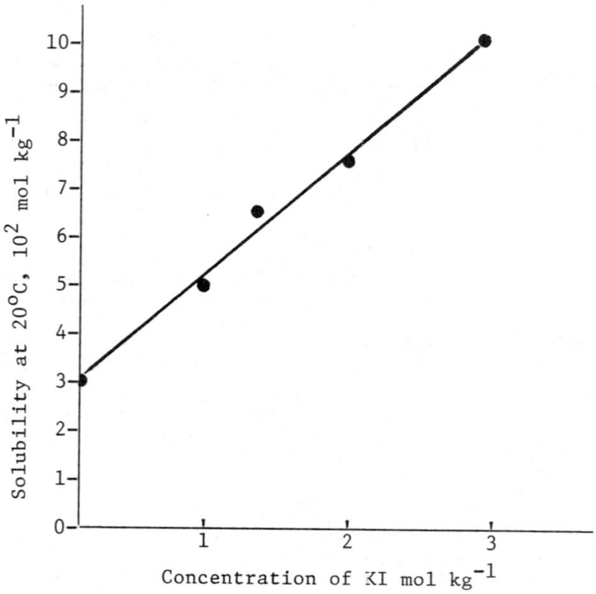

AUXILIARY INFORMATION

METHOD/APPARATUS/PROCEDURE:	SOURCE AND PURITY OF MATERIALS:
Satd solns of sulfanilamide were prepd in 50-ml tightly closed ampuls in which 20 ml of a KI soln was placed and a small excess of sulfanilamide. The mixts were equilibrated for 18 h at 20°C. Aliquots were pipetted out through a filter, dild, and assayed spectrophotometrically at 285 nm on a SF-IV spectrophotometer.	Sulfanilamide was recrystd. Its purity was 99.22%. KI was purified by a recommended procedure (1). Purity of the water was not specified.
	ESTIMATED ERROR: Soly: measurements were repeated several times (authors). Temp: ±0.1°C (authors).
	REFERENCES: 1. Karyakin, Ya. V.; Angelov, I. I. *Chistye khimicheskye reaktivy*, Moscow, 1955.

COMPONENTS:	ORIGINAL MEASUREMENTS:
(1) Benzenesulfonamide, 4-amino- (sulfanilamide); $C_6H_8N_2O_2S$; [63-74-1] (2) Potassium thiocyanate; KSCN; [333-20-0] (3) Water; H_2O; [7732-18-5]	Gusyakov, V.P.; Likhol'ot, N.M. *Farm. Zh. (Kiev)* 1960, *15(3)*, 21-4.
VARIABLES:	PREPARED BY:
Concentration of KSCN	R. Piekos

EXPERIMENTAL VALUES:

Concentration of KSCN	Solubility at 20°C	
Weight %	g/100 g water	10^2 mol kg^{-1} [a]
0.96	0.757	4.40
2.37	0.784	4.55
4.63	0.851	4.94
8.85	1.101	6.39

[a] calculated by compiler

AUXILIARY INFORMATION

METHOD/APPARATUS/PROCEDURE:

A small excess of sulfanilamide was equilibrated for 8 h in a 50-ml test tube with 20 ml of aqueous KSCN soln. Aliquots were taken with a pipet fitted with a filter. Sulfanilamide was detd in the filtrate at 285 nm using a SF-4 spectrophotometer.

SOURCE AND PURITY OF MATERIALS:

Sulfanilamide was crystd three times from hot water. Its purity conformed to the requirements of the State Pharmacopeia VIII. KSCN was doubly crystd. Purity of the water was not specified.

ESTIMATED ERROR:

Soly: the accuracy corresponded to that of colorimetric detns (authors).

Temp: not specified.

REFERENCES:

COMPONENTS:	ORIGINAL MEASUREMENTS:
(1) Benzenesulfonamide, 4-amino- (sulfanilamide); $C_6H_8N_2O_2S$; [63-74-1] (2) Thiocyanic acid, potassium salt; KCNS; [333-20-0] (3) Water; H_2O; [7732-18-5]	Likholet, M. N.; Gusyakov, V. P. *Med. Prom. SSSR* 1963, *17(5)*, 28-31.
VARIABLES: Concentration of KCNS	PREPARED BY: R. Piekos

EXPERIMENTAL VALUES:

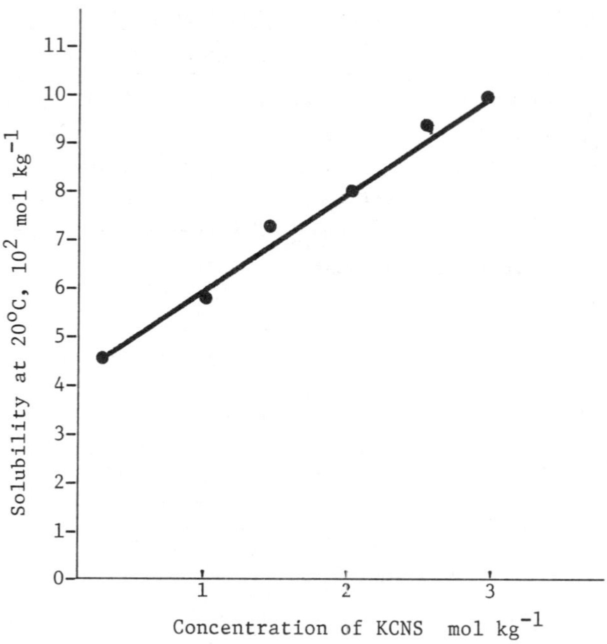

AUXILIARY INFORMATION

METHOD/APPARATUS/PROCEDURE:

Satd solns of sulfanilamide were prepd in 50-ml tightly closed ampuls in which 20 ml of a KCNS soln was placed and a small excess of sulfanilamide. The mixts were equilibrated for 18 h at 20°C. Aliquots were pipetted out through a filter, dild, and assayed spectrophotometrically at 285 nm on a SF-IV spectrophotometer.

SOURCE AND PURITY OF MATERIALS:

Sulfanilamide was recrystd. Its purity was 99.22%. KCNS was purified by a recommended procedure (1).

Purity of the water was not specified.

ESTIMATED ERROR:

Soly: measurements were repeated several times (authors).

Temp: ±0.1°C

REFERENCES:

1. Karyakin, Ya. V.; Angelov, I. I. *Chistye khimicheskye reaktivy*, Moscow, 1955.

COMPONENTS:

(1) Benzenesulfonamide, 4-amino- (sulfanilamide); $C_6H_8N_2O_2S$; [63-74-1]

(2) Phosphoric acid, monopotassium salt; KH_2PO_4; [7778-77-0]

(3) Water; H_2O; [7732-18-5]

ORIGINAL MEASUREMENTS:

Krüger-Thiemer, E.

Arch. Dermatol. Syphilis 1942, *183*, 90-116.

VARIABLES:

One temperature: ca 20°C; one pH: 4.37

PREPARED BY:

R. Piekos

EXPERIMENTAL VALUES:

Solubility of sulfanilamide in a 0.735 M (10%) KH_2PO_4 solution of pH 4.37, at room temperature (about 20°C) is 0.572 g% (3.32×10^{-2} mol dm^{-3} solution, compiler).

AUXILIARY INFORMATION

METHOD/APPARATUS/PROCEDURE:

Sulfanilamide (0.5 g) was dissolved in 10 cm^3 of a 0.735 M(10%) KH_2PO_4 soln, shaken for 2 h, and filtered. A 1-cm^3 aliquot of the filtrate was then withdrawn, cooled, acidified with 2 N HCl, and the sulfanilamide content was detd colorimetrically the method of Marshall modified by Kimmig (1) using an Authenreith colorimeter. The pH was detd on an ultraionograph using a glass electrode.

SOURCE AND PURITY OF MATERIALS:

Sulfanilamide was the product manufd by Schering AG. The source and purity of the remaining materials was not specified.

ESTIMATED ERROR:

Soly: precision $\pm$5% (author)

Temp: not specified

pH: $\pm$0.05 pH unit (author)

REFERENCES:

1. Kimmig, J. *Arch. Dermatol.* 1938, *176*, 722; *Erg. Hyg.* 1941, *24*, 398.

COMPONENTS:

(1) Benzenesulfonamide, 4-amino- (sulfanilamide); $C_6H_8N_2O_2S$; [63-74-1]

(2) Ammonium chloride; NH_4Cl; [12125-02-9]

(3) Water; H_2O; [7732-18-5]

ORIGINAL MEASUREMENTS:

Likholet, M. N.; Gusyakov, V. P.

Med. Prom. SSSR 1963, *17(5)*, 28-31.

VARIABLES:

Concentration of NH_4Cl

PREPARED BY:

R. Piekos

EXPERIMENTAL VALUES:

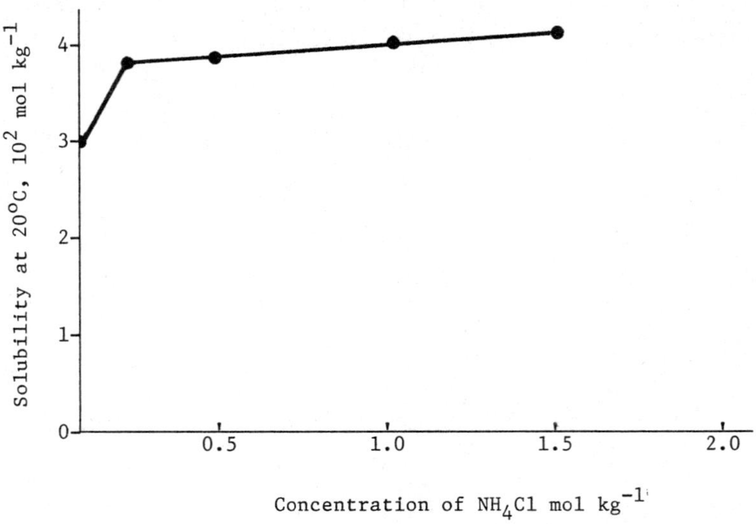

AUXILIARY INFORMATION

METHOD/APPARATUS/PROCEDURE:

Satd solns of sulfanilamide were prepd in 50-ml tightly closed ampuls in which 20 ml of a NH_4Cl soln was placed and a small excess of sulfanilamide. The mixts were equilibrated for 18 h at 20°C. Aliquots were pipetted out through a filter, dild, and assayed spectrophotometrically at 285 nm on a SF-IV spectrophotometer.

SOURCE AND PURITY OF MATERIALS:

Sulfanilamide was recrystd. Its purity was 99.22%. NH_4Cl was purified by a recommended procedure (1).

Purity of the water was not specified.

ESTIMATED ERROR:

Soly: measurements were repeated several times (authors).

Temp: ±0.1°C (authors).

REFERENCES:

1. Karyakin, Ya. V.; Angelov, I. I. *Chistye khimicheskye reaktivy*, Moscow, 1955.

COMPONENTS:

(1) Benzenesulfonamide, 4-amino- (sulfanilamide); $C_6H_8N_2O_2S$; [63-74-1]

(2) Magnesium chloride; $MgCl_2$; [7786-30-3]

(4) Water; H_2O; [7732-18-5]

ORIGINAL MEASUREMENTS:

Likholet, M. N.; Gusyakov, V. P.

Med. Prom. SSSR 1963, *17(5)*, 28-31.

VARIABLES:

Concentration of $MgCl_2$

PREPARED BY:

R. Piekos

EXPERIMENTAL VALUES:

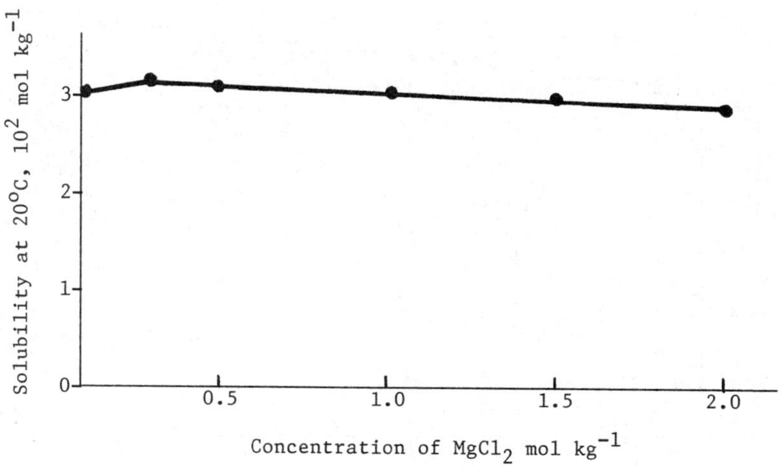

AUXILIARY INFORMATION

METHOD/APPARATUS/PROCEDURE:

Satd solns of sulfanilamide were prepd in 50-ml tightly closed ampuls in which 20 ml of a $MgCl_2$ soln was placed and a small excess of sulfanilamide. The mixts were equilibrated for 18 h at 20°C. Aliquots were pipetted out through a filter, dild, and assayed spectrophotometrically at 285 nm on a SF-IV spectrophotometer.

SOURCE AND PURITY OF MATERIALS:

Sulfanilamide was recrystd. Its purity was 99.22%. $MgCl_2$ was purified by a recommended procedure (1).

Purity of the water was not specified.

ESTIMATED ERROR:

Soly: measurements were repeated several times (authors).

Temp: ±0.1°C (authors).

REFERENCES:

1. Karyakin, Ya. V.; Angelov, I. I. *Chistye khimicheskye reaktivy*, Moscow, 1955.

COMPONENTS:	ORIGINAL MEASUREMENTS:
(1) Benzenesulfonamide, 4-amino- (sulfanilamide); $C_6H_8N_2O_2S$; [63-74-1] (2) Calcium chloride; $CaCl_2$; [10043-52-4] (3) Water; H_2O; [7732-18-5]	Likholet, M. N.; Gusyakov, V. P. *Med. Prom. SSSR* 1963, *17(5)*, 28-31.
VARIABLES:	PREPARED BY:
Concentration of $CaCl_2$	R. Piekos

EXPERIMENTAL VALUES:

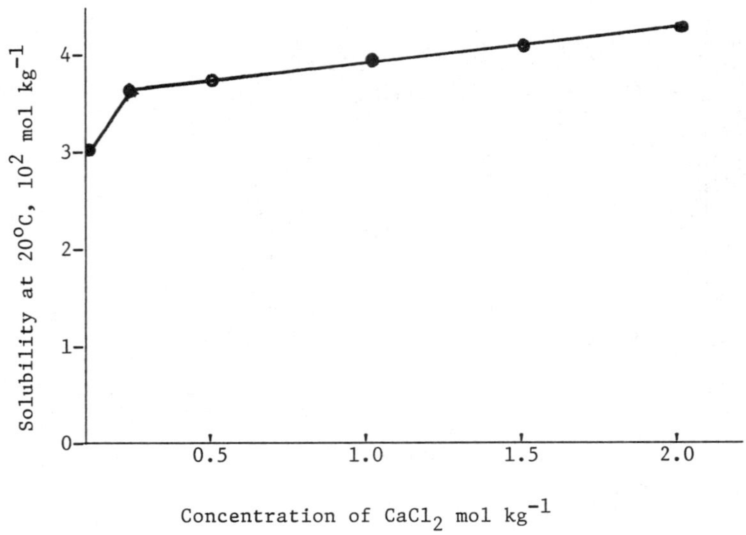

AUXILIARY INFORMATION

METHOD/APPARATUS/PROCEDURE:

Satd solns of sulfanilamide were prepd in 50-ml tightly closed ampuls in which 20 ml of a $CaCl_2$ soln was placed and a small excess of sulfanilamide. The mixts were equilibrated for 18 h at 20°C. Aliquots were pippeted out through a filter, dild, and assayed spectrophotometrically at 285 nm on a SF-IV spectrophotometer.

SOURCE AND PURITY OF MATERIALS:

Sulfanilamide was recrystd. Its purity was 99.22%. $CaCl_2$ was purified by a recommended procedure (1).
Purity of the water was not specified.

ESTIMATED ERROR:

Soly: measurements were repeated several times (authors).

Temp: ±0.1 (authors).

REFERENCES:

1. Karyakin, Ya. V.; Angelov, I. I. *Chistye khimicheskye reaktivy*, Moscow, 1955.

COMPONENTS:

(1) Benzenesulfonamide, 4-amino- (sulfanilamide); $C_6H_8N_2O_2S$; [63-74-1]

(2) Barium chloride; $BaCl_2$; [10361-37-2]

(3) Water; H_2O; [7732-18-5]

ORIGINAL MEASUREMENTS:

Likholet, M. N.; Gusyakov, V. P.

Med. Prom. SSSR 1963, *17(5)*, 28-31.

VARIABLES:

Concentration of $BaCl_2$

PREPARED BY:

R. Piekos

EXPERIMENTAL VALUES:

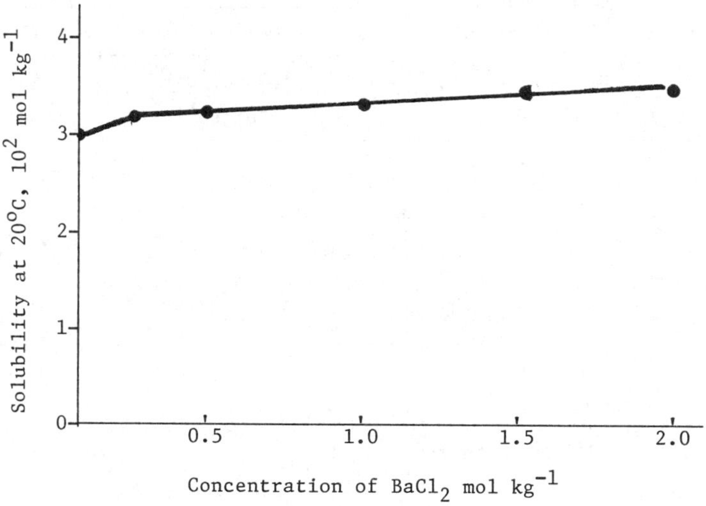

AUXILIARY INFORMATION

METHOD/APPARATUS/PROCEDURE:

Satd solns of sulfanilamide was prepd in 50-ml tightly closed ampuls in which 20 ml of a $BaCl_2$ soln was placed and a small excess of sulfanilamide. The mixts were equilibrated for 18 h at 20°C. Aliquots were pipetted out through a filter, dild, and assayed spectrophotometrically at 285 nm on a SF-IV spectrophotometer.

SOURCE AND PURITY OF MATERIALS:

Sulfanilamide was recrystd. Its purity was 99.22%. $BaCl_2$ was purified by a recommended procedure (1).

Purity of the water was not specified.

ESTIMATED ERROR:

Soly: measurements were repeated several times (authors).

Temp: ±0.1°C (authors).

REFERENCES:

1. Karyakin, Ya. V.; Angelov, I. I. *Chistye khimicheskye reaktivy* Moscow, 1955.

COMPONENTS:	ORIGINAL MEASUREMENTS:
(1) Benzenesulfonamide, 4-amino- (sulfanilamide); $C_6H_8N_2O_2S$; [63-74-1] (2) Phosphoric acid, disodium salt; Na_2HPO_4; [7558-94-4] (3) Phosphoric acid, monopotassium salt; KH_2PO_4; [7778-77-0] (4) Water; H_2O; [7732-18-5]	Krüger-Thiemer, E. *Arch. Dermatol. Syphilis* 1942, *183*, 90-116.
VARIABLES:	PREPARED BY:
Temperature, pH	R. Piekos

EXPERIMENTAL VALUES:

Composition of 1/15M phosphate buffer solutions			pH	Solubility			
				room temp (ca 20°C)		37°C	
Na_2HPO_4	KH_2PO_4	%Content		g%	10^2 mol dm^{-3} solution[a]	g%	10^2 mol dm^{-3} solution[a]
1.0	99.0	0.91	4.944	0.694	4.03	-	-
10.0	90.0	0.91	5.906	0.704	4.09	1.220	7.08
61.1	38.9	0.93	7.005	0.698	4.05	1.260	7.32
9.5	0.5	0.733[b]	7.51	0.573	3.33	-	-
94.7	5.3	0.95	8.018	0.694	4.03	-	-

[a]Calculated by compiler

[b]Molar content; 10% buffer solution

AUXILIARY INFORMATION

METHOD/APPARATUS/PROCEDURE:

Sulfanilamide (0.5 g) was dissolved in 10 cm^3 of a buffer soln, shaken for 2 h at 20°C (or left for 48 h at 37°C), and filtered at respective temp. A 1-cm^3 aliquot of the filtrate was then withdrawn, cooled (dild for expts at 37°C), acidified with 1 cm^3 of 2N HCl, and the sulfanilamide content was detd colorimetrically by the method of Marshall modified by Kimmig (1, 2) using an Autenrieth colorimeter. The pH was detd on an ultraionograph using a glass electrode.

SOURCE AND PURITY OF MATERIALS:

Sulfanilamide was the product manufd by Schering AG. The source and purity of the remaining materials was not specified.

ESTIMATED ERROR:

Soly: precision ±5% (author)

Temp: not specified

pH: ±0.05 pH unit (author)

REFERENCES:

1. Kimmig, J. *Arch. Dermatol.* 1938, *176*, 722.
2. Kimmig, J. *Erg. Hyg.* 1941, *24*, 398.

COMPONENTS:

(1) Benzenesulfonamide, 4-amino- (sulfanilamide); $C_6H_8N_2O_2S$; [63-74-1]

(2) Phosphoric acid, disodium salt; Na_2HPO_4; [7778-94-4]

(3) Phosphoric acid, monopotassium salt; KH_2PO_4; [7778-77-0]

(4) Water; H_2O; [7732-18-5]

ORIGINAL MEASUREMENTS:

Kienle, R.H.; Sayward, J.M.

J. Am. Chem. Soc. 1942, *64*, 2464-8.

VARIABLES:

One temperature: 37°C; one pH: 6.9

PREPARED BY:

R. Piekos

EXPERIMENTAL VALUES:

Solubility of sulfanilamide in a buffer solution prepared by mixing together 55.2 cm^3 of 1/15 M Na_2HPO_4 with 44.8 cm^3 of 1/15 M KH_2PO_4 (pH 6.9, ionic strength calculated from dissociation constants 0.03[a]) at 37.0°C is 1.44 g/100 cm^3 solution (8.36×10^{-2} mol dm^{-3}, compiler).

[a]Not specified for which reactions were the dissociation constants calculated - compiler

AUXILIARY INFORMATION

METHOD/APPARATUS/PROCEDURE:

An excess of sulfanilamide was rotated with the buffer soln usually overnight. Equilibrium was approached from above. Sampling was accomplished by forcing the soln through a filter into a pycnometer. From the pycnometer the contents were flushed into a volumetric flask. Duplicate aliquots were acidified, iced below 15°C and titrated with a 0.04 mol dm^{-3} $NaNO_2$ soln to first blue on a starch - iodide paper.

SOURCE AND PURITY OF MATERIALS:

Sulfanilamide (U.S.P.) from plant production was recrystd from alcohol and from hot water; mp 165.9°C. Titrn with nitrite indicated that the compd was 100.0±0.3% pure. Elemental analysis confirmed this value.

Source and purity of the remaining materials was not specified.

ESTIMATED ERROR:

Soly: ±0.01 g/100 g soln or $\pm 0.012 \times 10^{-3}$ in mole fraction (authors).

Temp: ±0.02°C (authors).

REFERENCES:

COMPONENTS:

(1) Benzenesulfonamide, 4-amino- (sulfanilamide) $C_6H_8N_2O_2S$; [63-74-1]

(2) Phosphoric acid, disodium salt; Na_2HPO_4; [7558-94-4]

(3) Phosphoric acid, monopotassium salt; KH_2HPO_4; [7778-77-0]

(4) Water; H_2O; [7732-18-5]

ORIGINAL MEASUREMENTS:

Langecker, H.

Arch. Exptl. Path. Pharmakol. 1948, *205*, 291-301.

VARIABLES:

pH

PREPARED BY:

R. Piekos

EXPERIMENTAL VALUES:

pH of the 1/15M phosphate buffer	Solubility at 37°C mg%	10^2 mol dm^{-3} [a]
4.9	1064	6.179
5.9	1050	6.097
6.9	1023	5.941
7.5	1167	6.777

[a] Calculated by compiler

AUXILIARY INFORMATION

METHOD/APPARATUS/PROCEDURE:

An excess of sulfanilamide was added to buffer solution and boiled for 1 h in a sealed ampul at 37°C. Sulfanilamide was assayed colorimetrically by the method of Bratton and Marshall (1) using a Havemann colorimeter (2), and by microanal detn of the solid residue.

SOURCE AND PURITY OF MATERIALS:

Source and purity of the materials was not specified.

ESTIMATED ERROR:

Nothing specified.

REFERENCES:

1. Bratton, A.G.; Marshall, E.K. *J. Biol. Chem.* 1939, *128*, 537.
2. Havemann, R. *Klin. Wochenschr.* 1940, p. 503.

COMPONENTS:	ORIGINAL MEASUREMENTS:
(1) Benzenesulfonamide, 4-amino- (sulfanilamide); $C_6H_8N_2O_2S$; [63-74-1] (2) Phosphoric acid, disodium salt; Na_2HPO_4; [7558-94-4] (3) Phosphoric acid, monopotassium salt; KH_2PO_4; [7778-77-0] (4) Water; H_2O; [7732-18-5]	Yamazaki, M.; Aoki, M.; Kamada, A; Yata, N. *Yakuzaigaku* 1967, *27(1)*, 37-40.
VARIABLES:	PREPARED BY:
One temperature: 30°C; one pH: 7.4	R. Piekos

EXPERIMENTAL VALUES:

Solubility of sulfanilamide in a phosphate buffer solution of pH 7.4 (μ = 0.17) at 30°C is 57.0 mmol/L (9.815 g dm^{-3}, compiler).

AUXILIARY INFORMATION

METHOD/APPARATUS/PROCEDURE:

Sulfanilamide (0.5 g) was placed in an L-shaped tube together with 20 ml of the buffer soln. The mixt was shaken in a thermostat until equilibrium was attained. The sulfanilamide was then assayed in the supernatant spectrophotometrically at 545 nm on a Beckmann DU spectrophotometer. The results were taken from a calibration graph.

SOURCE AND PURITY OF MATERIALS:

Nothing specified.

ESTIMATED ERROR:

Soly and pH: not specified.

Temp: $\pm$1°C (authors).

REFERENCES:

COMPONENTS:

(1) Benzenesulfonamide, 4-amino- (sulfanilamide); $C_6H_8N_2O_2S$; [63-74-1]

(2) Carbonic acid, disodium salt; Na_2CO_3; [497-19-8]

(3) Carbonic acid, monosodium salt; $NaHCO_3$; [144-55-8]

(4) Water; H_2O; [7732-18-5]

ORIGINAL MEASUREMENTS:

Takubo, T.; Matsumaru, H.; Tsuchiya, S.; Hiura, M. *Chem. Pharm. Bull.* 1973, *21(7)* 1440-5.

VARIABLES:

pH

PREPARED BY:

R. Piekos

EXPERIMENTAL VALUES:

Na_2CO_3	$NaHCO_3$	pH	Solubility at 37°C	
g/100 ml water	g/100 ml water		mg/ml soln[a]	10^2 mol dm^{-3} soln[b]
0.212	1.512	9.1	13.99	8.124
0.848	1.008	9.8	14.95	8.682
1.908	0.168	10.7	14.26	8.281

[a]Numerical values to the graphical data were given by one of the authors (S.T.) in personal communication.

[b]Calculated by compiler.

AUXILIARY INFORMATION

METHOD/APPARATUS/PROCEDURE:

Aliquots of the buffer solns were placed in glass-stoppered flasks with excess of sulfanilamide. The flasks were allowed to stand at 37$\pm$1°C and shaken vigorously for 4 h until equilibrium was established. One ml of the supernatant was removed by means of a filter pipet and the sulfanilamide was assayed by the previously reported method (1).

SOURCE AND PURITY OF MATERIALS:

The sulfanilamide was of pharmaceutical grade. The source and purity of Na_2CO_3 and $NaHCO_3$ was not specified. Distd water was used.

ESTIMATED ERROR:

Soly and pH: not specified.

Temp: $\pm$1°C (authors).

REFERENCES:

1. Takubo, T.; Tsuchiya, S.; Hiura, M. *Yakuzaigaku* 1971, *31*, 298.

COMPONENTS:	ORIGINAL MEASUREMENTS:
(1) Benzenesulfonamide, 4-amino- (sulfanilamide); $C_6H_8N_2O_2S$; [63-74-1] (2) Hydrochloric acid; HCl; [7647-14-5] (3) Sodium chloride; NaCl; [7647-14-5] (4) Water; H_2O; [7732-18-5]	Avico, U.; Cavazutti, G.; di Francesco, R.; Signoretti Ciranni, E.; Zuccaro, P. *Farmaco, Ed. Pratica* 1975, *30(1)*, 40-6.
VARIABLES:	PREPARED BY:
Temperature	R. Piekos

EXPERIMENTAL VALUES:

t/°C	Solubility of amorphous sulfanilamide in equimolal NaCl solutions containing a small excess of HCl	
	g/100 g water	10^2 mol kg^{-1} water[a]
25	4.72	2.74
35	6.50	3.78
40	7.40	4.30

[a]Calculated by compiler.

AUXILIARY INFORMATION

METHOD/APPARATUS/PROCEDURE:

A soln of sulfanilamide-HCl was added to an NaOH soln contg stoichiometric quantity of the base to neutralize the HCl salt. A small excess of HCl was used to dissolve the sulfanilamide. The neutralization was carried out in a thermostat and the pH of the mixt was maintained close to that of a satd aq sulfanilamide soln. The procedure was repeated using various initial concns of the reagents to find the max concn of sulfanilamide at which no pptn occurred.

SOURCE AND PURITY OF MATERIALS:

Source and purity of sulfanilamide was not specified. The mp of crystalline sulfanilamide was 164.5-6.5°C. Purity of the water was not specified.

ESTIMATED ERROR:

Nothing specified.

REFERENCES:

COMPONENTS:

(1) Benzenesulfonamide, 4-amino- (sulfanilamide); $C_6H_8N_2O_2S$; [63-74-1]

(2) Hydrochloric acid; HCl; [7647-01-0]

(3) Potassium chloride; KCl; [7447-40-7]

(4) Water; H_2O; [7732-18-5]

ORIGINAL MEASUREMENTS:

Kienle, R. H.; Sayward, J. M.

J. Am. Chem. Soc. 1942, *64*, 2464-8.

VARIABLES:

pH; ionic strength

PREPARED BY:

R. Piekos

EXPERIMENTAL VALUES:

pH of HCl-KCl	Ionic strength[a]	Solubility at 37.0 °C g/100 cm^3	mol dm^{-3} [b]
1.2	0.12	4.07	0.236
2.2	0.06	1.57	0.091

[a] calculated from dissociation constants.

[b] calculated by compiler

AUXILIARY INFORMATION

METHOD/APPARATUS/PROCEDURE:

An excess of sulfanilamide was rotated with a buffer soln usually overnight. Equilibrium was approached usually from above. Sampling was accomplished by forcing the soln through a filter into a pycnometer. From the pycnometer the contents were flushed into a volumetric flask. Duplicate aliquots were acidified, iced below 15 °C and titrated with a 0.04 mol dm^{-3} $NaNO_2$ soln to first blue on starch - iodide paper.

SOURCE AND PURITY OF MATERIALS:

Sulfanilamide (U.S.P.) from plant production was recrystd from alcohol and from hot water; mp 165.9 °C. Titrn with nitrite indicated that the compd was 100$\pm$0.3% pure. Elemental analysis and mixed mp detns confirmed this value. Purity of the remaining materials was not specified.

ESTIMATED ERROR:

Soly: $\pm$0.01 g/100 g soln or $\pm 0.012 \times 10^{-3}$ in mole fraction (authors).

Temp: $\pm$0.02 °C (authors).

REFERENCES:

COMPONENTS:

(1) Benzenesulfonamide, 4-amino- (sulfanilamide); $C_6H_8N_2O_2S$; [63-74-1]

(2) Boric acid, H_3BO_3; [10043-35-3]

(3) Potassium chloride; KCl; [7447-40-7]

(4) Sodium hydroxide; NaOH; [1310-73-2]

(5) Water; H_2O; [7732-18-5]

ORIGINAL MEASUREMENTS:

Kienle, R. H.; Sayward, J. M.

J. Am. Chem. Soc. 1942, *64*, 2464-8.

VARIABLES:

pH; ionic strength

PREPARED BY:

R. Piekos

EXPERIMENTAL VALUES:

pH of borate buffer	Ionic strength[a]	Solubility at 37.0°C	
		g/100 cm^3 solution	10^2 mol dm^{-3} [b]
9.4[c]	0.08	1.55	9.00
9.7[d]	0.09	1.60	9.29

[a] Calculated from dissociation constants (reactions not specified).

[b] Calculated by compiler.

[c] Obtained by mixing together 50 cm^3 of a 0.1 M solution in both H_3BO_3 and KCl with 32.1 cm^3 of 0.1 M NaOH and diluting with water up to 100 cm^3.

[c] Obtained by mixing together 50 cm^3 of a 0.1 M solution in both H_3BO_3 and KCl with 38.75 cm^3 of a 0.1 M NaOH and diluting with water up to 100 cm^3.

AUXILIARY INFORMATION

METHOD/APPARATUS/PROCEDURE:

An excess of sulfanilamide was rotated with the buffer soln usually overnight. Equilibrium was approached from above. Sampling was accomplished by forcing the soln through a filter into a pycnometer. From the pycnometer the contents were flushed into a volumetric flask. Duplicate aliquots were acidified, iced below 15°C and titrated with a 0.04 mol dm^{-3} $NaNO_2$ soln to first blue on a starch - iodide paper.

SOURCE AND PURITY OF MATERIALS:

Sulfanilamide (U.S.P.) from plant production was recrystd from alcohol and from hot water; mp 165.9°C. Titrn with nitrite indicated that the compd was 100.0±0.3% pure. Elemental analysis confirmed this value.

Source and purity of the remaining materials was not specified.

ESTIMATED ERROR:

Soly: ±0.01 g/100 g soln or $\pm 0.012 \times 10^{-3}$ in mole fraction (authors).

Temp: ±0.02°C (authors).

REFERENCES:

COMPONENTS:	ORIGINAL MEASUREMENTS:
(1) Benzenesulfonamide, 4-amino (sulfanilamide); $C_6H_8N_2O_2S$; [63-74-1] (2) Phosphoric acid, disodium salt; Na_2HPO_4; [7558-94-4] (3) 1,2,3-Propanetricarboxylic acid, 2-hydroxy- (citric acid); $C_6H_8O_7$; [77-92-9] (4) Water; H_2O; [7732-18-5]	Kienle, R. H.; Sayward, J. M. *J. Am. Chem. Soc.* 1942, *64*, 2464-8.
VARIABLES:	PREPARED BY:
One temperature: 37°C; one pH: 4.2	R. Piekos

EXPERIMENTAL VALUES:

Solubility of sulfanilamide in a solution prepared by mixing together 41.4 cm^3 of 0.2 M Na_2HPO_4 with 58.6 cm^3 of 0.1M citric acid (pH 4.2, ionic strength calculated from dissociation constants 0.84[a]) at 37.0°C is 1.40 g/100 cm^3 solution (8.13 x 10^{-2} mol dm^{-3}, compiler).

[a]Not specified for which reactions were the dissociation constants calculated - compiler.

AUXILIARY INFORMATION

METHOD/APPARATUS/PROCEDURE:

An excess of sulfanilamide was rotated with the buffer soln usually overnight. Equilibrium was approached from above. Sampling was accomplished by forcing the soln through a filter into a pycnometer. From the pycnometer the contents were flushed into a volumetric flask. Duplicate aliquots were acidified, iced below 15°C and titrated with a 0.04 mol dm^{-3} $NaNO_2$ soln to first blue on a starch - iodide paper.

SOURCE AND PURITY OF MATERIALS:

Sulfanilamide (U.S.P.) from plant production was recrystd from alcohol and from hot water; mp 165.9°C.

Titrn with nitrite indicated that the compd was 100.0±0.3% pure. Elemental analysis confirmed this value. Source and purity of the remaining materials was not specified.

ESTIMATED ERROR:

Soly: ±0.01 g/100 g soln or ±0.012 x 10^{-3} in mole fraction (authors).

Temp: ±0.02°C (authors).

REFERENCES:

COMPONENTS:

(1) Benzenesulfonamide, 4-amino- (sulfanilamide); $C_6H_8N_2O_2S$; [63-74-1]

(2) Phosphoric acid, disodium salt; Na_2HPO_4; [7558-94-4]

(3) 1,2,3-Propanetricarboxylic acid, 2-hydroxy- (citric acid); $C_6H_8O_7$; [77-92-9]

(4) Water; H_2O; [7732-18-5]

ORIGINAL MEASUREMENTS:

Likhol'ot, N. M. *Farm. Zh. (Kiev)* 1965, *20(5)*, 44-6.

VARIABLES:

pH

PREPARED BY:

R. Piekos

EXPERIMENTAL VALUES:

pH of McIlvaine's buffer solution	Solubility at 20°C	
	g/100 ml	10^2 mol dm^{-3} [a]
4.1	0.525	3.05
5.1	0.504	2.93
5.9	0.488	2.83
6.5	0.475	2.76
6.9	0.467	2.71
7.5	0.458	2.66

[a] calculated by compiler.

AUXILIARY INFORMATION

METHOD/APPARATUS/PROCEDURE:

An earlier described method was employed (1) whereby a small excess of sulfanilamide was equilibrated with the McIlvaine's buffer soln for 8 h in a 50-ml test tube. Aliquots were removed through a filter and sulfanilamide was assayed bromatometrically.

SOURCE AND PURITY OF MATERIALS:

Sulfanilamide; not specified.
McIlvaine's buffer solns were prepd from a 0.2M Na_2HPO_4 and a 0.1M citric acid solns. Source and purity of the buffer components were not specified.

ESTIMATED ERROR:

Soly: not specified.
Temp: ±0.1°C (authors).
pH: not specified.

REFERENCES:

1. Gusyakov, V. P.; Likhol'ot, N. M. *Farm. Zh. (Kiev)* 1960, *15(8)*, 21.

COMPONENTS:	ORIGINAL MEASUREMENTS:
(1) Benzenesulfonamide, 4-amino- (sulfanilamide); $C_6H_8N_2O_2S$; [63-74-1] (2) Phosphoric acid, disodium salt; Na_2HPO_4; [7558-94-4] (3) 1,2,3-Propanetricarboxylic acid, 2-hydroxy- (citric acid); $C_6H_8O_7$; [77-92-9] (4) Water; H_2O; [7732-18-5]	Takubo, T.; Matsumaru, H.; Tsuchiya, S.; Hiura, M. *Chem. Pharm. Bull.* 1973, *21(7)* 1440-5.
VARIABLES:	PREPARED BY:
pH	R. Piekos

EXPERIMENTAL VALUES:

Citric acid	Na_2HPO_4	pH	Solubility at 37°C	
g/100 ml water	g/100 ml water		mg/ml soln[a]	10^2mol dm^{-3} soln[b]
1.680	0.572	3.1	16.03	9.309
1.260	1.144	4.2	14.40	8.362
0.840	1.716	5.8	13.00	7.549
0.420	2.228	6.8	12.60	7.317

[a]Numerical values to the graphical data were given by one of the authors (S.T.) in personal communication.

[b]Calculated by compiler.

AUXILIARY INFORMATION

METHOD/APPARATUS/PROCEDURE:

Aliquots of the buffer solns were placed in glass-stoppered flasks with excess of sulfanilamide. The flasks were allowed to stand at 37±1°C and shaken vigorously for 4 h until equilibrium was established. One ml of the supernatant was removed by means of a filter pipet and the sulfanilamide was assayed by the previously reported method (1).

SOURCE AND PURITY OF MATERIALS:

The sulfanilamide was of pharmaceutical grade. The source and purity of citric acid and Na_2HPO_4 was not specified. Distd water was used.

ESTIMATED ERROR:

Soly and pH: not specified.

Temp: ±1°C (authors).

REFERENCES:

1. Takubo, T.; Tsuchiya, S.; Hiura, M. *Yakuzaigaku* 1971, *31*, 298.

COMPONENTS:	ORIGINAL MEASUREMENTS:
(1) Benzenesulfonamide, 4-amino- (sulfanilamide); $C_6H_8N_2O_2S$; [63-74-1] (2) 1,2-Benzenedicarboxylic acid, monopotassium salt; $C_8H_5KO_4$; [877-24-7] (3) Hydrochloric acid; HCl; [7647-01-0] (4) Water; H_2O; [7732-18-5]	Kienle, R. H.; Sayward, J. M. *J. Am. Chem. Soc.* 1942, *64*, 2464-8.
VARIABLES:	PREPARED BY:
One temperature: 37.0°C; one pH: 2.2	R. Piekos

EXPERIMENTAL VALUES:

Solubility of sulfanilamide in a buffer solution prepared by mixing together 50 cm^3 of 0.1M monopotassium 1,2-benzenedicarboxylate with 49.5 cm^3 of 0.1M HCl and diluting up to 100 cm^3 with water (pH 2.2, ionic strength calculated from dissociation constants 0.06[a]) at 37.0°C is 1.79 g/100 cm^3 solution (0.104 mol dm^{-3}, compiler).

[a]Not specified for which reactions were the dissociation constants calculated - compiler.

AUXILIARY INFORMATION

METHOD/APPARATUS/PROCEDURE:	SOURCE AND PURITY OF MATERIALS:
An excess of sulfanilamide was rotated with the buffer soln usually overnight. Equilibrium was approached from above. Sampling was accomplished by forcing the soln through a filter into a pycnometer. From the pycnometer the contents were flushed into a volumetric flask. Duplicate aliquots were acidified, iced below 15°C and titrated with a 0.04 mol dm^{-3} $NaNO_2$ soln to first blue on a starch - iodide paper.	Sulfanilamide (U.S.P.) from plant production was recrystd from alcohol and from hot water; mp 165.9°C. Titrn with nitrite indicated that the compd was 100±0.3% pure. Elemental analysis confirmed this value. Purity of the remaining materials was not specified. ESTIMATED ERROR: Soly: ±0.01 g/100 g soln or ±0.012 x 10^{-3} in mole fraction (authors); Temp: ±0.02°C (authors). REFERENCES:

COMPONENTS:	ORIGINAL MEASUREMENTS:
(1) Benzenesulfonamide, 4-amino- (sulfanilamide); $C_6H_8N_2O_2S$; [63-74-1] (2) Ethanamine, N-ethyl-, (diethylamine); $C_4H_{11}N$; [109-89-7] (3) Sodium chloride; NaCl; [7647-14-5] (4) Water; H_2O; [7732-18-5]	Meyer, E. J. E., *Pharm. Weekblad* 1939, *76*, 977-9.

VARIABLES:	PREPARED BY:
One temperature: 20°C	R. Piekos

EXPERIMENTAL VALUES:

Solubility of sulfanilamide in a 6.5% diethylamine solution in physiological saline (0.9% aqueous NaCl) solution at 20°C is 6% (0.35 mol kg^{-1} solution, compiler).

[diethylamine] = 0.89 mol kg^{-1}, compiler
[NaCl] = 0.15 mol kg^{-1}, compiler

AUXILIARY INFORMATION

METHOD/APPARATUS/PROCEDURE:	SOURCE AND PURITY OF MATERIALS:
Nothing specified.	Nothing specified.
	ESTIMATED ERROR:
	Nothing specified.
	REFERENCES:

COMPONENTS:

(1) Benzenesulfonamide, 4-amino- (sulfanilamide) $C_6H_8N_2O_2S$; [63-74-1]

(2) Formic acid, sodium salt; $CHNaO_2$; [141-53-7]

(3) Water; H_2O; [7732-18-5]

ORIGINAL MEASUREMENTS:

Likhol'ot, N. M.; Gusyakov, V. P.

Farm. Zh. (Kiev) 1964, *19(1)*, 52-5.

VARIABLES:

Concentration of $CHNaO_2$

PREPARED BY:

R. Piekos

EXPERIMENTAL VALUES:

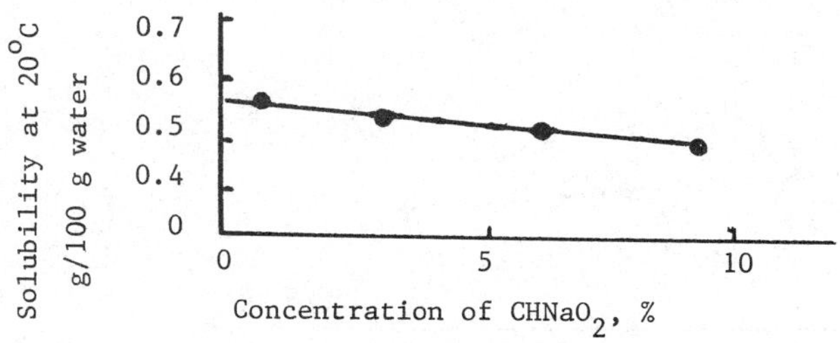

The solubility in a 1 molal (mol kg^{-1} water, compiler) $CHNaO_2$ solution at 20°C is 0.497 g/100 g water (2.89×10^{-2} mol kg^{-1} water, compiler).

AUXILIARY INFORMATION

METHOD/APPARATUS/PROCEDURE:

The previously described method (1) was employed whereby a small excess of sulfanilamide was equilibrated in a 50-ml test tube with 20 ml of a $CHNaO_2$ soln for 8 h. Aliquots of the satd soln were removed through a filter and sulfanilamide was assayed colorimetrically (2).

SOURCE AND PURITY OF MATERIALS:

Sulfanilamide (source not specified) was recrystd from water.

$CHNaO_2$ (source not specified) was also recrystd from water.

Purity of the water was not specified.

ESTIMATED ERROR:

Nothing specified.

REFERENCES:

1. Gusyakov, V. P.; Likhol'ot, N. M. *Farm. Zh. (Kiev)* 1960, *15(3)*, 21.
2. Predtechenskii, B. E.; Borovskaya, V.M.; Margolina, L. T., *Laboratornye metody issledovanya, Medgiz, Moscow*, 1950, p. 371.

COMPONENTS:	ORIGINAL MEASUREMENTS:
(1) Benzenesulfonamide, 4-amino- (sulfanilamide); $C_6H_8N_2O_2S$; [63-74-1] (2) Acetic acid, sodium salt (Na acetate); $C_2H_3NaO_2$; [127-09-3] (3) Water; H_2O; [7732-18-5]	Likhol'ot, N. M.; Gusyakov, V. P. *Farm. Zh. (Kiev)* 1964, *19(1)*, 52-5.
VARIABLES:	PREPARED BY:
Concentration of Na acetate	R. Piekos

EXPERIMENTAL VALUES:

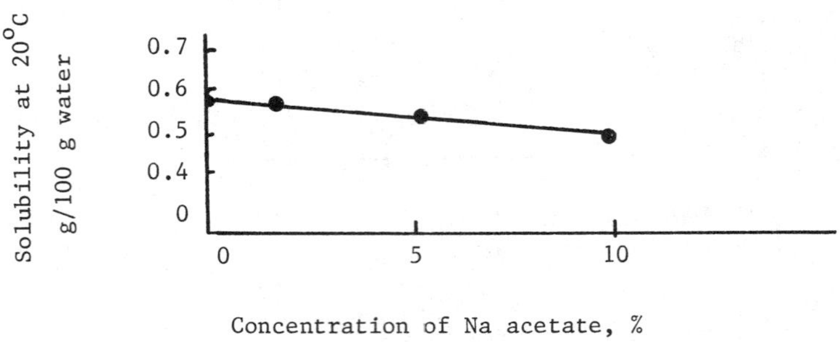

The solubility in a 1 molal (mol kg^{-1} water, compiler) Na acetate solution at 20°C is 0.511 g/100 g water (2.97×10^{-2} mol kg^{-1} water, compiler).

AUXILIARY INFORMATION

METHOD/APPARATUS/PROCEDURE:

The previously described method (1) was employed whereby a small excess of sulfanilamide was equilibrated in a 50-ml test tube with 20 ml of a Na acetate soln for 8 h. Aliquots of the satd soln were removed through a filter and sulfanilamide was assayed bromatometrically (2).

SOURCE AND PURITY OF MATERIALS:

Sulfanilamide (source not specified) was recrystd from water. Na acetate (source not specified) was also recrystd from water.

Purity of the water was not specified.

ESTIMATED ERROR:

Nothing specified.

REFERENCES:

1. Gusyakov, V. P.; Likhol'ot, N. M. *Farm. Zh. (Kiev)* 1960, *15(3)*, 21.
2. *The Extra Pharmacopeia (Martindale)* 1955, *2(23)*, 353 and 389.

COMPONENTS:

(1) Benzenesulfonamide, 4-amino- (sulfanilamide); $C_6H_8N_2O_2S$; [63-74-1]

(2) Propanoic acid, sodium salt (Na propionate); $C_3H_5NaO_2$; [137-40-6]

(3) Water; H_2O; [7732-18-5]

ORIGINAL MEASUREMENTS:

Likhol'ot, N. M.; Gusyakov, V. P.

Farm. Zh. (Kiev) 1964, *19(1)* 52-5.

VARIABLES:

Concentration of Na propionate

PREPARED BY:

R. Piekos

EXPERIMENTAL VALUES:

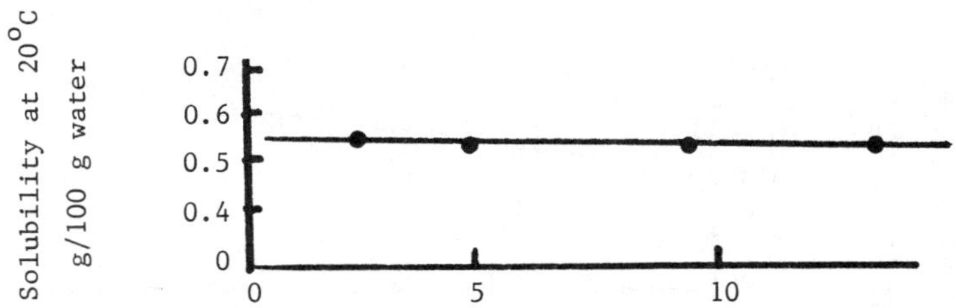

The solubility in a 1 molal (mol kg^{-1} water, compiler) Na propionate solution at 20°C is 0.541 g/100 g water (3.14 x 10^{-2} mol kg^{-1} water, compiler).

AUXILIARY INFORMATION

METHOD/APPARATUS/PROCEDURE:

The previously described method (1) was used whereby a small excess of sulfanilamide was equilibrated in a 50-ml test tube with 20 ml of a Na propionate soln for 8 h. Aliquots of the satd soln were removed through a filter and sulfanilamide was assayed bromatometrically (2).

SOURCE AND PURITY OF MATERIALS:

Sulfanilamide was recrystd from water. Its source was not specified. Na propionate was prepd by neutralization of propionic acid (source not specified) with Na_2CO_3 or NaOH.

Purity of the water was not specified.

ESTIMATED ERROR:

Nothing specified.

REFERENCES:

1. Gusyakov, V. P.; Likhol'ot, N. M. *Farm. Zh. (Kiev)* 1960, *15(3)* 21.
2. *The Extra Pharmacopeia (Martindale)* 1955, *2(23)*, 353 and 389

COMPONENTS:	ORIGINAL MEASUREMENTS:
(1) Benzenesulfonamide, 4-amino- (sulfanilamide); $C_6H_8N_2O_2S$; [63-74-1] (2) Pentanoic acid, sodium salt (Na valerate); $C_5H_9NaO_2$;[6106-41-8] (3) Water; H_2O; [7732-18-5]	Likhol'ot, N. M.; Gusyakov, V. P. *Farm. Zh. (Kiev)* 1964, *19(1)* 52-5.
VARIABLES:	PREPARED BY:
Concentration of Na valerate	R. Piekos

EXPERIMENTAL VALUES:

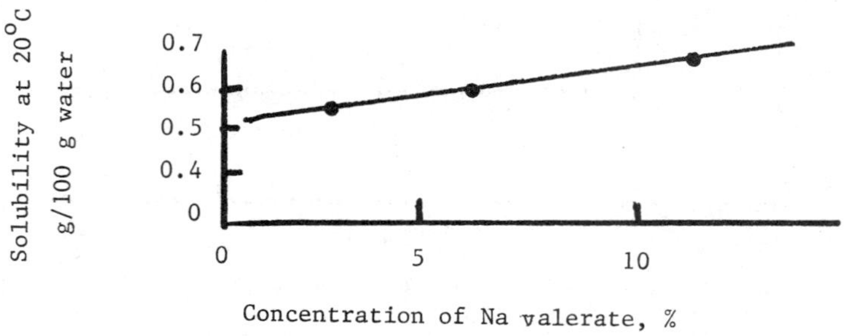

The solubility in a 1 molal (mol kg^{-1} water, compiler) Na valerate solution at 20°C is 0.678 g/100 g water (3.94×10^{-2} mol kg^{-1} water, compiler).

AUXILIARY INFORMATION

METHOD/APPARATUS/PROCEDURE:

The previously described method (1) was used whereby a small excess of sulfanilamide was equilibrated in a 50-ml test tube with 20 ml of a Na valerate soln for 8 h. Aliquots of the satd soln were removed through a filter and sulfanilamide was assayed bromatometrically (2).

SOURCE AND PURITY OF MATERIALS:

Sulfanilamide was recrystd from water. Na valerate was prepd by neutralization of valeric acid with Na_2CO_3 or NaOH. The source and purity of the materials was not specified.

ESTIMATED ERROR:

Nothing specified.

REFERENCES:

1. Gusyakov, V. P.; Likhol'ot, N. M. *Farm. Zh. (Kiev)* 1960, *15(3)*, 21.

The Extra Pharmacopeia (Martindale) 1955, *2(23)*, 353 and 389.

COMPONENTS:

(1) Benzenesulfonamide, 4-amino- sulfanilamide); $C_6H_8N_2O_2S$; [63-74-1]

(2) Hexanoic acid, sodium salt (Na caproate); $C_6H_{11}NaO_2$; [10051-44-2]

(3) Water; H_2O; [7732-18-5]

ORIGINAL MEASUREMENTS:

Likhol'ot, N. M.; Gusyakov, V. P.

Farm. Zh. (Kiev) 1964, *19(1)* 52-5.

VARIABLES:

Concentration of Na caproate

PREPARED BY:

R. Piekos

EXPERIMENTAL VALUES:

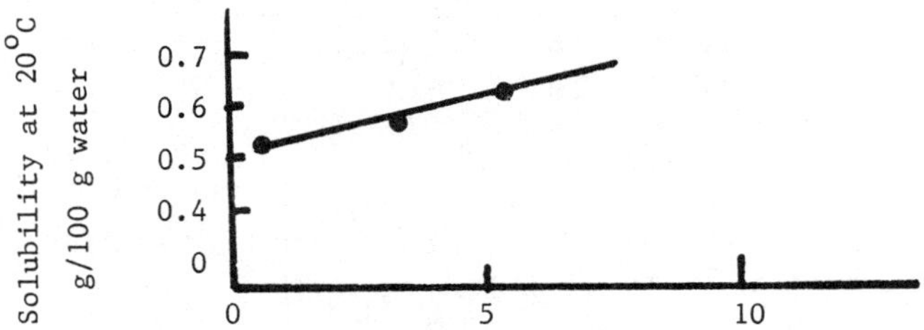

Concentration of Na caproate, %

The solubility in a 0.4 molal (mol kg^{-1} water, compiler) Na caproate solution at 20°C is 0.651 g/100 g water (3.78×10^{-2} mol/kg water, compiler).

AUXILIARY INFORMATION

METHOD/APPARATUS/PROCEDURE:

The previously described method (1) was used whereby a small excess of sulfanilamide was equilibrated in a 50-ml test tube with 20 ml of a Na caproate soln for 8 h. Aliquots of the satd soln were removed through a filter and sulfanilamide was assayed bromatometrically (2).

SOURCE AND PURITY OF MATERIALS:

Sulfanilamide was recrystd from water.

Na caproate was prepd by neutralization of caproic acid with Na_2CO_3 or NaOH.

The source and purity of the materials was not specified.

ESTIMATED ERROR:

Nothing specified.

REFERENCES:

1. Gusyakov, V. P.; Likhol'ot, N. M. *Farm. Zh. (Kiev)* 1960, *15(3)*, 21.

2. *The Extra Pharmacopeia (Martindale)* 1955, *2(23)* 353 and 389.

COMPONENTS:	ORIGINAL MEASUREMENTS:
(1) Benzenesulfonamide, 4-amino- (sulfanilamide); $C_6H_8N_2O_2S$; [63-74-1] (2) Hexanoic acid, sodium salt, (Na caproate); $C_6H_{11}NaO_2$; [10051-44-2] (3) Water; H_2O; [7732-18-5]	Likhol'ot, N. M.; Gusyakov, V. P., *Farm. Zh. (Kiev)* 1964, *19(1)*, 52-5.
VARIABLES:	**PREPARED BY:**
One temperature: 20°C	R. Piekos

EXPERIMENTAL VALUES:

Solubility of sulfanilamide in a 0.4 mol kg^{-1} (molal) sodium caproate solution at 20°C is 0.651 g/100 g water (3.78×10^{-2} mol kg^{-1} water, compiler).

AUXILIARY INFORMATION

METHOD/APPARATUS/PROCEDURE:

A small excess of sulfanilamide was equilibrated for 8 hr in a 50-ml open test tube with 20 ml of Na caproate soln. Aliquots were removed through a filter, and sulfanilamide was assayed bromatometrically (1).

SOURCE AND PURITY OF MATERIALS:

Sulfanilamide (source not specified) was recrystd from water (purity not specified). Na caproate was prepd by neutralization of caproic acid (source and purity not specified) with Na carbonate or hydroxide (source and purity not specified).

ESTIMATED ERROR:

Nothing specified.

REFERENCES:

1. *The Extra Pharmacopeia (Martindale)* 1955, *2(23)*, 353 and 389.

COMPONENTS:	ORIGINAL MEASUREMENTS:
(1) Benzenesulfonamide, 4-amino- (sulfanilamide); $C_6H_8N_2O_2S$; [63-74-1] (2) Aminoacetic acid (glycine); $C_2H_5NO_2$; [56-40-6] (3) Water; H_2O; [7732-18-5]	Kienle, R. H.; Sayward, J. M. *J. Am. Chem. Soc.* 1942, *64*, 2464-8.
VARIABLES:	PREPARED BY:
One temperature: 37.0°C; one pH: 11.8	R. Piekos

EXPERIMENTAL VALUES:

Solubility of sulfanilamide in an aqueous solution of glycine (pH 11.8, ionic strength calculated from dissociation constants 0.11[a]) at 37.0°C is 1.93 g/100 cm^3 solution (0.112 mol dm^{-3}, compiler).

[a]Not specified for which reactions were the dissociation constants calculated - compiler.

AUXILIARY INFORMATION

METHOD/APPARATUS/PROCEDURE:	SOURCE AND PURITY OF MATERIALS:
An excess of sulfanilamide was rotated with the glycine soln usually overnight. Equilibrium was approached from above. Sampling was accomplished by forcing the soln through a filter into a pycnometer. From the pycnometer the contents were flushed into a volumetric flask. Duplicate aliquots were acidified, iced below 15°C and titrated with a 0.04 mol dm^{-3} $NaNO_2$ soln to first blue on a starch - iodide paper.	Sulfanilamide (U.S.P.) from plant production was recrystd from alcohol and from hot water; mp 165.9°C. Titrn with nitrite indicated that the compd was 100.0 ±0.3% pure. Elemental analysis confirmed this value. Source and purity of the remaining materials was not specified.
	ESTIMATED ERROR: Soly: ±0.01 g/100 g soln or ±0.012 x 10^{-3} in mole fraction (authors). Temp: ±0.02°C (authors).
	REFERENCES:

COMPONENTS:	ORIGINAL MEASUREMENTS:
(1) Benzenesulfonamide, 4-amino- (sulfanilamide); $C_6H_8N_2O_2S$; [63-74-1] (2) Pentanoic acid, sodium salt, (Na valerate); $C_5H_9NaO_2$; [6106-41-8] (3) Water; H_2O; [7732-18-5]	Likhol'ot, N. M.; Gusyakov, V. P., *Farm. Zh. (Kiev)* 1964, *19(1)*, 52-5.
VARIABLES:	PREPARED BY:
One temperature: 20°C	R. Piekos

EXPERIMENTAL VALUES:

Solubility of sulfanilamide in a 1 mol kg^{-1} (molal) sodium valerate solution at 20°C is 0.678 g/100 g water (3.94 x 10^{-2} mol kg^{-1} water, compiler).

AUXILIARY INFORMATION

METHOD/APPARATUS/PROCEDURE:	SOURCE AND PURITY OF MATERIALS:
A small excess of sulfanilamide was equilibrated for 8 hr in a 50-ml test tube with 20 ml of Na valerate soln. Aliquots were taken through a filter and sulfanilamide was assayed bromatometrically (1).	Sulfanilamide (source not specified) was recrystd from water (purity not specified). Na valerate was prepd by neutralization of valeric acid (source and purity not specified) with Na carbonate or hydroxide (source and purity not specified).
	ESTIMATED ERROR:
	Nothing specified.
	REFERENCES:
	1. *The Extra Pharmacopeia (Martindale)* 1955, *2(23)*, 353 and 389.

COMPONENTS:	ORIGINAL MEASUREMENTS:
(1) Benzenesulfonamide, 4-amino- (sulfanilamide); $C_6H_8N_2O_2S$; [63-74-1] (2) Propanoic acid, 2-hydroxy-, monosodium salt (Na hydroxypropionate) $C_3H_5NaO_3$; [72-17-3] (3) Water; H_2O; [7732-18-5]	Likhol'ot, N. M.; Gusyakov, V. P. *Farm. Zh. (Kiev)* 1964, *19(1)*, 52-5.
VARIABLES:	PREPARED BY:
Concentration of Na hydroxypropionate	R. Piekos

EXPERIMENTAL VALUES:

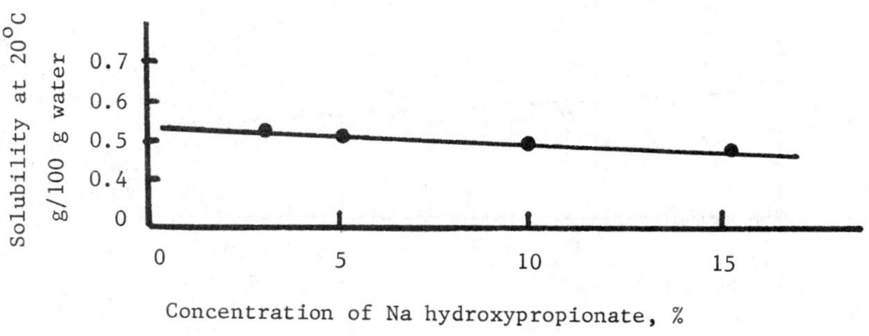

The solubility in a 1 molal (mol kg^{-1} water, compiler) Na hydroxypropionate solution at 20°C is 0.493 g/100 g water (2.86×10^{-2} mol kg^{-1} water, compiler).

AUXILIARY INFORMATION

METHOD/APPARATUS/PROCEDURE:

The previously described method (1) was used whereby a small excess of sulfanilamide was equilibrated in a 50-ml test tube with 20 ml of a Na hydroxypropionate soln for 8 h. Aliquots of the satd soln were removed through a filter and sulfanilamide was assayed bromatometrically (2).

SOURCE AND PURITY OF MATERIALS:

Sulfanilamide was recrystd from water. Its source was not specified. The soln of Na 2-hydroxypropionate was of the Czechoslovak origin (purity not specified).

Purity of the water was not specified.

ESTIMATED ERROR:

Nothing specified.

REFERENCES:

1. Gusyakov, V. P.; Likhol'ot, N. M. *Farm. Zh. (Kiev)* 1960, *15(3)* 21.
2. *The Extra Pharmacopoia (Martindale)* 1955, *2(23)*, 353 and 389.

COMPONENTS:	ORIGINAL MEASUREMENTS:
(1) Benezenesulfonamide, 4-amino- (sulfanilamide); $C_6H_8N_2O_2S$; [63-74-1] (2) Propanoic acid, 2-amino-, monosodium salt (Na aminopropionate); $C_3H_6NNaO_2$; [23388-69-4] (3) Water; H_2O; [7732-18-5]	Likhol'ot, N. M.; Gusyakov, V. P. *Farm. Zh. (Kiev)* 1964, *19(1)* 52-5.
VARIABLES:	PREPARED BY:
Concentration of Na aminopropionate	R. Piekos

EXPERIMENTAL VALUES:

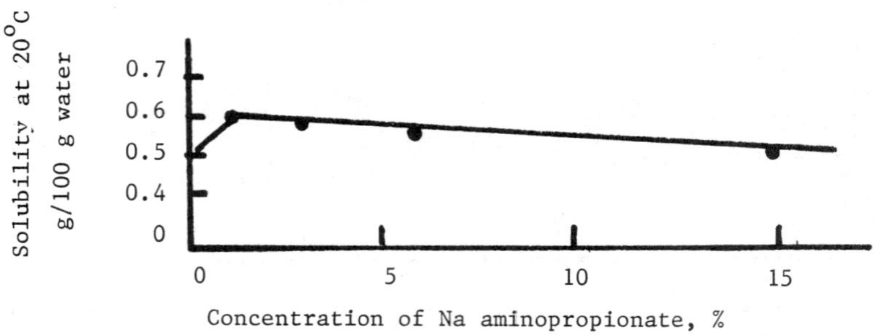

The solubility in a 1 molal (mol kg^{-1} water, compiler) Na aminopropionate solution at 20°C is 0.533 g/100 g water (3.10×10^{-2} mol kg^{-1} water, compiler).

AUXILIARY INFORMATION

METHOD/APPARATUS/PROCEDURE:

The previously described method was used (1) whereby a small excess of sulfanilamide was equilibrated in a 50-ml test tube with 20 ml of a Na aminopropionate soln for 8 h. Aliquots of the satd soln were removed through a filter and sulfanilamide was assayed bromatometrically (2).

SOURCE AND PURITY OF MATERIALS:

Sulfanilamide was recrystd from water. Na aminopropionate was prepd by neutralization of 2-aminopropionic acid with Na_2CO_3 or NaOH. The source and purity of the materials was not specified.

ESTIMATED ERROR:

Nothing specified.

REFERENCES:

1. Gusyakov, V. P.; Likhol'ot, N. M. *Farm. Zh. (Kiev)* 1960, *15(3)* 21.
2. *The Extra Pharmacopeia (Martindale)* 1955, *2(23)*, 353 and 389.

COMPONENTS:	ORIGINAL MEASUREMENTS:
(1) Benzenesulfonamide, 4-amino- (sulfanilamide); $C_6H_8N_2O_2S$; [63-74-1] (2) Butanedioic acid, 2,3-dihydroxy- disodium salt (di-Na tartrate); $C_4H_4Na_2O_6$; [868-18-8] (3) Water; H_2O; [7732-18-5]	Likhol'ot, N. M.; Gusyakov, V. P. *Farm. Zh. (Kiev)* 1964, *19(1)*, 52-5.
VARIABLES:	PREPARED BY:
Concentration of di-Na tartrate	R. Piekos

EXPERIMENTAL VALUES:

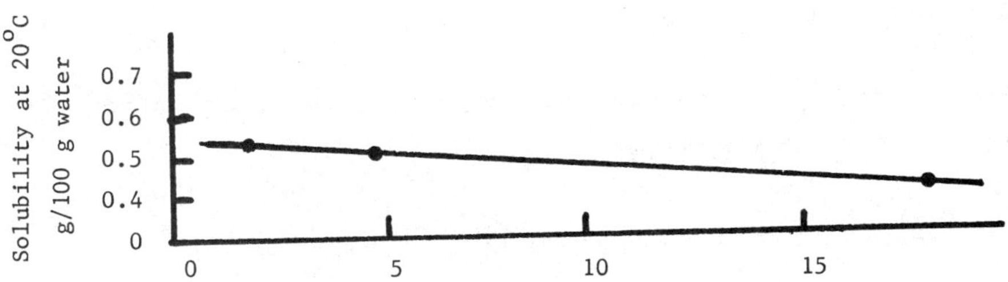

Concentration of di-Na tartrate, %

The solubility in a 1 molal (mol kg^{-1} water, compiler) di-Na tartrate solution at 20°C is 0.426 g/100 g water (2.47×10^{-2} mol kg^{-1} water, compiler).

AUXILIARY INFORMATION

METHOD/APPARATUS/PROCEDURE:	SOURCE AND PURITY OF MATERIALS:
The previously described method (1) was used whereby a small excess of sulfanilimide was equilibrated in a 50-ml test tube with 20 ml of a di-Na tartrate soln for 8 h. Aliquots of the satd soln were removed through a filter and sulfanilamide was assayed bromatometrically (2).	Sulfanilamide and di-Na tartrate (source not specified) were purified by recrystn from water. Purity of the water was not specified.
	ESTIMATED ERROR:
	Nothing specified.
	REFERENCES:
	1. Gusyakov, V. P.; Likhol'ot, N. M. *Farm. Zh. (Kiev)* 1960, *15(3)*, 21. *The Extra Pharmacopeia (Martindale)* 1955, *2(23)*, 353 and 389.

COMPONENTS:

(1) Benzenesulfonamide, 4-amino- (sulfanilamide); $C_6H_8N_2O_2S$; [63-74-1]

(2) 1,2,3-Propanetricarboxylic acid, 2-hydroxy- (citric acid); $C_6H_8O_7$; [77-92-9]

(3) Water; H_2O; [7732-18-5]

ORIGINAL MEASUREMENTS:

Takubo, T.; Matsumaru, H.; Tsuchiya, S.; Hiura, M. *Chem. Pharm. Bull.* 1973, *21(7)*, 1440-5.

VARIABLES:

One temperature: 37°C; one pH: 2.1

PREPARED BY:

R. Piekos

EXPERIMENTAL VALUES:

Solubility of sulfanilamide in a citric acid solution (2.100 g citric acid per 100 ml water) of pH 2.1 at 37°C is 19.24 mg/ml solution[a] (0.1117 mol dm^{-3} solution, compiler).

[a]Numerical value to the graphical one was given by one of the authors (S.T.) in personal communication.

AUXILIARY INFORMATION

METHOD/APPARATUS/PROCEDURE:

Aliquots of the citric acid soln were placed in glass-stoppered flasks with excess of sulfanilamide. The flasks were allowed to stand at 37±1°C and shaken vigorously for 4 h until equilibrium was established. One ml of the supernatant was removed by means of a filter pipet and the sulfanilamide was assayed by the previously reported method (1).

SOURCE AND PURITY OF MATERIALS:

The sulfanilamide was of pharmaceutical grade. Source and purity of the citric acid was not specified.

Distd water was used.

ESTIMATED ERROR:

Soly: not specified.
pH: not specified.
Temp: ±1°C (authors).

REFERENCES:

1. Takubo, T.; Tsuchiya, S.; Hiura, M. *Yakuzaigaku* 1971, *31*, 298.

COMPONENTS:

(1) Benzenesulfonamide, 4-amino- (sulfanilamide); $C_6H_8N_2O_2S$; [63-74-1]

(2) 1,2,3-Propane tricarboxylic acid, 2-hydroxy-, trisodium salt: (tri-Na citrate); $C_6H_5Na_3O_7$; [68-04-2]

(3) Water; H_2O; [7732-18-5]

ORIGINAL MEASUREMENTS:

Likhol'ot, N. M.; Gusyakov, V. P.
Farm. Zh. (Kiev) 1964, *19(1)*, 52-5.

VARIABLES:

Concentration of tri-Na citrate

PREPARED BY:

R. Piekos

EXPERIMENTAL VALUES:

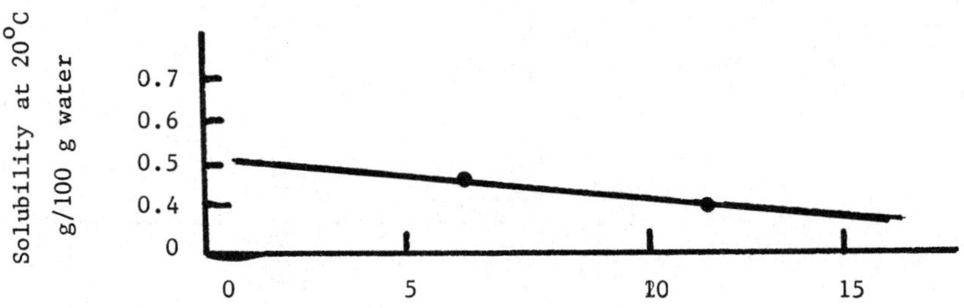

The solubility in a 1 molal (mol kg^{-1} water, compiler) tri-Na citrate solution at 20°C is 0.281 g/100 g water (1.63×10^{-2} mol kg^{-1} water, compiler).

AUXILIARY INFORMATION

METHOD/APPARATUS/PROCEDURE:

The previously described method (1) was used whereby a small excess of sulfanilamide was equilibrated in a 50-ml test tube with 20 ml of a tri-Na citrate soln for 8 h. Aliquots of the satd soln were removed through a filter and assayed bromatrometrically (2).

SOURCE AND PURITY OF MATERIALS:

Sulfanilamide and tri-Na citrate (source not specified) were purified by crystn from water.

Purity of the water was not specified.

ESTIMATED ERROR:

Nothing specified.

REFERENCES:

1. Gusyakov, V. P.; Likhol'ot, N. M. *Farm. Zh. (Kiev)* 1960. *15(3)*, 21.
2. *The Extra Pharmacopeia (Martindale)* 1955, *2(23)*, 353 and 389.

COMPONENTS:	ORIGINAL MEASUREMENTS:
(1) Benzenesulfonamide, 4-amino- (sulfanilamide); $C_6H_8N_2O_2S$; [63-74-1] (2) Benzoic acid, sodium salt; $C_7H_5NaO_2$; [532-32-1] (3) Water; H_2O; [7732-18-5]	Gusyakov, V. P.; Sukmans'ka, I. V. *Farm. Zh. (Kiev)* 1960, *15(1)*, 20-23.
VARIABLES:	PREPARED BY:
One temperature: 20°C	R. Piekos

EXPERIMENTAL VALUES:

Solubility of sulfanilamide in a 1 mol kg^{-1} water sodium benzoate solution at 20°C is 0.94 g/100 ml solution (5.5 x 10^{-2} mol dm^{-3}, compiler).

AUXILIARY INFORMATION

METHOD/APPARATUS/PROCEDURE:

To 50-ml tightly stoppered test tubes contg 25 ml of a 1 mol kg^{-1} water Na benzoate soln, placed in a thermostat, accurately weighed 0.02-0.002-g portions of sulfanilamde were added under agitation until satn was attained.

SOURCE AND PURITY OF MATERIALS:

Both sulfanilamide and Na benzoate conformed to the requirements of the State Pharmacopeia VIII.

Distilled water was used.

ESTIMATED ERROR:

Temp: ±0.1°C (authors).

Soly: the accuracy of the detn of the concn was similar to that attained by volumetric method (authors).

REFERENCES:

COMPONENTS:

(1) Benzenesulfonamide, 4-amino- (sulfanilamide); $C_6H_8N_2O_2S$; [63-74-1]

(2) Benzoic acid, sodium salt; $C_7H_5NaO_2$; [532-32-1]

(3) Water; H_2O; [7732-18-5]

ORIGINAL MEASUREMENTS:

Kňažko, L. *Farm. Obzor* 1966, *35*, 298-311.

VARIABLES:

One temperature: 20°C

PREPARED BY:

R. Piekos

EXPERIMENTAL VALUES:

Solubility of sulfanilamide in a 0.5 mol dm^{-3} sodium benzoate solution at 20°C is 0.81 g/100 ml solution (4.7 x 10^{-2} mol dm^{-3}, compiler) or 0.79/100 g solution.

AUXILIARY INFORMATION

METHOD/APPARATUS/PROCEDURE:

Sulfanilamide was added in small portions to a known volume of Na benzoate soln until reaching satn. The equilibration time was 3-4 h under stirring. The temp was held const by means of the Höppler ultrathermostat.

SOURCE AND PURITY OF MATERIALS:

Sulfanilamide and Na benzoate conformed to the requirements of the Czechoslovak Pharmacopeia 2, Suppl. 1959.
Purity of the water was not specified.

ESTIMATED ERROR:

Nothing specified.

REFERENCES:

COMPONENTS:

(1) Benzenesulfonamide, 4-amino- (sulfanilamide); $C_6H_8N_2O_2S$; [63-74-1]

(2) Benzoic acid, 2-hydroxy-, monosodium salt (Na salicylate); $C_7H_5NaO_3$; [54-21-7]

(3) Water; H_2O; [7732-18-5]

ORIGINAL MEASUREMENTS:

Kňažko, L. *Farm. Obzor* 1966, *35*, 298-311.

VARIABLES:

Concentration of Na salicylate

PREPARED BY:

R. Piekos

EXPERIMENTAL VALUES:

Concentration of Na salicylate ($mol\ dm^{-3}$)	Solubility at 20 °C		
	g/100 ml	g/100 g	10^2 mol dm^{-3} [a]
0.5	1.05	1.02	6.10
1	1.50	1.415	8.71
1.5	1.83	1.674	10.63
2	2.79	2.48	16.20
2.5	3.78	3.27	21.95
3	4.238	3.57	24.61

[a] of Na salicylate solution, calculated by compiler.

AUXILIARY INFORMATION

METHOD/APPARATUS/PROCEDURE:

Sulfanilamide was added in small portions to a known volume of Na salicylate solns until reaching satn. The equilibration time was 3-4 h under stirring. The temp was held const by means of the Höppler ultrathermostat.

SOURCE AND PURITY OF MATERIALS:

Sulfanilamide and Na salicylate conformed to the requirements of the Czechoslovak Pharmacopeia 2, Suppl. 1959.
Purity of the water was not specified.

ESTIMATED ERROR:

Nothing specified.

REFERENCES:

COMPONENTS:

(1) Benzenesulfonamide, 4-amino- (sulfanilamide); $C_6H_8N_2O_2S$; [63-74-1]

(2) Benzoic acid, 2-hydroxy-, monosodium salt (Na salicylate); $C_7H_5NaO_3$; [54-21-7]

(3) Water; H_2O; [7732-18-5]

ORIGINAL MEASUREMENTS:

Gusyakov, V. P.; Sukmans'ka, I. V.

Farm. Zh. (Kiev) 1960, *15(1)*, 20-23.

VARIABLES:

One temperature: 20°C

PREPARED BY:

R. Piekos

EXPERIMENTAL VALUES:

Solubility of sulfanilamide in a 1 mol kg^{-1} water Na salicylate solution at 20°C is 1.29 g/100 ml solution (7.49×10^{-2} mol dm^{-3}, compiler).

AUXILIARY INFORMATION

METHOD/APPARATUS/PROCEDURE:

To 50-ml tightly stoppered test tubes contg 25 ml of a 1 mol kg^{-1} water Na salicylate soln, placed in a thermostat, accurately weighed 0.02-0.002-g portions of sulfanilamide were added under agitation until satn was attained.

SOURCE AND PURITY OF MATERIALS:

Both sulfanilamide and Na salicylate conformed to the requirements of the State Pharmacopeia VIII.

Distilled water was used.

ESTIMATED ERROR:

Temp: ± 0.1°C (authors).

Soly: the accuracy of the detn of the concn was similar to that attained by volumetric method (authors).

REFERENCES:

COMPONENTS:	ORIGINAL MEASUREMENTS:
(1) Benzenesulfonamide, 4-amino- (sulfanilamide); $C_6H_8N_2O_2S$; [63-74-1] (2) Benzoic acid, 2-hydroxy-, monosodium salt; $C_7H_5NaO_3$; [54-21-7] (3) Water; H_2O; [7732-18-5]	Gusyakov, V. P.; Sukmans'ka, I. V. *Farm. Zhur. (Kiev)* 1961, *16*, 25-8.
VARIABLES:	PREPARED BY:
One temperature: 20°C	R. Piekos

EXPERIMENTAL VALUES:

Solubility of sulfanilamide in a 1 molal monosodium 2-hydroxybenzoate solution at 20°C is 1.29 g/100 ml monosodium 2-hydroxybenzoate solution (7.49×10^{-2} mol dm^{-3}, compiler).

AUXILIARY INFORMATION

METHOD/APPARATUS/PROCEDURE:	SOURCE AND PURITY OF MATERIALS:
Sulfanilamide was added in small portions (0.02 - 0.002 g) to a known volume of 1 molal monosodium 2-hydroxybenzoate soln, held on a water bath, until satn was attained. Moreover, the concn of sulfanilamide was assessed by means of a FEK-M photoelectrocolorimeter.	Sulfanilamide and Na 2-hydroxybenzoate were pharmacopeial products. Purity of the water was not specified.
	ESTIMATED ERROR:
	Soly: results of the colorimetric and gravimetric runs differed by 1-3% (authors). Temp: ±0.1°C (authors).
	REFERENCES:

COMPONENTS:

(1) Benzenesulfonamide, 4-amino- (sulfanilamide); $C_6H_8N_2O_2S$; [63-74-1]

(2) Benzoic acid, 4-hyroxy-, monosodium salt; $C_7H_5NaO_3$; [114-63-6]

(3) Water; H_2O; [7732-18-5]

ORIGINAL MEASUREMENTS:

Gusyakov, V. P.; Sukmans'ka, I. V.

Farm. Zh. (Kiev) 1961, *16*, 25-8.

VARIABLES:

One temperature: 20°C

PREPARED BY:

R. Piekos

EXPERIMENTAL VALUES:

Solubility of sulfanilamide in a 1 molal monosodium 4-hydroxybenzoate solution at 20°C is 1.04 g/100 ml monosodium 4-hydroxybenzoate solution (6.04 x 10^{-2} mol dm^{-3}, compiler).

AUXILIARY INFORMATION

METHOD/APPARATUS/PROCEDURE:

Sulfanilamide was added in small portions (0.02 - 0.002 g) to a known volume of 1 molal monosodium 4-hydroxybenzoate soln, held on a water bath, until satn was attained. Moreover, the concn of sulfanilamide was detd by means of a FEK-M photoelectrocolorimeter.

SOURCE AND PURITY OF MATERIALS:

Sulfanilamide was a pharmacopeial product. Na 4-hydroxybenzoate was obtained by the authors by neutralization of 4-hydroxybenzoic acid which was 99.7% pure.

ESTIMATED ERROR:

Soly: results of colorimetric and gravimetric runs differed by 1-3% (authors).

Temp: ±0.1°C (authors).

REFERENCES:

COMPONENTS:

(1) Benzenesulfonamide, 4-amino- (sulfanilamide); $C_6H_8N_2O_2S$; [63-74-1]

(2) Benzoic acid, 4-amino-2-hydroxy-, monosodium salt; $C_7H_6NNaO_3$; [133-10-8]

(3) Water; H_2O; [7732-18-5]

ORIGINAL MEASUREMENTS:

Gusyakov, V. P.; Sukmans'ka, I. V.

Farm. Zh. (Kiev) 1960, *15(1)* 20-23.

VARIABLES:

One temperature: 20°C

PREPARED BY:

R. Piekos

EXPERIMENTAL VALUES:

Solubility of sulfanilamide in a 1 mol kg^{-1} water Na 4-amino-2-hydroxybenzoate solution at 20°C is 1.42 g/100 ml solution (8.25×10^{-2} mol dm^{-3}, compiler).

AUXILIARY INFORMATION

METHOD/APPARATUS/PROCEDURE:

To 50-ml tightly stoppered test tubes contg 25 ml of a 1 mol kg^{-1} water Na 4-amino-2-hydroxybenzoate soln, placed in a thermostat, accurately weighed 0.02-0.002-g quantities of sulfanilamide were added under agitation until satn was attained.

SOURCE AND PURITY OF MATERIALS:

Both sulfanilamide and Na 4-amino-2-hydroxybenzoate conformed to the requirements of the State Pharmacopeia VIII.

Distilled water was used.

ESTIMATED ERROR:

Temp. ±0.1°C (authors).

Soly: the accuracy of the detn of the concn was similar to that attained by volumetric method (authors).

REFERENCES:

COMPONENTS:

(1) Benzenesulfonamide, 4-amino- (sulfanilamide); $C_6H_8N_2O_2S$; [63-74-1]

(2) Benzenesulfonic acid, 4-methyl-, sodium salt (Na 4-toluenesulfonate) $C_7H_7NaO_3S$; [657-84-1]

(3) Water; H_2O; [7732-18-4]

ORIGINAL MEASUREMENTS:

Kňažko, L. *Farm. Obzor* 1966, *35*, 298-311.

VARIABLES:

Concentration of Na 4-toluenesulfonate

PREPARED BY:

R. Piekos

EXPERIMENTAL VALUES:

Concentration of Na 4-toluenesulfonate ($mol\ dm^{-3}$)	Solubility at 20°C		
	g/100 ml	g/100 g	$10^2 mol\ dm^{-3}$ [a]
0.5	0.95	0.92	5.52
1	1.25	1.165	7.26
1.5	1.56	1.405	9.06

[a] of Na 4-toluenesulfonate solution, calculated by compiler.

AUXILIARY INFORMATION

METHOD/APPARATUS/PROCEDURE:

Sulfanilamide was added in small portions to a known volume of Na 4-toluenesulfonate solns until reaching satn. The equilibration time was 3-4 h under stirring. The temp was held const by means of the Höppler ultrathermostat.

SOURCE AND PURITY OF MATERIALS:

Sulfanilamide and Na 4-tolenesulfonate conformed to the requirements of the Czechoslovak Pharmacopeia 2, Suppl. 1959.

Purity of the water was not specified.

ESTIMATED ERROR:

Nothing specified.

REFERENCES:

COMPONENTS:	ORIGINAL MEASUREMENTS:
(1) Benzenesulfonamide, 4-amino- (sulfanilamide); $C_6H_8N_2O_2S$; [63-74-1] (2) Benzoic acid, 3-hydroxy-, monosodium salt; $C_7H_5NaO_3$; [7720-19-6] (3) Water; H_2O; [7732-18-5]	Gusyakov, V. P.; Sukmans'ka, I. V. *Farm. Zh. (Kiev)* 1961, *16*, 25-8.
VARIABLES: One temperature: 20°C	PREPARED BY: R. Piekos

EXPERIMENTAL VALUES:

Solubility of sulfanilamide in a 1 molal monosodium 3-hydroxybenzoate solution at 20°C is 1.09 g/100 ml monosodium 3-hydroxybenzoate solution (6.33×10^{-2} mol dm^{-3}, compiler).

AUXILIARY INFORMATION

METHOD/APPARATUS/PROCEDURE:	SOURCE AND PURITY OF MATERIALS:
Sulfanilamide was added in small portions (0.02 - 0.002 g) to a known volume of 1 molal monosodium 3-hydroxybenzoate soln, held on a water bath, until satn was attained. Moreover, the concn of sulfanilamide was detd by means of a FEK-M photoelectrocolorimeter.	Sulfanilamide was a pharmacopeial product. Na 3-hydroxybenzoate was obtained by neutralization of a comm 3-hydroxybenzoic acid with Na_2CO_3. The purity of the product was 97.7% (authors). Purity of the water was not specified.
	ESTIMATED ERROR: Soly: results of the colorimetric and gravimetric runs differed by 1-3% (authors). Temp: ± 0.1°C (authors).
	REFERENCES:

COMPONENTS:

(1) Benzenesulfonamide, 4-amino-, (sulfanilamide); $C_6H_8N_2O_2S$; [63-74-1]

(2) Pectin; $(C_{13}H_{18}O_{12})_n$; [9000-69-5]

(3) Water; H_2O; [7732-18-5]

ORIGINAL MEASUREMENTS:

Becher, R.; Leya, S., *Experientia* 1946, *2*, 459-60.

VARIABLES:

One temperature: 18-19°C

PREPARED BY:

R. Piekos

EXPERIMENTAL VALUES:

Solubility of sulfanilamide in a 2.5% pectin solution ([pectin] = 6.8×10^{-2} mol kg^{-1}, compiler), of pH about 2.6, at room temperature (18 - 19°C) is 866 mg% (5.00×10^{-2} mol dm^{-3}, compiler).

AUXILIARY INFORMATION

METHOD/APPARATUS/PROCEDURE:

The soln was allowed to stand for more than 2 days at room temp. The soln was then filtered, and sulfanilamide assayed in the filtrate colorimetrically by the method of Druey and Oesterheld (1).

SOURCE AND PURITY OF MATERIALS:

A high quality apple pectin was used: the rel viscosity of a 0.5% soln was 6.2, and for neutralization of 1 g of the pectin, 1.67 cm^3 of 1 mol dm^{-3} NaOH was used. The source and purity of sulfanilamide and water was not specified.

ESTIMATED ERROR:

Nothing specified.

REFERENCES:

1. Druey, J.; Oesterheld, G., *Helv. Chim. Acta* 1942, *25*, 753.

COMPONENTS:	ORIGINAL MEASUREMENTS:
(1) Benzenesulfonamide, 4-amino-, (sulfanilamide); $C_6H_6N_2O_2S$; [63-74-1] (2) Pectinic acid, sodium salt; $(C_{13}H_{17}NaO_{12})_n$; [9049-37-0] (3) Water; H_2O; [7732-18-5]	Becher, R., Leya, S., *Experientia* 1946, *2*, 459-60.
VARIABLES:	**PREPARED BY:**
One temperature: 18-19°C	R. Piekos

EXPERIMENTAL VALUES:

Solubility of sulfonamide in a 2.6% neutral sodium pectinate solution at room temperature (18-19°C) is 757 mg% (4.40×10^{-2} mol dm^{-3}, compiler).

[Na pectinate] = 6.7×10^{-2} mol kg^{-1}, compiler.

AUXILIARY INFORMATION

METHOD/APPARATUS/PROCEDURE:

The soln was allowed to stand for two days at room temp. The soln was then filtered, and sulfanilamide assayed in the filtrate colorimetrically by the method of Druey and Oesterheld (1).

SOURCE AND PURITY OF MATERIALS:

Nothing specified.

ESTIMATED ERROR:

Nothing specified.

REFERENCES:

1. Druey, J.; Oesterheld, G., *Helv. Chim. Acta* 1942, *25*, 753.

COMPONENTS:	ORIGINAL MEASUREMENTS:
(1) Benzenesulfonamide, 4-amino- (sulfanilamide); $C_6H_8N_2O_2S$; [63-74-1] (2) Ethanol; C_2H_6O; [64-17-5] (3) Water; H_2O; [7732-18-5]	Sapozhnikova, N. V.; Postovskii, I. Ya. *Zh. Prikl. Khim.* 1944, *17*, 527-34.
VARIABLES: Concentration of ethanol	PREPARED BY: R. Piekos

EXPERIMENTAL VALUES:

Concentration of ethanol	Solubility at 37°C	
Weight%	Weight%	mol kg^{-1} solvent[a]
0	1.4	0.84
38.3	2.4	0.14
57.6	3.8	0.23
67.2	4.9	0.30
76.4	7.0	0.44
86	4.8	0.29
96	3.9	0.24

[a]calculated by compiler.

AUXILIARY INFORMATION

METHOD/APPARATUS/PROCEDURE:

Sulfanilamide was dissolved in EtOH-water mixts to form satd solns which were occasionally agitated in glass vessels immersed in a thermostat. The equilibrium was usually attained after 1 h. Five- to 100-cm^3 samples of the satd soln were placed in Pt crucibles or dishes and evapd to dryness at temps lower than 110-115°C. The residue was dried to const wt at 105-110°C and weighed.

SOURCE AND PURITY OF MATERIALS:

Pure, recrystd. sulfanilamide was used. Its mp conformed to that reported in the literature.

The purity of ethanol and water was not specified.

ESTIMATED ERROR:

Soly: quite reliable results were obtained (authors).

Temp: ±0.05°C (authors).

REFERENCES:

COMPONENTS:

(1) Benzenesulfonamide, 4-amino- (sulfanilamide); $C_6H_8N_2O_2S$; [63-74-1]

(2) Ethanol; C_2H_6O; [64-17-5]

(3) Water; H_2O; [7732-18-5]

ORIGINAL MEASUREMENTS:

Shkadova, A. I.

Farm. Zh. (Kiev) 1969, *24(3)*, 39-41.

VARIABLES:

Concentration of ethanol

PREPARED BY:

R. Piekos

EXPERIMENTAL VALUES:

Concentration of ethanol		Solubility at 20°C	
mole %	weight %	10^2 mol kg^{-1}	g/100 g[a]
0	0	3.06	0.527
10	22.14	7.54	1.298
20	39.01	17.15	2.953
30	52.31	21.67	3.732
40	63.04	25.25	4.348
50	71.90	26.89	4.630
60	79.33	25.09	4.320
70	85.65	23.01	3.962
80	91.10	20.77	3.577
90	95.83	14.50	2.496

[a] calculated by compiler.

AUXILIARY INFORMATION

METHOD/APPARATUS/PROCEDURE:

Satd solns of sulfanilamide were equilibrated with the solvent in a water thermostat at 20±0.1°C. Sulfanilamide was detd bromatometrically.

SOURCE AND PURITY OF MATERIALS:

Purity of sulfanilamide conformed to the requirements of the State Pharmacopoeia IX. The EtOH-water mixtures were prepd from abs EtOH (purity and source not specified) and distd water.

ESTIMATED ERROR:

Soly: not specified.

Temp: ±0.1°C (author).

REFERENCES:

COMPONENTS:	ORIGINAL MEASUREMENTS:
(1) Benzenesulfonamide, 4-amino- (sulfanilamide); $C_6H_8N_2O_2S$; [63-74-1] (2) Ethanol; C_2H_6O; [64-17-5] (3) Water; H_2O; [7732-18-5]	Burger, A. *Pharm. Ind.* 1973, *35*, 626-33.
VARIABLES:	PREPARED BY:
Concentration of ethanol	R. Piekos

EXPERIMENTAL VALUES:

Concentration of ethanol	Saturation solubility, C_s,[a] of crystalline form II of sulfanilamide at 20.0°C	
Volume %	mg/100 ml soln	mol dm^{-3}[b]
96	2680	0.1556
90	3279	0.1904
80	3735	0.2169
70	3931	0.2283
65	4024	0.2337
60	3843	0.2332
55	3409	0.1980
47.5	2615	0.1519
45	2475	0.1437
40	2285	0.1327

[a] C_s = [HA] + [A^-], where [HA] is the molar concentration of the dissolved, undissociated molecules of sulfanilamide and [A^-] is the concentration of the dissolved anion of the sulfanilamide.

[b] Calculated by compiler.

AUXILIARY INFORMATION

METHOD/APPARATUS/PROCEDURE:

Sulfanilamide was assayed spectrophotometrically at 258.5 nm using a Zeiss PMQ II spectrophotometer, in 1/15 M phosphate buffer of pH 7.00 ($E^{1\%}_{1\ cm}$ = 945).

SOURCE AND PURITY OF MATERIALS:

Form II of sulfanilamide was obtained by the common method (1). Its purity was not specified.

Purity of the water and EtOH was not specified.

ESTIMATED ERROR:

Soly: not specified.

Temp: $\pm$0.1°C (author)

REFERENCES:

1. Burger, A. *Sci. Pharm.* 1973, *41*, 290 and 303.

COMPONENTS:	ORIGINAL MEASUREMENTS:
(1) Benzenesulfonamide, 4-amino- (sulfanilamide); $C_6H_8N_2O_2S$; [63-74-1] (2) 1,2-Ethanediol; $C_2H_6O_2$; [107-21-1] (3) Water; H_2O; [7732-18-5]	Mingoia, Q. *Ann. Chim. Farm. (Suppl. to Farm. Ital.) Apr.*, 1939, 48-58.
VARIABLES:	PREPARED BY:
Concentration of 1,2-ethanediol	R. Piekos

EXPERIMENTAL VALUES:

% Water	% 1,2-Ethanediol	Solubility of sulfanilamide at 20°C	
		wt.%	10 mol kg^{-1} solution[a]
90	10	0.81	0.47
75	25	1.40	0.81
50	50	3.22	1.87
40	60	6.49	3.77
25	75	9.07	5.27

[a] Calculated by compiler.

AUXILIARY INFORMATION

METHOD/APPARATUS/PROCEDURE:

Five cm^3 of a sulfanilamide soln in aq 1,2-ethanediol was evapd to const wt on a boiling water bath. The residue was dissolved in distd water and its volume was adjusted to 50 cm^3 with the water. In 25 cm^3 of this soln sulfanilamide was assayed by known methods (probably colorimetric).

SOURCE AND PURITY OF MATERIALS:

Sulfanilamide: source and purity not specified.

1,2-Ethanediol was from Merck (purity not specified.

Distilled water was used.

ESTIMATED ERROR:

Nothing specified.

REFERENCES:

COMPONENTS:	ORIGINAL MEASUREMENTS:
(1) Benzenesulfonamide, 4-amino- (sulfanilamide); $C_6H_8N_2O_2S$; [63-74-1] (2) 1,2-Propanediol; $C_3H_8O_2$; [57-55-6] (3) Water; H_2O; [7732-18-5]	Mingoia, Q. *Ann. Chim. Farm. (Suppl. to Farm. Ital.)* Apr., 1939, 48-58.
VARIABLES:	PREPARED BY:
Concentration of 1,2-propanediol	R. Piekos

EXPERIMENTAL VALUES:

% Water	% 1,2-Propanediol	Solubility of sulfanilamide at 20°C wt%	Solubility of sulfanilamide at 20°C $10\ mol\ kg^{-1}$ solution[a]
90	10	1.05	0.61
75	25	1.88	1.09
50	50	3.85	2.24
40	60	5.58	3.24
25	75	7.90	4.59

[a]Calculated by compiler.

AUXILIARY INFORMATION

METHOD/APPARATUS/PROCEDURE:

Five cm^3 of a sulfanilamide soln in aq 1,2-propanediol was evapd to const wt on a boiling water bath. The residue was dissolved in distd water and its volume was adjusted to 50 cm^3 with the water. In 25 cm^3 of this soln sulfanilamide was assayed by known methods (probably colorimetric).

SOURCE AND PURITY OF MATERIALS:

Sulfanilamide: source and purity not specified.

1,2-Propanediol was from Merck (purity not specified).

Distilled water was used.

ESTIMATED ERROR:

Nothing specified.

REFERENCES:

COMPONENTS:	ORIGINAL MEASUREMENTS:
(1) Benzenesulfonamide, 4-amino- (sulfanilamide); $C_6H_8N_2O_2S$; [63-74-1] (2) Ethanol; C_2H_6O; [64-17-5] (3) 1,2,3-Propanetriol; $C_3H_8O_3$; [56-81-5[(4) Water; H_2O; [7732-18-5]	Dolique, R.; Foucault, J. *Trav. soc. pharm. Montpellier* 1952, *12*, 145-53.
VARIABLES:	PREPARED BY:
One temperature: 26-28 °C	R. Piekos

EXPERIMENTAL VALUES:

Solubility of sulfanilamide in a mixture of 1,2,3-propanetriol and 95° ethanol (2:1 by wt) at 26-28 °C is 6.23% (0.386 mol kg^{-1} solvent, compiler).

AUXILIARY INFORMATION

METHOD/APPARATUS/PROCEDURE:	SOURCE AND PURITY OF MATERIALS:
The sulfanilamide content was detd by diazotization of the amine group in a cold acidified 0.1N KNO_2 soln. An excess of KNO_2 was detected by using iodinated starch.	Nothing specified.
	ESTIMATED ERROR:
	Nothing specified.
	REFERENCES:

COMPONENTS:

(1) Benzenesulfonamide, 4-amino- (sulfanilamide); $C_6H_8N_2O_2S$; [63-74-1]

(2) Ethanol, 2,2'-oxybis- (diethylene glycol); $C_4H_{10}O_3$; [111-46-6]

(3) Water; H_2O; [7732-18-5]

ORIGINAL MEASUREMENTS:

Mingoia, Q.

Ann. Chim. Farm. (Suppl. to Farm. Ital.) Apr., 1939, 48-58.

VARIABLES:

Concentration of diethylene glycol

PREPARED BY:

R. Piekos

EXPERIMENTAL VALUES:

% Water	% Diethylene glycol	Solubility of sulfanilamide at 20°C: wt%	$10\ mol\ kg^{-1}$ solution[a]
90	10	1.38	0.801
75	25	2.98	1.73
50	50	6.48	3.76
40	60	12.60	7.317
25	75	26.75	15.53

[a]Calculated by compiler.

AUXILIARY INFORMATION

METHOD/APPARATUS/PROCEDURE:

Five cm^3 of a sulfanilamide soln in aq diethylene glycol was evapd to const wt on a boiling water bath. The residue was dissolved in distd water and its volume was adjusted to 50 cm^3 with the water. In 25 cm^3 of this soln sulfanilamide was assayed by known methods (probably colorimetric).

SOURCE AND PURITY OF MATERIALS:

Sulfanilamide: source and purity not specified.

Diethylene glycol was from Carbide and Carbon Co. (purity was not specified).

Distilled water was used.

ESTIMATED ERROR:

Nothing specified.

REFERENCES:

COMPONENTS:

(1) Benzenesulfonamide, 4-amino- (sulfanilamide); $C_6H_8N_2O_2S$; [63-74-1]

(2) Urea; CH_4N_2O; [57-13-6]

(3) Water; H_2O; [7732-18-5]

ORIGINAL MEASUREMENTS:

Dolique, R.; Foucault, J.

Trav. soc. pharm. Montpellier 1952, *12* 145-53.

VARIABLES:

Concentration of urea

PREPARED BY:

R. Piekos

EXPERIMENTAL VALUES:

Concentration of urea	Solubility of sulfanilamide at 26 °C	
g/100 g water	g/100 g water	mol kg^{-1} water[a]
0	0.8	.05
20	1.9_3	.11
40	3.3_5	.20
80	5.9	.34
120	9.9	.58

[a]Calculated by compiler.

AUXILIARY INFORMATION

METHOD/APPARATUS/PROCEDURE:

The satd soln was agitated for 12 h at 26 °C and filtered. The filtrate was evapd at 100-110 °C and the residue was weighed. Measurements were carried out in test tubes containing 25 cm^3 of water.

SOURCE AND PURITY OF MATERIALS:

Nothing specified.

ESTIMATED ERROR:

Nothing specified.

REFERENCES:

COMPONENTS:

(1) Benzenesulfonamide, 4-amino- (sulfanilamide); $C_6H_8N_2O_2S$; [63-74-1]

(2) Urea; CH_4N_2O; [57-13-6]

(3) Water; H_2O; [7732-18-5]

ORIGINAL MEASUREMENTS:

Rohdewald, P.

Pharmazie 1975, *30(7)*, 460-3.

VARIABLES:

Concentration of urea

PREPARED BY:

R. Piekos

EXPERIMENTAL VALUES:

Concentration of urea	Solubility at 20°C	
mol/l[a]	g/100 ml	10^2 mol dm^{-3}[b]
0.100	0.572	3.32
0.200	0.610	3.54
0.300	0.628	3.65
0.489	0.604	3.51
0.700	0.660	3.83
0.957	0.676	3.93
1.551	0.814	4.73

[a]Numerical values given by the author in personal communication.

[b]Calculated by compiler.

AUXILIARY INFORMATION

METHOD/APPARATUS/PROCEDURE:

The previously employed method (1) was used whereby the solns (50 cm^3) were placed in 100-cm^3 flasks with glass stoppers and rotated in a thermostated bath for 2 h. They were then filtered through a G3 glass filter and the undissolved sulfanilamide was dried at 90°C to const wt and weighed.

SOURCE AND PURITY OF MATERIALS:

Sulfanilamide (source not specified) conformed to the requirements of the DAB 7-BRD. Urea (Schuchardt) was recrystd from aq MeOH. Purity of the water was not specified.

ESTIMATED ERROR:

Soly: not specified.

Temp: ±0.05°C (author).

REFERENCES:

1. Schulte, K. E.; Rohdewald, R.; Weinhold, P. *Pharmazie* 1968, *23(5)*, 252.

COMPONENTS:	ORIGINAL MEASUREMENTS:
(1) Benzenesulfonamide, 4-amino- (sulfanilamide); $C_6H_8N_2O_2S$; [63-74-1] (2) 3-Pyridinecarboxamide; $C_6H_6N_2O$; [98-92-0] (3) Water; H_2O; [7732-18-5]	Rohdewald, P. *Pharm. Ztg.* 1971, *No. 38*, 1342-4.
VARIABLES:	**PREPARED BY:**
Concentration of 3-pyridinecarboxamide	R. Piekos

EXPERIMENTAL VALUES:

$$-k_s = \log \frac{L_{H_2O}}{L_s \; c_s} = 0.66 \text{ l/mol},$$

where L_{H_2O} (0.318_8 g/50 ml = $2.975_8 \times 10^{-2}$ mol dm^{-3}, compiler) and L_s are solubilities of sulfanilamide in water and in aqueous 3-pyridinecarboxamide solution, respectively, and c_s is the concentration of 3-pyridinecarboxamide. L_s values were supplied by the author in personal communication and are shown below.

Concentration of 3-pyridinecarboxamide	L_s at 20°C	
mol/l	g/100 ml	10^2 mol dm^{-3} [a]
0.030	0.542	3.15
0.082	0.612	3.55
0.164	0.686	3.98
0.328	0.854	4.96
0.500	1.090	6.33

[a] Calculated by compiler

AUXILIARY INFORMATION

METHOD/APPARATUS/PROCEDURE:

The solns were equilibrated by agitation for 2 h at 20°C and the sulfanilamide was assayed by differential gravimetric analysis. No details were given.

SOURCE AND PURITY OF MATERIALS:

The source and purity of sulfanilamide and water was not specified. Anal reagent grade 3-pyridinecarboxamide (source not specified) dried over mol sieve was used.

ESTIMATED ERROR:

Soly: not specified.

Temp: ±0.05°C (author).

REFERENCES:

COMPONENTS:	ORIGINAL MEASUREMENTS:
(1) Benzenesulfonamide, 4-amino- (sulfanilamide); $C_6H_8N_2O_2S$; [63-74-1] (2) 3-Pyridinecarboxamide, N,N-diethyl- (nicetamide); $C_{10}H_{14}N_2O$; [59-26-7] (3) Water; H_2O; [7732-18-5]	Rohdewald, P. *Pharm. Ztg.* **1971**, *No. 38* 1342-4.
VARIABLES:	PREPARED BY:
Concentration of nicetamide	R. Piekos

EXPERIMENTAL VALUES:

$$-k_x = \log \frac{L_{H_2O}}{L_s \; c_s} = 0.64 \text{ l/mol},$$

where L_{H_2O} (0.318_8 g/40 ml = $2.975_8 \times 10^{-2}$ mol dm^{-3}, compiler) and L_s are solubilities of sulfanilamide in water and in aqueous nicetamide solutions, respectively, and c_s is the concentration of nicetamide.

L_s values were supplied by the author in personal communication and are shown below.

Concentration of nicetamide	L_s at 20°C	
mol/l	g/100 ml	10^2 mol dm^{-3} [a]
0.100	0.674	3.91
0.200	0.696	4.04
0.300	0.932	5.41
0.500	1.246	7.24

[a] Calculated by compiler.

AUXILIARY INFORMATION

METHOD/APPARATUS/PROCEDURE:

The solns were equilibrated by agitation for 2 h at 20°C and the sulfanilamide was assayed by differential gravimetric analysis. No details were given.

SOURCE AND PURITY OF MATERIALS:

The source and purity of sulfanilamide and water was not specified.
Anal reagent grade nicetamide (source not specified) dried over mol sieve was used.

ESTIMATED ERROR:

Soly: not specified.

Temp: ±0.05°C (author).

REFERENCES:

COMPONENTS:	ORIGINAL MEASUREMENTS:
(1) Benzenesulfonamide, 4-amino- (sulfanilamide); $C_6H_8N_2O_2S$; [63-74-1] (2) Urea, methyl-; $C_2H_6N_2O$; [598-50-5] (3) Water; H_2O; [7732-18-5]	Rohdewald, P. *Pharmazie* 1975, *30(7)* 460-3.
VARIABLES:	PREPARED BY:
Concentration of methylurea	R. Piekos

EXPERIMENTAL VALUES:

Concentration of methylurea mol/l[a]	Solubility at 20°C	
	g/100 ml	10^2 mol cm^{-3} [b]
0.050	0.576	3.34
0.100	0.586	3.40
0.200	0.602	3.50
0.400	0.614	3.57
0.600	0.676	3.93
0.700	0.732	4.25
0.800	0.740	4.30
1.000	0.790	4.59

[a]Numerical values given by the author in personal communication.

[b]Calculated by compiler.

AUXILIARY INFORMATION

METHOD/APPARATUS/PROCEDURE:

The previously employed method (1) was used whereby the solns (50 cm^3) were placed in 100-cm^3 flasks with glass stoppers and rotated in a thermostated bath for 2 h. They were then filtered through a G3 glass filter and the undissolved sulfanilamide was dried at 90°C to const wt and weighed.

SOURCE AND PURITY OF MATERIALS:

Sulfanilamide (source not specified) conformed to the requirements of the DAB 7-BRD. Methylurea (Schuchardt) was recrystd from aq MeOH. Purity of the water was not specified.

ESTIMATED ERROR:

Soly: not specified.

Temp: $\pm$0.05°C (author).

REFERENCES:

1. Schulte, K. E.; Rohdewald, P.; Weinhold, P. *Pharmazie* 1968, *23(5)*, 252.

COMPONENTS:

(1) Benzenesulfonamide, 4-amino- (sulfanilamide); $C_6H_8N_2O_2S$; [63-74-1]

(2) Urea, ethyl-; $C_3H_8N_2O$; [625-52-5]

(3) Water; H_2O; [7732-18-5]

ORIGINAL MEASUREMENTS:

Rohdewald, P.

Pharmazie 1975, *30(7)*, 460-3.

VARIABLES:

Concentration of ethylurea

PREPARED BY:

R. Piekos

EXPERIMENTAL VALUES:

Concentration of ethylurea mol/l[a]	Solubility at 20°C g/100 ml	10^2 mol dm^{-3} [b]
0.200	0.620	3.60
0.300	0.664	3.86
0.500	0.738	4.29
0.700	0.826	4.80

[a] Numerical values given by the author in personal communication.

[b] Calculated by compiler.

AUXILIARY INFORMATION

METHOD/APPARATUS/PROCEDURE:

The previously employed method (1) was used whereby the solns (50 cm^3) were placed in 100-cm^3 flasks with glass stoppers and rotated in a thermostated bath for 2 h. They were then filtered through a G3 glass filter and the undissolved sulfanilamide was dried at 90°C to const wt and weighed.

SOURCE AND PURITY OF MATERIALS:

Sulfanilamide (source not specified) conformed to the requirements of the DAB 7-BRD. Ethylurea (Schuchardt) was recrystd from aq MeOH. Purity of the water was not specified.

ESTIMATED ERROR:

Soly: not specified.

Temp: ±0.05°C (author).

REFERENCES:

1. Schulte, K. E.; Rohdewald, P.; Weinhold, P. *Pharmazie* 1968, *23(5)*, 252.

COMPONENTS:

(1) Benzenesulfonamide, 4-amino- (sulfanilamide); $C_6H_8N_2O_2S$; [63-74-1]

(2) Urea, N,N'-dimethyl-; $C_3H_8N_2O$; [96-31-1]

(3) Water; H_2O; [7732-18-5]

ORIGINAL MEASUREMENTS:

Rohdewald, P.

Pharmazie 1975, *30(7)*. 460-3.

VARIABLES:

Concentration of N,N'-dimethylurea

PREPARED BY:

R. Piekos

EXPERIMENTAL VALUES:

Concentration of N,N'-dimethylurea mol/l[a]	Solubility at 20°C g/100 ml	10^2 mol dm^{-3} [b]
0.200	0.624	3.62
0.300	0.660	3.83
0.400	0.678	3.94
0.500	0.726	4.22
0.600	0.750	4.35
0.700	0.792	4.60
0.800	0.842	4.89

[a]Numerical values given by the author in personal communication.

[b]Calculated by compiler.

AUXILIARY INFORMATION

METHOD/APPARATUS/PROCEDURE:

The previously employed method (1) was used whereby the solns (50 cm^3) were placed in 100-cm^3 flasks with glass stoppers and rotated in a thermostated bath for 2 h. They were then filtered through a G3 glass filter and the undissolved sulfanilamide was dried at 90°C to const wt and weighed.

SOURCE AND PURITY OF MATERIALS:

Sulfanilamide (source not specified) conformed to the requirements of the DAB 7-BRD. N,N'-dimethylurea (Schuchardt) was recrystd from aq MeOH. Purity of the water was not specified.

ESTIMATED ERROR:

Soly: not specified.

Temp: $\pm$0.05°C (author).

REFERENCES:

1. Schulte, K. E.; Rohdewald, P.; Weinhold, P. *Pharmazie* 1968, *23(5)*, 252.

COMPONENTS:

(1) Benzenesulfonamide, 4-amino- (sulfanilamide); $C_6H_8N_2O_2S$; [63-74-1]

(2) Urea, N,N-dimethyl-; $C_3H_8N_2O$; [598-94-7]

(3) Water; H_2O; [7732-18-5]

ORIGINAL MEASUREMENTS:

Rohdewald, P.

Pharmazie 1975, *30(7)* 460-3.

VARIABLES:

Concentration of N,N-dimethylurea

PREPARED BY:

R. Piekos

EXPERIMENTAL VALUES:

Concentration of N,N-dimethylurea	Solubility at 20°C	
mol/l[a]	g/100 ml	10^2 mol dm^{-3} [b]
0.200	0.656	3.81
0.300	0.678	3.94
0.500	0.708	4.11
0.700	0.838	4.87
0.800	0.870	5.05
0.927	1.016	5.90

[a] Numerical values given by the author in personal communication.

[b] Calculated by compiler.

AUXILIARY INFORMATION

METHOD/APPARATUS/PROCEDURE:

The previously employed method (1) was used whereby the solns (50 cm^3) were placed in 100-cm^3 flasks with glass stoppers and rotated in a thermostated bath for 2 h. They were then filtered through a G3 glass filter and the undissolved sulfanilamide was dried at 90°C to const wt and weighed.

SOURCE AND PURITY OF MATERIALS:

Sulfanilamide (source not specified) conformed to the requirements of DAB 7-BRD. N,N-dimethylurea (Schuchardt) was recrystd from aq MeOH. Purity of the water was not specified.

ESTIMATED ERROR:

Soly: not specified.

Temp: ±0.05°C (author).

REFERENCES:

1. Schulte, K. E.; Rohdewald, P.; Weinhold, P. *Pharmazie* 1968, *23(5)*, 252.

COMPONENTS:

(1) Benzenesulfonamide, 4-amino- (sulfanilamide); $C_6H_8N_2O_2S$; [63-74-1]

(2) Urea, tetramethyl-; $C_5H_{12}N_2O$; [632-22-4]

(3) Water; H_2O; [7732-18-5]

ORIGINAL MEASUREMENTS:

Rohdewald, P.

Pharmazie 1975, *30(7)*, 460-3.

VARIABLES:

Concentration of tetramethylurea

PREPARED BY:

R. Piekos

EXPERIMENTAL VALUES:

Concentration of tetramethylurea	Solubility at 20 °C	
mol/l[a]	g/100 ml	10^2 mol dm^{-3} [b]
0.200	0.850	4.936
0.400	1.004	5.830
0.600	1.196	6.945
0.800	1.412	8.200
1.000	1.668	9.686

[a]Numerical values given by the author in personal communication.

[b]Calculated by compiler.

AUXILIARY INFORMATION

METHOD/APPARATUS/PROCEDURE:

The previously employed method (1) was used wereby the solns (50 cm^3) were placed in 100-cm^3 flasks with glass stoppers and rotated in a thermostated bath for 2 h. They were then filtered through a G3 glass filter and the undissolved sulfanilamide was dried at 30 °C to const wt and weighed.

SOURCE AND PURITY OF MATERIALS:

Sulfanilamide (source not specified) conformed to the requirements of the DAB 7-BRD. Tetramethylurea (Schuchardt) was recrystd from aq MeOH. Purity of the water was not specified.

ESTIMATED ERROR:

Soly: not specified.

Temp: $\pm$0.05 °C (author).

REFERENCES:

1. Schulte, K. E.; Rohdewald, P.; Weinhold, P. *Pharmazie* 1968, *23(5)*, 252.

COMPONENTS:	ORIGINAL MEASUREMENTS:
(1) Benzenesulfonamide, 4-amino- (sulfanilamide); $C_6H_8N_2O_2S$; [63-74-1] (2) Thiourea; CH_4N_2S; [62-56-6] (3) Water; H_2O; [7732-18-5]	Rohdewald, P. *Pharmazie* 1975, *30(7)*, 460-3.
VARIABLES: Concentration of thiourea	PREPARED BY: R. Piekos

EXPERIMENTAL VALUES:

Concentration of thiourea mol/l[a]	Solubility at 20°C g/100 ml	10^2 mol dm^{-3} [b]
0.300	0.646	3.75
0.400	0.664	3.86
0.600	0.706	4.10
0.800	0.784	4.55
1.000	0.880	5.11
1.200	0.952	5.53

[a] Numerical values given by the author in personal communication.

[b] Calculated by compiler.

AUXILIARY INFORMATION

METHOD/APPARATUS/PROCEDURE:

The previously employed method (1) was used whereby the solns (50 cm^3) were placed in 100-cm^3 flasks with glass stoppers and rotated in a thermostated bath for 2 h. They were then filtered through a G3 glass filter and the undissolved sulfanilamide was dried at 90°C to const wt and weighed.

SOURCE AND PURITY OF MATERIALS:

Sulfanilamide (source not specified) conformed to the requirements of the DAB 7-BRD. Thiourea (Schuchardt) was recrystd from aq MeOH. Purity of the water was not specified.

ESTIMATED ERROR:

Soly: not specified.

Temp: ±0.05°C (author).

REFERENCES:

1. Schulte, K. E.; Rohdewald, P.; Weinhold, P. *Pharmazie* 1968, *23(5)*, 252.

COMPONENTS:	ORIGINAL MEASUREMENTS:
(1) Benzenesulfonamide, 4-amino-, (sulfanilamide); $C_6H_8N_2O_2S$; [63-74-1] (2) Carbamic acid, ethyl ester (urethane); $C_3H_7NO_2$; [51-79-6] (3) Water; H_2O; [7732-18-5]	Weinstein, L.; McDonald, A., *Science* 1945, *101*, 44-5.
VARIABLES: One temperature: 20°C	PREPARED BY: R. Piekos

EXPERIMENTAL VALUES:

Solubility of sulfanilamide in a 10% aqueous urethane solution at 20°C
1000 mg/100 cm^3 urethane solution (6 x 10^{-2} mol dm^{-3}, compiler).

AUXILIARY INFORMATION

METHOD/APPARATUS/PROCEDURE:	SOURCE AND PURITY OF MATERIALS:
Nothing specified.	Nothing specified.
	ESTIMATED ERROR: Nothing specified.
	REFERENCES:

COMPONENTS:	ORIGINAL MEASUREMENTS:
(1) Benzenesulfonamide, 4-amino- (sulfanilamide); $C_6H_8N_2O_2S$; [63-74-1] (2) Ethanol; C_2H_6O; [64-17-5] (3) 1,2,3-Propanetriol; $C_3H_8O_3$; [56-81-5] (4) Urea; CH_4N_2O; [57-13-6] (5) Water; H_2O; [7732-18-5]	Dolique, R.; Foucault, J. *Trav. soc. pharm. Montpellier* 1952, *12*, 145-53.
VARIABLES:	PREPARED BY:
One temperature: 26-28 °C	R. Piekos

EXPERIMENTAL VALUES:

Solubility of sulfanilamide at 26-28 °C in a saturated solution of urea in a mixture of 1,2,3-propanetriol and 95° ethanol (2:1 by wt), containing 54.5 g of urea per 100 g of the mixture, is 8.22% (0.52 mol kg^{-1} solvent, compiler).

AUXILIARY INFORMATION

METHOD/APPARATUS/PROCEDURE:	SOURCE AND PURITY OF MATERIALS:
The sulfanilamide content was detd by diazotization of the amine group in a cold acidified 0.1N KNO_2 soln. An excess of KNO_2 was detected by using iodinated starch.	Nothing specified. ESTIMATED ERROR: Nothing specified. REFERENCES:

COMPONENTS:	ORIGINAL MEASUREMENTS:
(1) Benzenesulfonamide, 4-amino- (sulfanilamide); $C_6H_8N_2O_2S$; [63-74-1] (2) Poly (oxy-1,2-ethanediyl), α-hydro-ω-hydroxy- (PEG 200); $(C_2H_4O)_nH_2O$; [25322-68-3] 200 (3) Water; H_2O; [7732-18-5]	Kawam, M. N.; Tawashi, R.; Czetsch-Lindenwald, H. v. *Sci. Pharm.* 1965, *33*, 151-61.
VARIABLES:	**PREPARED BY:**
Concentration of PEG 200; temperature	R. Piekos

EXPERIMENTAL VALUES:

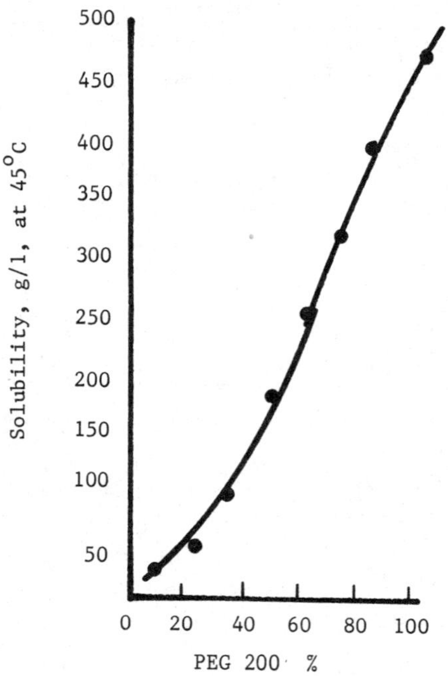

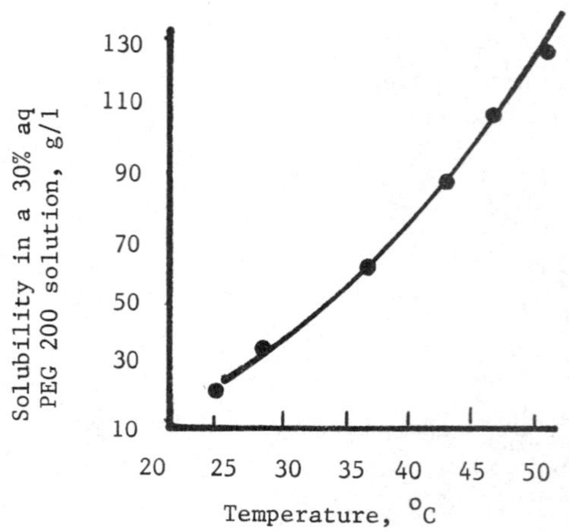

AUXILIARY INFORMATION

METHOD/APPARATUS/PROCEDURE:

An earlier developed method was employed (1) whereby a 100-ml conical flask contg a PEG 200 soln was placed in a drying cabinet at a given temp and an excess of sulfanilamide was added under stirring for 1 h. After 12 h the soln was filtered or decanted and the solute was assayed in the filtrate spectrophotometrically at 260 mμ using a Unicam SP 500 spectrophotometer and 1-ml quartz cuvets. Results were taken from a calibration graph.

SOURCE AND PURITY OF MATERIALS:

Sulfanilamide conformed with the requirements of BP 1953 or USP XIV. It was recrystd, dried, powdered, and again dried at 105°C to const wt. PEG 200 was a product of Farbwerke Hoechst. It was kept over concd H_2SO_4 in a desiccator.

Purity of the water was not specified.

ESTIMATED ERROR:

Nothing specified.

REFERENCES:

1. Khawam, M. N.; Tawashi, R.; Czetsch-Lindenwald, H. v. *Sci. Pharm.* 1965, *33*, 90.

COMPONENTS:	ORIGINAL MEASUREMENTS:
(1) Benzenesulfonamide, 4-amino- (sulfanilamide); $C_6H_8N_2O_2S$; [63-74-1] (2) Poly(oxy-1,2-ethanediyl), α-hydro-ω-hydroxy- (PEG 400); $(C_2H_4O)_nH_2O$; [25322-68-3] 400 (3) Water; H_2O; [7732-18-5]	Khawam, M. N.; Tawashi, R.; Czetsch-Lindenwald, H. v. *Sci. Pharm.* 1964, *32*, 271-9.
VARIABLES:	**PREPARED BY:**
One temperature: 45°C	R. Piekos

EXPERIMENTAL VALUES:

Solubility of sulfanilamide in a 10% aqueous solution of PEG 400 at 45°C is 5.57 g/100 g PEG 400 solution (0.323 mol kg^{-1}, compiler).

AUXILIARY INFORMATION

METHOD/APPARATUS/PROCEDURE:

Sulfanilamide was added in excess to an aq PEG 400 soln and the mixt was stirred for 30 min. The soln was then placed in a drying cabinet for 24 h and occasionally shaken. After filtration the sulfanilamide was assayed in the filtrate by the USP XVI method based on diazotization. The end point was detected by means of a starch paste as an indicator. Corrections were made for consumption of the 0.1 N $NaNO_2$ soln by PEG 400.

SOURCE AND PURITY OF MATERIALS:

Sulfanilamide conformed to the requirements of USP XVI.

Purity of the remaining materials was not specified.

ESTIMATED ERROR:

Soly: measurements were made in duplicate (authors).

Temp: ±1°C (authors).

REFERENCES:

COMPONENTS:	ORIGINAL MEASUREMENTS:
(1) Benzenesulfonamide, 4-amino- (sulfanilamide); $C_6H_8N_2O_2S$; [63-74-1] (2) Poly(oxy-1,2-ethanediyl), α-hydro- ω-hydroxy- (PEG 400); $(C_2H_4O)_nH_2O$; [25322-68-3] 400 (3) Water; H_2O; [7732-18-5]	Kawam, M. N.; Tawashi, R.; Czetsch-Lindenwald, H. v. *Sci. Pharm.* 1965, *33*, 151-61.
VARIABLES:	PREPARED BY:
Concentration of PEG 400	R. Piekos

EXPERIMENTAL VALUES:

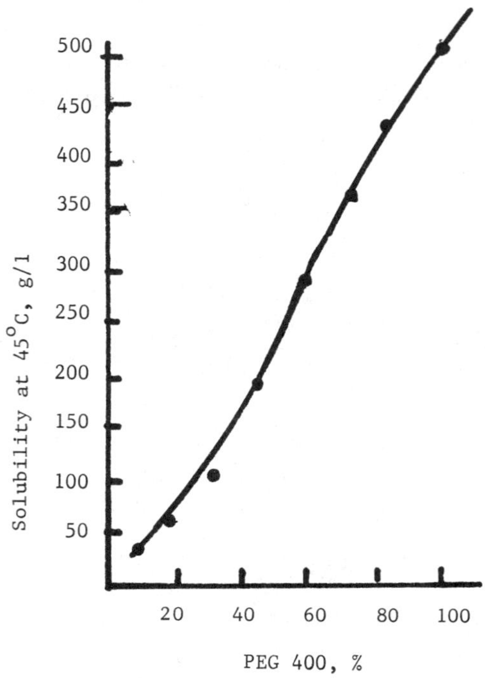

AUXILIARY INFORMATION

METHOD/APPARATUS/PROCEDURE:

An earlier developed method was employed (1) whereby a 100-ml conical flask contg a PEG 400 soln was placed in a drying cabinet at 45°C and an excess of sulfanilamide was added under stirring for 1 h. After 12 h the soln was filtered or decanted and the solute was assayed in the filtrate spectrophotometrically at 260 mμ using a Unicam SP 500 spectrophotometer and 1-ml quartz cuvets. Results were taken from a calibration graph.

SOURCE AND PURITY OF MATERIALS:

Sulfanilamide conformed with the requirements of BP 1953 or USP XIV. It was recrystd, dried, powdered, and again dried at 105°C to const wt. PEG 400 was a product of Farbwerke Hoechst. It was kept over concd H_2SO_4 in a desiccator.

Purity of the water was not specified.

ESTIMATED ERROR:

Nothing specified.

REFERENCES:

1. Khawam, M. N.; Tawashi, R.; Czetsch-Lindenwald, H. v. *Sci. Pharm.* 1965, *33*, 90.

COMPONENTS:	ORIGINAL MEASUREMENTS:
(1) Benzenesulfonamide, 4-amino- (sulfanilamide); $C_6H_8N_2O_2S$; [63-74-1] (2) Poly(oxy-1,2-ethanediyl), α -hydro- ω -hydroxy- (PEG 400); $(C_2H_4O)_nH_2O$; [25322-68-3] 400	Gusyakov, V. P.; Likhol'ot, N. M.; Kutna, I. M. *Farm. Zh. (Kiev)* 1968, *23(6)*, 56-61.
VARIABLES:	PREPARED BY:
One temperature: 21-25°C	R. Piekos

EXPERIMENTAL VALUES:

Solubility of sulfanilamide in α-hydro-ω -hydroxypoly(oxy-1,2-ethanediyl) 400 at room temperature (21-25°C) is 87.5% by weight (40.7 mol kg^{-1} PEG 400, compiler).

AUXILIARY INFORMATION

METHOD/APPARATUS/PROCEDURE:	SOURCE AND PURITY OF MATERIALS:
Small quantities (2-4 mg) of sulfanilamide were added to a known quantity of PEG 400 under stirring until satn was attained.	Sulfanilamide: neither source nor purity were specified. PEG 400: source not specified; sp gr 1.127 g cm^{-3}; temp of solidification approx 6°C; refractive index 1.466 (temp not indicated).
	ESTIMATED ERROR:
	Nothing specified.
	REFERENCES:

COMPONENTS:

(1) Benzenesulfonamide, 4-amino- (sulfanilamide); $C_6H_8N_2O_2S$; [63-74-1]

(2) Poly(oxy-1,2-ethanediyl), α-hydro-ω-hydroxy- (PEG 400); $(C_2H_4O)_nH_2O$; [25322-68-3] 400

ORIGINAL MEASUREMENTS:

Khawan, M. N.; Tawashi, R.; Czetsch-Lindenwald, H. v. *Sci. Pharm.* 1964, *32*, 271-9.

VARIABLES:

One temperature: 45°C

PREPARED BY:

R. Piekos

EXPERIMENTAL VALUES:

Solubility of sulfanilamide in PEG 400 at 45°C is 15.25 g/100 g PEG 400 (0.8856 mol kg^{-1}, compiler).

AUXILIARY INFORMATION

METHOD/APPARATUS/PROCEDURE:

Small weighed samples of sulfanilamide were added to PEG 400 under stirring until dissoln occurred.

SOURCE AND PURITY OF MATERIALS:

Sulfanilamide conformed to the requirements of USP XVI.

Source and purity of PEG 400 was not specified.

ESTIMATED ERROR:

Soly: measurements were made in duplicate (authors).

Temp: ±1°C (authors).

REFERENCES:

COMPONENTS:	ORIGINAL MEASUREMENTS:
(1) Benzenesulfonamide, 4-amino- (sulfanilamide); $C_6H_8N_2O_2S$; [63-74-1] (2) Poly(oxy-1,2-ethanediyl), α-hydro-ω-hydroxy- (PEG 600); $(C_2H_4O)_nH_2O$; [25322-68-3] 600 (3) Water; H_2O; [7732-18-5]	Kawam, M. N.; Tawashi, R.; Czetsch-Lindenwald, H. v. *Sci. Pharm.* 1965, *33*, 153-61.
VARIABLES:	PREPARED BY:
Concentration of PEG 600	R. Piekos

EXPERIMENTAL VALUES:

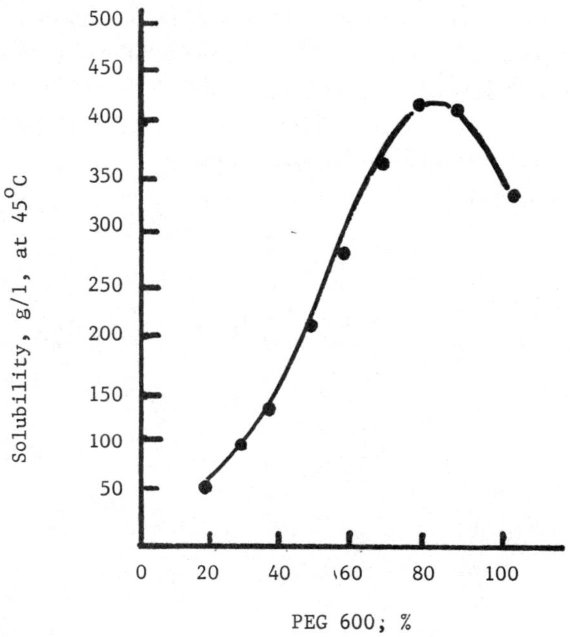

AUXILIARY INFORMATION

METHOD/APPARATUS/PROCEDURE:

An earlier developed method was employed (1) whereby a 100-ml conical flask contg a PEG 600 soln was placed in a drying cabinet at a given temp and an excess of sulfanilamide was added under stirring for 1 h. After 12 h the soln was filtered or decanted and the solute was assayed in the filtrate spectrophotometrically at 260 mμ using a Unicam SP 500 spectrophotometer and 1-ml quartz cuvets. Results were taken from a calibration graph.

SOURCE AND PURITY OF MATERIALS:

Sulfanilamide conformed to the requirements of BP 1953 or USP XIV. It was recrystd, dried, powdered, and again dried at 105°C to const wt. PEG 600 was a product of Farbwerke Hoechst. It was kept over concd H_2SO_4 in a desiccator.

Purity of the water was not specified.

ESTIMATED ERROR:

Nothing specified.

REFERENCES:

1. Khawam, M. N.; Tawashi, R.; Czetsch-Lindenwald, H. v. *Sci. Pharm.* 1965, *33*, 90.

COMPONENTS:	ORIGINAL MEASUREMENTS:
(1) Benzenesulfonamide, 4-amino- (sulfanilamide); $C_6H_8N_2O_2S$; [63-74-1] (2) Poly(oxy-1,2-ethanediyl), α-hydro-ω-hydroxy- (PEG 600); $(C_2H_4O)_nH_2O$; [25322-68-3] 600 (3) Water; H_2O; [7732-18-5]	Gusyakov, V. P.; Likhol'ot, N. M.; Kutna, I. M. *Farm. Zh. (Kiev)* 1968, *23(6)*, 56-61.
VARIABLES:	PREPARED BY:
One temperature: 21-25°C	R. Piekos

EXPERIMENTAL VALUES:

Solubility of sulfanilamide in a 5% (by weight) aqueous α-hydro-ω-hydroxypoly(oxy-1,2-ethanediyl) 600 at room temperature (21-25°C) is 0.702 g/100 ml (4.08×10^{-2} mol dm^{-3}, compiler).

AUXILIARY INFORMATION

METHOD/APPARATUS/PROCEDURE:

A small excess of sulfanilamide was added to a 5% (by wt) aq PEG 600 soln, the mixt was sealed in an ampul and agitated for 24 h (1). The concn of sulfanilamide was detd colorimetrically (2).

SOURCE AND PURITY OF MATERIALS:

Sulfanilamide: neither source nor purity was specified. PEG 600 was of the Austrian or West German origin; purity not specified.

Purity of the water was not specified.

ESTIMATED ERROR:

Nothing specified.

REFERENCES:

1. Gusyakov, V. P.; Likhol'ot, N. M.; Kutna, I. M. *Farm. Zh. (Kiev)* 1967, *22(3)*, 34.
2. Predchetenskii, B. E.; Borovskaya, V. M.; Morgolina, L. T. *Laboratornye metody issledovaniya, Medgiz, Moscow* 1950, *p.* 371

COMPONENTS:	ORIGINAL MEASUREMENTS:
(1) Benzenesulfonamide, 4-amino- (sulfanilamide); $C_6H_8N_2O_2S$; [63-74-1] (2) Poly(oxy-1,2-ethanediyl), α-hydro-ω-hydroxy- (PEG 1500); $(C_2H_4O)_nH_2O$; [25322-68-3] 1500 (3) Water; H_2O; [7732-18-5]	Kawam, M. N.; Tawashi, R.; Czetsch-Lindenwald, H. v. *Sci. Pharm.* 1965, *33*, 151-61.
VARIABLES:	PREPARED BY:
Temperature	R. Piekos

EXPERIMENTAL VALUES:

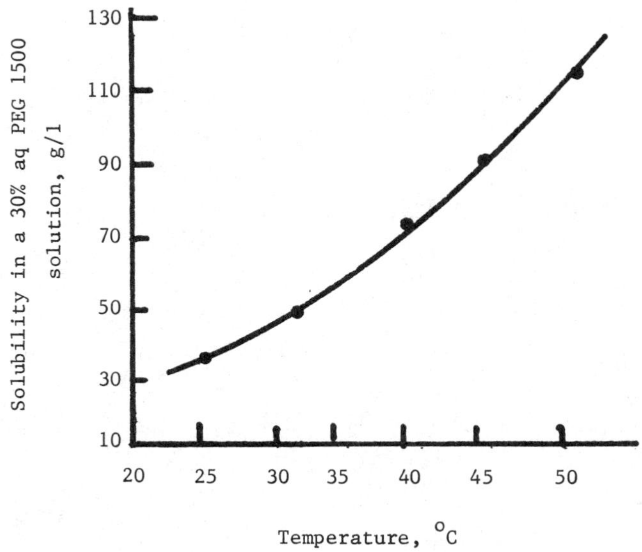

AUXILIARY INFORMATION

METHOD/APPARATUS/PROCEDURE:	SOURCE AND PURITY OF MATERIALS:
An earlier developed method was employed (1) whereby a 100-ml conical flask contg a PEG 1500 soln was placed in a drying cabinet at a given temp and an excess of sulfanilamide was added under stirring for 1 h. After 12 h the soln was filtered or decanted and the solute was assayed in the filtrate spectrophotometrically at 260 mμ using a Unicam SP 500 spectrophotometer and 1-ml quartz cuvets. Results were taken from a calibration graph.	Sulfanilamide conformed with the requirements of BP 1953 and USP XIV. It was recrystd, dried, powdered, and again dried at 105°C to const wt. PEG 1500 was a product of Farbwerke Hoechst. It was kept over concd H_2SO_4 in a desiccator. Purity of the water was not specified.
	ESTIMATED ERROR:
	Nothing specified.
	REFERENCES:
	1. Khawam, M. N.; Tawashi, R.; Czetsch-Lindenwald, H. v. *Sci. Pharm.* 1965, *33*, 90.

COMPONENTS:	ORIGINAL MEASUREMENTS:
(1) Benzenesulfonamide, 4-amino- (sulfanilamide); $C_6H_8N_2O_2S$; [63-74-1] (2) Poly(oxy-1,2-ethanediyl), α-hydro-ω-hydroxy- (PEG 4000); $(C_2H_4O)_nH_2O$; [25322-68-3] 4000 (3) Water; H_2O; [7732-18-5]	Khawam, M. N.; Tawashi, R.; Czetsch-Lindenwald, H. v. *Sci. Pharm.* 1964, *32*, 271-9.
VARIABLES:	PREPARED BY:
One temperature: 45°C	R. Piekos

EXPERIMENTAL VALUES:

Solubility of sulfanilamide in a 10% aqueous solution of PEG 4000 at 45°C is 20.5 g/100 g PEG 4000 solution (1.19 mol kg^{-1}, compiler).

AUXILIARY INFORMATION

METHOD/APPARATUS/PROCEDURE:

Sulfanilamide was added in excess to an aq PEG 4000 soln and the mixt was stirred for 30 min. The soln was then placed in a drying cabinet for 24 h and occasionally shaken. After filtration the sulfonamide was assayed in the filtrate by the USP XVI method based on diazotization. The end point was detected by means of a starch paste as an indicator. Corrections were made for consumption of the 0.1 N $NaNO_2$ soln by PEG 4000.

SOURCE AND PURITY OF MATERIALS:

Sulfanilamide conformed to the requirements of USP XVI.

Purity of the remaining materials was not specified.

ESTIMATED ERROR:

Soly: measurements were made in duplicate (authors).

Temp: ±1°C (authors).

REFERENCES:

COMPONENTS:

(1) Benzenesulfonamide, 4-amino- (sulfanilamide); $C_6H_8N_2O_2S$; [63-74-1]

(2) Poly(oxy-1,2-ethanediyl), α-hydro-ω-hydroxy- (PEG 4000); $(C_2H_4O)_nH_2O$; [25322-68-3] 4000

(3) Water; H_2O; [7732-18-5]

ORIGINAL MEASUREMENTS:

Gusyakov, V. P.; Likhol'ot, N. M.; Kutna, I. M. *Farm. Zh. (Kiev)* 1968, *23(6)*, 56-61.

VARIABLES:

One temperature: 21-25 °C

PREPARED BY:

R. Piekos

EXPERIMENTAL VALUES:

Solubility of sulfanilamide in a 5% (by weight) aqueous α-hydro-ω-hydroxypoly(oxy-1,2-ethanediyl) 4000 at room temperature (21-25 °C) is 0.766 g/100 ml (4.45×10^{-2} mol dm^{-3}, compiler).

AUXILIARY INFORMATION

METHOD/APPARATUS/PROCEDURE:

A small excess of sulfanilamide was added to a 5% (by wt) aq PEG 4000 soln, the mixture was sealed in an ampul and agitated for 24 h (1). The concn of sulfanilamide was detd colorimetrically (2).

SOURCE AND PURITY OF MATERIALS:

Sulfanilamide: neither source nor purity was specified. PEG 4000 was of the Austrian or West German origin. Its purity was not specified.

Purity of the water was not specified.

ESTIMATED ERROR:

Nothing specified.

REFERENCES:

1. Gusyakov, V. P.; Likhol'ot, N. M.; Kutna, I. M. *Farm. Zh. (Kiev)* 1967, *22(3)*, 34.
2. Predchetenskii, B. E.; Borovskaya, V. M.; Morgolina, L. T. *Laboratornye metody issledovaniya, Medgiz, Moscow* 1950, *p.* 371.

COMPONENTS:	ORIGINAL MEASUREMENTS:
(1) Benzenesulfonamide, 4-amino- (sulfanilamide); $C_6H_8N_2O_2S$; [63-74-1] (2) Poly(oxy-1,2-ethanediyl), α-hydro-ω-hydroxy- (PEG 6000); $(C_2H_4O)_nH_2O$; [25322-68-3] 6000 (3) Water; H_2O; [7732-18-5]	Khawam, M. N.; Tawashi, R.; Czetsch-Lindenwald, H. v. *Sci. Pharm.* 1964, *32*, 271-9.
VARIABLES:	PREPARED BY:
One temperature: 45°C	R. Piekos

EXPERIMENTAL VALUES:

Solubility of sulfanilamide in a 10% aqueous PEG 6000 solution at 45°C is 21.8 g/100 g PEG 6000 solution (1.27 mol kg^{-1}, compiler).

AUXILIARY INFORMATION

METHOD/APPARATUS/PROCEDURE:	SOURCE AND PURITY OF MATERIALS:
Sulfanilamide was added in excess to an aq PEG 6000 soln and the mixt was stirred for 30 min. The soln was then placed in a drying cabinet for 24 h and occasionally shaken. After filtration the sulfonamide was assayed in the filtrate by the USP XVI method based on diazotization. The end point was detected by means of a starch paste as an indicator. Corrections were made for consumption of the 0.1 N $NaNO_2$ soln by PEG 6000.	Sulfanilamide conformed to the requirements of USP XVI. Purity of the remaining materials was not specified.
	ESTIMATED ERROR:
	Soly: measurements were made in duplicate (authors). Temp: ±1°C (authors).
	REFERENCES:

COMPONENTS:

(1) Benzenesulfonamide, 4-amino- (sulfanilamide); $C_6H_8N_2O_2S$; [63-74-1]

(2) Poly(oxy-1,2-ethanediyl), α-hydro-ω-hydroxy- (PEG 6000); $(C_2H_4O)_nH_2O$; [25322-68-3] 6000

(3) Water; H_2O; [7732-18-5]

ORIGINAL MEASUREMENTS:

Kawam, M. N.; Tawashi, R.; Czetsch-Lindenwald, H. v. *Sci. Pharm.* 1965, *33*, 153-61.

VARIABLES:

Concentration of PEG 6000; temperature

PREPARED BY:

R. Piekos

EXPERIMENTAL VALUES:

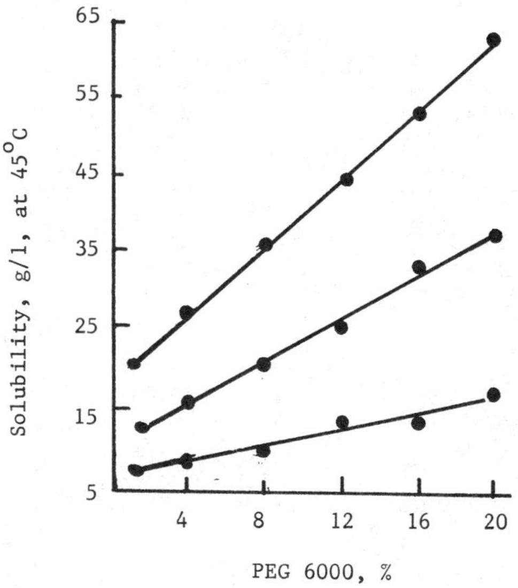

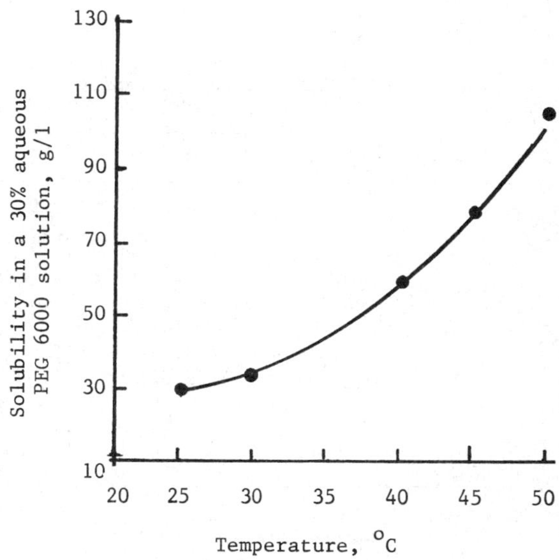

AUXILIARY INFORMATION

METHOD/APPARATUS/PROCEDURE:

An earlier developed method was employed (1) whereby a 100-ml conical flask contg a PEG 6000 soln was placed in a drying cabinet at a given temp and an excess of sulfanilamide was added under stirring for 1 h. After 12 h the soln was filtered or decanted and the solute was assayed in the filtrate spectrophotometrically at 260 mμ using a Unicam SP 500 spectrophotometer and 1-ml quartz cuvets. Results were taken from a calibration graph.

SOURCE AND PURITY OF MATERIALS:

Sulfanilamide conformed with the requirements of BP 1953 or USP XIV. It was recrystd, dried, powdered, and again dried at 105°C to const wt. PEG 6000 was a product of Farbwerke Hoechst. It was kept over concd H_2SO_4 in a desiccator.

Purity of the water was not specified.

ESTIMATED ERROR:

Nothing specified.

REFERENCES:

1. Khawam, M. N.; Tawashi, R.; Czetsch-Lindenwald, H. v. *Sci. Pharm.* 1965, *33*, 90.

COMPONENTS:	ORIGINAL MEASUREMENTS:
(1) Benzenesulfonamide, 4-amino- (sulfanilamide); $C_6H_8N_2O_2S$; [63-74-1] (2) Sorbitan monolaurate, polyoxyethylene derivatives (Tween 20); [9005-64-5] (3) Water; H_2O; [7732-18-5]	Khawam, M. N.; Tawashi, R.; Czetsch-Lindenwald, H. v. *Sci. Pharm.* 1964, *32*, 271-9.
VARIABLES:	PREPARED BY:
Temperature; concentration of Tween 20	R. Piekos

EXPERIMENTAL VALUES:

t/°C	Solubility of sulfanilamide in g/l (10^2 mol dm^{-3} [a]) at various Tween 20 concentrations					
	0.1%	0.5%	1%	2%	3%	4% (w/v)
16	4.43 (2.57)	4.71 (2.73)	4.91 (2.85)	5.62 (3.26)	6.10 (3.54)	6.75 (3.92)
24	6.98 (4.05)	7.6 (4.41)	8.20 (4.76)	9.18 (5.33)	9.81 (5.70)	10.33 (6.00)
34	12.85 (7.46)	13.47 (7.82)	13.9 (8.07)	15.54 (9.02)	16.14 (9.37)	16.88 (9.80)
44	21.22 (12.32)	21.98 (12.76)	21.22 (12.32)	23.81 (13.83)	25.05 (14.55)	27.72 (16.10)

[a] Calculated by compiler.

AUXILIARY INFORMATION

METHOD/APPARATUS/PROCEDURE:

Sulfanilamide was added in excess to an aq Tween 20 soln and the mixture was stirred for 30 min. The soln was then placed in a drying cabinet for 24 h and occasionally shaken. After filtration the sulfanilamide was assayed in the filtrate by the USP XIV method based on diazotization. The end point was detected by means of a starch paste as an indicator. Corrections were made for consumption of the 0.1 N $NaNO_2$ soln by Tween 20.

SOURCE AND PURITY OF MATERIALS:

Sulfanilamide conformed to the requirements of the USP XVI.

Tween 20 was a commercially available reagent with a HLB value of 16.7, manufd by Atlas Goldschmidt A. G., Essen, West Germany.

Purity of the water was not specified.

ESTIMATED ERROR:

Soly: Measurements were made in duplicate (authors).

Tem;: ± 1°C (authors).

REFERENCES:

COMPONENTS:

(1) Benzenesulfonamide, 4-amino- (sulfanilamide); $C_6H_8N_2O_2S$; [63-74-1]

(2) Sorbitan monolaurate, polyoxyethylene derivatives (Tween 20); [9005-64-5]

(3) Water; H_2O; [7732-18-5]

ORIGINAL MEASUREMENTS:

Khawam, M. N.; Tawashi, R.; Czetsch-Lindenwald, H. v. *Sci. Pharm.* 1965, *33*, 90-101.

VARIABLES:

Concentration of Tween 20

PREPARED BY:

R. Piekos

EXPERIMENTAL VALUES:

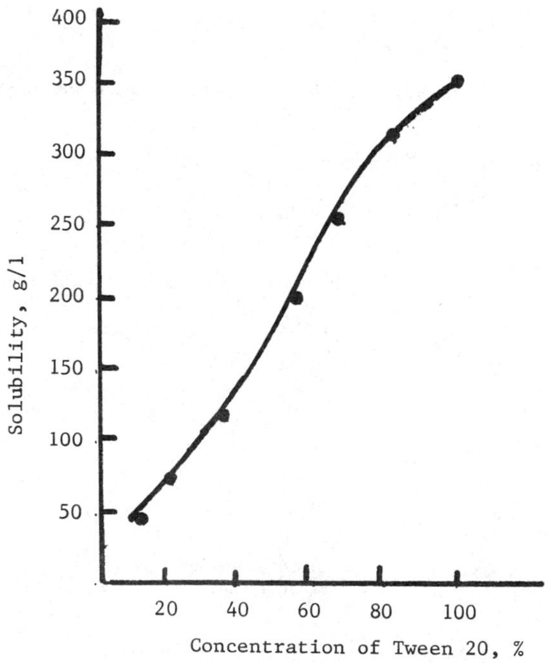

AUXILIARY INFORMATION

METHOD/APPARATUS/PROCEDURE:

A 100-ml conical flask contg a Tween 20 soln was placed in a drying cabinet at 25°C and sulfanilamide was added in excess under stirring for 1 h. After 12 h the soln was filtered or decanted and sulfanilamide was assayed in the filtrate spectrophotometrically at 260 mμ using a Unicam SP 500 spectrophotometer and 1-ml quartz cuvets. Results were taken from a calibration graph.

SOURCE AND PURITY OF MATERIALS:

Sulfanilamide (BP 1953 and USP XIV) was recrystd, dried, powdered, and dried again at 105°C to const wt. Tween 20 was an Atlas-Goldschmidt product with HLB = 16.7 and dielec const 9.89.

Distd water was used.

ESTIMATED ERROR:

Soly: not specified.

Temp: ±1°C (authors).

REFERENCES:

COMPONENTS:	ORIGINAL MEASUREMENTS:
(1) Benzenesulfonamide, 4-amino- (sulfanilamide); $C_6H_8N_2O_2S$; [63-74-1] (2) Sorbitan monolaurate, polyoxyethylene derivatives (Tween 20); [9005-64-5] (3) Water; H_2O; [7732-18-5]	Gusyakov, V. P.; Likhol'ot, N. M.; Kutna, I. M. *Farm. Zh. (Kiev)* 1967, *22(3)*, 34-9.
VARIABLES:	PREPARED BY:
One temperature: 20°C	R. Piekos

EXPERIMENTAL VALUES:

$$S/S_o = 1.23 \text{ at } 20^\circ C,$$

where S is the solubility of sulfanilamide in a 2% by weight aqueous Tween 20 solution, and

S_o is the solubility of sulfanilamide in water (0.53 g/100 ml).

Hence S = 0.65 g/100 ml (3.8×10^{-2} mol dm^{-3}) - compiler.

AUXILIARY INFORMATION

METHOD/APPARATUS/PROCEDURE:	SOURCE AND PURITY OF MATERIALS:
An excess of sulfanilamide in a 2% by wt aq Tween 20 soln was equilibrated for 24 h in an ampul immersed in a water thermostat. Aliquots of the satd soln were withdrawn through a filter and the sulfanilamide content was assayed in the filtrate photometrically.	Sulfanilamide conformed to the requirements of the State Pharmacopeia IX. Tween 20 was a product of Gee Lawson, England. Purity of the water was not specified.
	ESTIMATED ERROR:
	Soly: not specified. Temp: ±0.1°C (authors).
	REFERENCES:

COMPONENTS:	ORIGINAL MEASUREMENTS:
(1) Benzenesulfonamide, 4-amino- (sulfanilamide); $C_6H_8N_2O_2S$; [63-74-1] (2) Sorbitan monopalmitate, polyoxyethylene derivatives (Tween 40); [9005-66-7] (3) Water; H_2O; [7732-18-5]	Gusyakov, V. P.; Likhol'ot, N. M.; Kutna, I. M. *Farm. Zh. (Kiev)* 1967, *22(3)* 34-9.
VARIABLES:	PREPARED BY:
One temperature: 20°C	R. Piekos

EXPERIMENTAL VALUES:

$$S/S_o = 1.1 \text{ at } 20^{\circ}C$$

where S is the solubility of sulfanilamide in a 2% by weight aqueous Tween 40 solution, and

S_o is the solubility of sulfanilamide in water (0.53 g/100 ml).

Hence S = 0.58 g/100 ml (3.4×10^{-2} mol dm^{-3}) - compiler.

AUXILIARY INFORMATION

METHOD/APPARATUS/PROCEDURE:	SOURCE AND PURITY OF MATERIALS:
An excess of sulfanilamide in a 2% by wt aq Tween 40 soln was equilibrated for 24 h in an ampul immersed in a water thermostat. Aliquots of the satd soln were withdrawn through a filter and the sulfanilamide content was assayed in the filtrate photometrically.	Sulfanilamide conformed to the requirements of the State Pharmacopeia IX. Tween 40 was a product of Gee Lawson, England. Purity of the water was not specified.
	ESTIMATED ERROR: Soly: not specified. Temp: $\pm 0.1^{\circ}C$ (authors).
	REFERENCES:

COMPONENTS:	ORIGINAL MEASUREMENTS:
(1) Benzenesulfonamide, 4-amino- (sulfanilamide); $C_6H_8N_2O_2S$; [63-74-1] (2) Sorbitan monostearate, polyoxyethylene derivatives (Tween 60); [9005-67-8] (3) Water; H_2O; [7732-18-5]	Khawam, M. N.; Tawashi, R.; Czetsch-Lindenwald, H. v. *Sci. Pharm.* 1964, *32*, 271-9.
VARIABLES: Temperature; concentration of Tween 60	PREPARED BY: R. Piekos

EXPERIMENTAL VALUES:

t/°C	Solubility of sulfanilamide in g/l (10^2 mol dm^{-3} [a]) at various Tween 60 concentrations					
	0.1%	0.5%	1%	2%	3%	4% (w/v)
15	4.04 (2.35)	4.32 (2.51)	4.59 (2.66)	5.22 (3.03)	5.70 (3.31)	6.31 (3.66)
24	7.25 (4.21)	7.67 (4.45)	8.67 (5.03)	9.23 (5.36)	9.92 (5.76)	10.7 (6.21)
34	12.71 (7.38)	13.50 (7.84)	14.49 (8.41)	15.28 (8.87)	16.51 (9.59)	17.05 (9.90)
44	20.79 (12.07)	21.35 (12.40)	21.63 (12.56)	23.20 (13.47)	24.65 (14.31	26.26 (15.25)

[a] Calculated by compiler.

AUXILIARY INFORMATION

METHOD/APPARATUS/PROCEDURE:

Sulfanilamide was added in excess to an aq Tween 60 soln and the mixture was stirred for 30 min. The soln was then placed in a drying cabinet for 24 h and occasionally shaken. After filtration the sulfanilamide was assayed in the filtrate by the USP XVI method based on diazotization. The end point was detected by means of a starch paste as an indicator. Corrections were made for consumption of the 0.1N $NaNO_2$ soln by Tween 60.

SOURCE AND PURITY OF MATERIALS:

Sulfanilamide conformed to the requirements of USP XVI.

Tween 60 was a commercially available reagent (source and purity not specified).

Purity of the water was not specified.

ESTIMATED ERROR:

Soly: Measurements were made in duplicate (authors).

Temp: ± 1 °C (authors).

REFERENCES:

COMPONENTS:	ORIGINAL MEASUREMENTS:
(1) Benzenesulfonamide, 4-amino- (sulfanilamide); $C_6H_8N_2O_2S$; [63-74-1] (2) Sorbitan monolaurate, polyoxyethylene derivatives (Tween 60); [9005-67-8] (3) Water; H_2O; [7732-18-5]	Khawam, M. N.; Tawashi, R.; Czetsch-Lindenwald, H. v. *Sci. Pharm.* 1965, *33*, 90-101.
VARIABLES:	PREPARED BY:
Concentration of Tween 60	R. Piekos

EXPERIMENTAL VALUES:

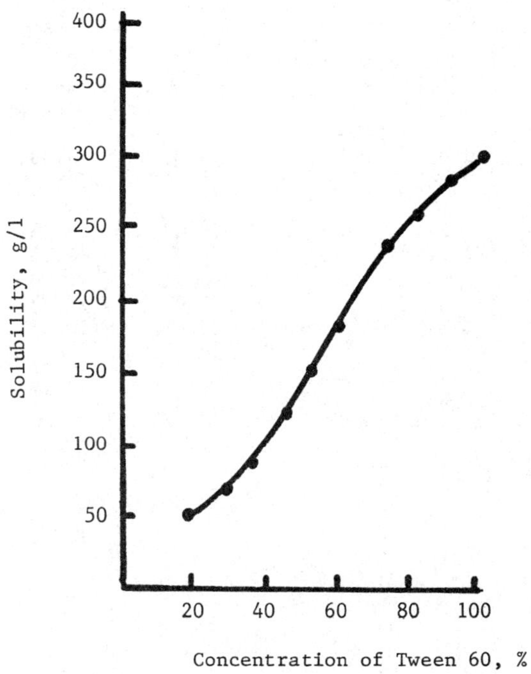

AUXILIARY INFORMATION

METHOD/APPARATUS/PROCEDURE:	SOURCE AND PURITY OF MATERIALS:
A 100-ml conical flask contg Tween 60 soln was placed in a drying cabinet at 25°C and sulfanilamide was added in excess under stirring for 1 h. After 12 h the soln was filtered or decanted and sulfanilamide was assayed in the filtrate spectrophotometrically at 260 mμ using a Unicam SP 500 spectrophotometer and 1-ml quartz cuvets. Results were taken from a calibration graph.	Sulfanilamide (BP 1953 and USP XIV) was recrystd, dried, powdered, and dried again at 105°C to const wt. Tween 20 was an Atlas-Goldschmidt product with HLB = 14.9 and dielec const 8.27. Distd water was used. ESTIMATED ERROR: Soly: not specified. Temp: ±1°C (authors). REFERENCES:

COMPONENTS:

(1) Benzenesulfonamide, 4-amino- (sulfanilamide); $C_6H_8N_2O_2S$; [63-74-1]

(2) Sorbitan monooleate, polyoxyethylene derivatives (Tween 80); [9005-65-6]

(3) Water; H_2O; [7732-18-5]

ORIGINAL MEASUREMENTS:

Khawam, M. N.; Tawashi, R.; Czetsch-Lindenwald, H. v. *Sci. Pharm.* 1964, *32*, 271-9.

VARIABLES:

Temperature; concentration of Tween 60

PREPARED BY:

R. Piekos

EXPERIMENTAL VALUES:

t/°C	Solubility of sulfanilamide in g/l (10^2 mol dm^{-3} [a]) at various Tween 80 concentrations					
	0.1%	0.5%	1%	2%	3%	4% (w/v)
13	3.62 (2.10)	3.86 (2.24)	4.06 (2.36)	4.68 (2.72)	5.30 (3.08)	5.88 (3.41)
24	7.40 (4.30)	7.76 (4.51)	8.21 (4.77)	8.90 (5.17)	10.37 (6.02)	10.85 (6.30)
34	12.47 (7.24)	13.47 (7.82)	13.50 (7.84)	13.7 (7.96)	15.2 (8.83)	17.69 (10.27)
44	21.05 (12.22)	22.02 (12.79)	21.42 (12.44)	24.15 (14.02)	25.79 (14.98)	26.71 (15.51)

[a] Calculated by compiler.

AUXILIARY INFORMATION

METHOD/APPARATUS/PROCEDURE:

Sulfanilamide was added in excess to an aq Tween 80 soln and the mixt was stirred for 30 min. The soln was then placed in a drying cabinet for 24 h and occasionally shaken. After filtration the sulfanilamide was assayed in the filtrate by the USP XVI method based on diazotization. The end point was detected by means of a starch paste as an indicator. Corrections were made for consumption of the 0.1N $NaNO_2$ soln by Tween 80.

SOURCE AND PURITY OF MATERIALS:

Sulfanilamide conformed to the requirements of USP XVI.

Tween 80 was a commercially available reagent (source and purity not specified).

Purity of the water was not specified.

ESTIMATED ERROR:

Soly: Measurements were made in duplicate (authors).

Temp: ± 1°C (authors).

REFERENCES:

COMPONENTS:

(1) Benzenesulfonamide, 4-amino- (sulfanilamide); $C_6H_8N_2O_2S$; [63-74-1]

(2) Sorbitan monolaurate, polyoxyethylene derivatives (Tween 80); [9005-65-6]

(3) Water; H_2O; [7732-18-5]

ORIGINAL MEASUREMENTS:

Khawam, M. N.; Tawashi, R.; Czetsch-Lindenwald, H. v. *Sci. Pharm.* 1965, *33*, 90-101.

VARIABLES:

Concentration of Tween 80

PREPARED BY:

R. Piekos

EXPERIMENTAL VALUES:

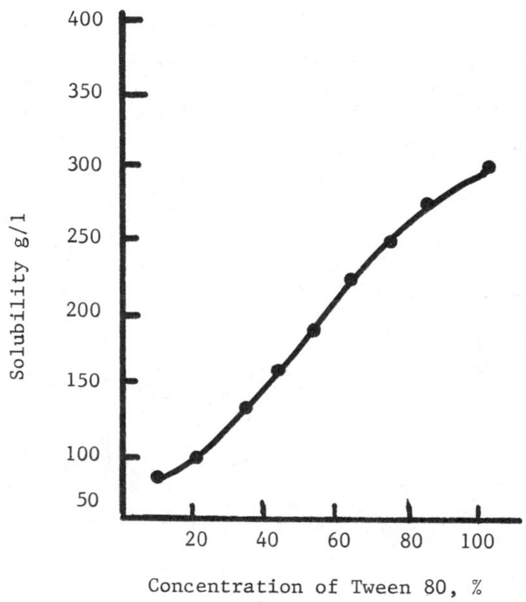

AUXILIARY INFORMATION

METHOD/APPARATUS/PROCEDURE:

A 100-ml conical flask contg a Tween 80 soln was placed in a drying cabinet at 25°C and sulfanilamide was added in excess under stirring for 1 h. After 12 h the soln was filtered or decanted and sulfanilamide was assayed in the filtrate spectrophotometrically at 260 mμ using a Unicam SP 500 spectrophotometer and 1-ml quartz cuvets. Results were taken from a calibration graph.

SOURCE AND PURITY OF MATERIALS:

Sulfanilamide (BP 1953 and USP XIV) was recrystd, dried, powdered, and dried again at 105°C to const wt. Tween 80 was an Atlas-Goldschmidt product with HLB = 15.0 and dielec const 8.75.

Distd water was used.

ESTIMATED ERROR:

Soly: not specified.

Temp: ±1°C (authors).

REFERENCES:

COMPONENTS:	ORIGINAL MEASUREMENTS:
(1) Benzenesulfonamide, 4-amino- (sulfanilamide); $C_6H_8N_2O_2S$; [63-74-1] (2) Sorbitan monooleate, polyoxyethylene derivatives (Tween 80); [9005-65-6] (3) Water; H_2O; [7732-18-5]	Gusyakov, V. P.; Likhol'ot, N. M.; Kutna, I. M. *Farm. Zh. (Kiev)* 1967, *22(3)*, 34-9.
VARIABLES:	PREPARED BY:
One temperature: 20°C	R. Piekos

EXPERIMENTAL VALUES:

$$S/S_o = 1.19 \text{ at } 20^\circ C$$

where S is the solubility of sulfanilamide in a 2% by weight aqueous Tween 80 solution, and

S_o is the solubility of sulfanilamide in water (0.53 g/100 ml).

Hence S = 0.63 g/100 ml (3.7×10^{-2} mol dm^{-3}) - compiler.

AUXILIARY INFORMATION

METHOD/APPARATUS/PROCEDURE:

An excess of sulfanilamide in a 2% by wt aq Tween 80 soln was equilibrated for 24 h in an ampul immersed in a water thermostat. Aliquots of the satd soln were withdrawn through a filter and the sulfanilamide content was assayed in the filtrate photometrically.

SOURCE AND PURITY OF MATERIALS:

Sulfanilamide conformed to the requirements of the State Pharmacopeia IX.

Tween 80 was a product of Gee Lawson, England.

Purity of the water was not specified.

ESTIMATED ERROR:

Soly: not specified.

Temp: ±0.1°C (authors).

REFERENCES:

COMPONENTS:	ORIGINAL MEASUREMENTS:
(1) Benzenesulfonamide, 4-amino- (sulfanilamide); $C_6H_8N_2O_2S$; [63-74-1] (2) D-Glucose; $C_6H_{12}O_6$; [50-99-7] (3) Water; H_2O; [7732-18-5]	Becher, R.; Leya, S. *Experientia* 1946, *2*, 459-60.
VARIABLES:	PREPARED BY:
One temperature: 18-19°C	R. Piekos

EXPERIMENTAL VALUES:

Solubility of sulfanilamide in a 10% D-glucose solution at room temperature (18-19°C) is 654 mg% (3.80×10^{-2} mol dm^{-3}, compiler).

AUXILIARY INFORMATION

METHOD/APPARATUS/PROCEDURE:	SOURCE AND PURITY OF MATERIALS:
After standing for more than two days the soln of sulfanilamide was filtered and the sulfonamide was assayed in the filtrate colorimetrically by the method of Druey and Oesterheld (1).	Nothing specified.
	ESTIMATED ERROR:
	Nothing specified.
	REFERENCES:
	1. Druey, J.; Oesterheld, G. *Helv. Chim. Acta* 1942, *25*, 753.

COMPONENTS:	ORIGINAL MEASUREMENTS:
(1) Benzenesulfonamide, 4-amino- (sulfanilamide); $C_6H_8N_2O_2S$; [63-74-1] (2) 2-Propanol; C_3H_8O; [67-63-0]	Burlage, H. M. *J. Am. Pharm. Assoc., Sci. Ed.* 1948, *37* 345.
VARIABLES:	**PREPARED BY:**
One temperature: 25°C	R. Piekos

EXPERIMENTAL VALUES:

Solubility of sulfanilamide in 2-propanol at 25°C is 0.7970 g/100 cm^3 solution (4.628×10^{-2} mol dm^{-3}, compiler).

AUXILIARY INFORMATION

METHOD/APPARATUS/PROCEDURE:	SOURCE AND PURITY OF MATERIALS:
Satd solns of sulfanilamide in 2-propanol were prepd at 25°C and definite vols of the solns were measured into tared dishes by means of standard pipets. The alcohol was allowed to evap at room temp and the residue was dried at 105°C. In the case of losses due to apparent decompn, the residue was dried in a dessicator (1).	The sulfonamide was manufd by Gane and Ingram and was of the U.S.P. purity. The source and purity of 2-propanol was not specified. **ESTIMATED ERROR:** Nothing specified. **REFERENCES:** 1. Burlage, H. M. *J. Am. Pharm. Assoc., Sci. Ed.* 1947, *36(1)*, 16.

COMPONENTS:

(1) Benzenesulfonamide, 4-amino- (sulfanilamide); $C_6H_8N_2O_2S$; [63-74-1]

(2) 1-Butanol; $C_4H_{10}O$; [71-36-3]

ORIGINAL MEASUREMENTS:

Burger, A.

Sci. Pharm. 1973, *4*, 303-14.

VARIABLES:

Temperature

PREPARED BY:

R. Piekos

EXPERIMENTAL VALUES:

t/°C	Saturation solubility, C_s, of crystalline form II: mg/100 ml soln	10^2 mol dm^{-3} [a]
10.4	192	1.12
15.3	226	1.31
20.0	272	1.58
26.0	336	1.95
30.5	392	2.28
35.5	462	2.68
40.9	552	3.21
45.8	648	3.76
50.5	780	4.53
55.8	937	5.44
60.5	1096	6.36
66.0	1325	7.69
70.5	1596	9.27
75.5	1963	11.4
80.3	2261	13.1

[a] Calculated by compiler.

The following equation was derived based on the above data (C_s is in mg/100 ml solution): $C_s = 156.79 + 4643.79\ T^{-1} + 25.80\ \ln T$,

where T is absolute temperature.

AUXILIARY INFORMATION

METHOD/APPARATUS/PROCEDURE:

Satn soly was detd by the earlier developed method (1). Sulfanilamide was assayed photometrically at 263 nm ($E^{1\%}_{1\ cm}$ = 1130) on a Zeiss - PMQ II spectrophotometer.

SOURCE AND PURITY OF MATERIALS:

Cryst form II of sulfanilamide mp 156°C, was obtained by crystn from 96% EtOH (2).

Source and purity of the BuOH was not specified.

ESTIMATED ERROR:

Soly: not specified.

Temp: ±0.1°C (author).

REFERENCES:

1. Kuhnert-Brandstätter, M.; Burger, A. *Pharm. Ind.* 1972, *34*, 187.
2. Burger, A. *Sci. Pharm.* 1973, *41*, 290.

COMPONENTS:

(1) Benzenesulfonamide, 4-amino- (sulfanilamide); $C_6H_8N_2O_2S$; [63-74-1]

(2) 1-Pentanol, (amyl alcohol); $C_5H_{12}O$; [71-41-0]

ORIGINAL MEASUREMENTS:

Lin, O. H.; Ph.D. Dissertation; The University of Iowa; Iowa City, IA; 1971; p. 76.

VARIABLES:

Four temperatures; four crystalline forms.

PREPARED BY:

J. K. Guillory

EXPERIMENTAL VALUES:

Solubilities of Sulfanilamide Polymorphic Forms:

t/°C	mol dm^{-3} α	β	γ	δ
30.0	1.36×10^{-2}	1.26×10^{-2}	1.35×10^{-2}	1.49×10^{-2}
37.0	1.74×10^{-2}	1.73×10^{-2}	1.72×10^{-2}	1.93×10^{-2}
45.0	2.32×10^{-2}	2.34×10^{-2}	2.33×10^{-2}	2.50×10^{-2}
50.0	2.90×10^{-2}	2.90×10^{-2}	2.86×10^{-2}	3.04×10^{-2}

AUXILIARY INFORMATION

METHOD/APPARATUS/PROCEDURE:

An excess of sulfanilamide was placed in 20-ml screw-capped vials with 20 ml of pentanol. Vials were rotated end-over-end in a bath whose temperature was controlled to ±0.1°C. Equilibrium was attained after 36 h or less. Supernatant was filtered through sintered glass, diluted 1:24 with 95% ethanol, and analyzed spectrophotometrically at 262 nm. Measurements were performed in duplicate. Residual crystals were analyzed by differential thermal analysis to detect change in crystal form. None was observed.

SOURCE AND PURITY OF MATERIALS:

Sulfanilamide (Mallinckrodt) was recrystd from 95% ethanol, methanol, n-pentanol, or n-butanol. Crystal forms were identified using densities, refractive indexes, x-ray diffraction measurements (1) and by infrared spectrophotometry and differential thermal analysis (2).

1-Pentanol (Fisher Certified).

ESTIMATED ERROR:

Uncertainty of temperature (±0.1°C).
Uncertainty of solubility measurements probably (1-2%) based on agreement of duplicate absorbance measurements.

REFERENCES:

1. Lin, H. O.; Baenziger, N. C.; and Guillory, J. K., *J. Pharm. Sci.* 1974, *63*, 145-6.
2. Lin, H. O.; Guillory, J. K., *J. Pharm. Sci.* 1970, *59*, 972-5.

COMPONENTS:	ORIGINAL MEASUREMENTS:
(1) Benzenesulfonamide, 4-amino- (sulfanilamide); $C_6H_8N_2O_2S$; [63-74-1] (2) 1,2-Ethanediol; $C_2H_6O_2$; [107-21-1]	Mingoia, Q. *Ann. Chim. Farm. (Suppl. to Farm. Ital.) Apr.*, 1939, 48-58.
VARIABLES:	PREPARED BY:
One temperature: 20 °C	R. Piekos

EXPERIMENTAL VALUES:

Solubility of sulfanilamide in 1,2-ethanediol at 20 °C is 9.80 wt.% (0.569 mol kg^{-1} solution, compiler).

AUXILIARY INFORMATION

METHOD/APPARATUS/PROCEDURE:	SOURCE AND PURITY OF MATERIALS:
Five cm^3 of a sulfanilamide soln in 1,2-ethanediol were diluted with distilled water to 50 cm^3 and sulfanilamide was assayed in this solution by known methods (probably colorimetric).	Source and purity of sulfanilamide were not specified. 1,2-Ethanediol was from Merck (purity not specified).
	ESTIMATED ERROR: Nothing specified.
	REFERENCES:

COMPONENTS:	ORIGINAL MEASUREMENTS:
(1) Benzenesulfonamide, 4-amino- (sulfanilamide); $C_6H_8N_2O_2S$; [63-74-1] (2) 1,2-Propanediol; $C_3H_8O_2$; [57-55-6]	Mingoia, Q. *Ann. Chim. Farm. (Suppl. to Farm. Ital.)* *Apr.*, 1939, 48-58.
VARIABLES:	PREPARED BY:
One temperature: 20°C	R. Piekos

EXPERIMENTAL VALUES:

Solubility of sulfanilamide in 1,2-propanediol at 20°C is 11.46 wt% (0.6655 mol kg^{-1} solution, compiler).

AUXILIARY INFORMATION

METHOD/APPARATUS/PROCEDURE:

Five cm^3 of a sulfanilamide soln in 1,2-propanediol were diluted with distilled water to 50 cm^3 and sulfanilamide was assayed in this solution by known methods (probably colorimetric).

SOURCE AND PURITY OF MATERIALS:

Source and purity of sulfanilamide were not specified.

1,2-Propanediol was from Merck (purity not specified).

ESTIMATED ERROR:

Nothing specified.

REFERENCES:

COMPONENTS:

(1) Benzenesulfonamide, 4-amino- (sulfanilamide); $C_6H_8N_2O_2S$; [63-74-1]

(2) Poly(oxy-1,2-ethanediyl), α-hydro-ω-hydroxy- (PEG 400); $(C_2H_4O)_nH_2O$; [25322-68-3] 400

ORIGINAL MEASUREMENTS:

Wahlgren, S., *Svensk farm. tidskr.* 1962, *66*, 585-91.

VARIABLES:

Temperature

PREPARED BY:

R. Piekos

EXPERIMENTAL VALUES:

t/°C	Solubility in PEG 4000	
	weight%	mol kg^{-1} [a]
20	21	1.5
60	37	3.4

[a] Calculated by compiler.

AUXILIARY INFORMATION

METHOD/APPARATUS/PROCEDURE:

The soly detns were made in 100-cm^3 Erlenmeyer flasks immersed in a const-temp bath. The suspension was stirred with an electrically driven propeller stirrer for at least 4 h.

SOURCE AND PURITY OF MATERIALS:

The source and purity of sulfanilamide was not specified. PEG 400: pH 4.7 (1.00 g in 20.0 g of water), ash content 0.030%, free acid: 0.30 cm^3 of 0.1 N NaOH soln was required to neutralize free acids in 5.0 g of PEG 400 dissolved in 20 cm^3 of EtOH; average mol wt 400; water content 0.2%.

ESTIMATED ERROR:

Temp: ±0.5°C (author).

Soly: duplicate tests were made of concns on both sides of the borderline value (author).

REFERENCES:

COMPONENTS:

(1) Benzenesulfonamide, 4-amino- (sulfanilamide); $C_6H_8N_2O_2S$; [63-74-1]

(2) Poly(oxy-1,2-ethanediyl), α-hydro-ω-hydroxy- (poly(ethylene glycol) (PEG 3000); $(C_2H_4O)_nH_2O$; [25322-68-3] 3000

ORIGINAL MEASUREMENTS:

Wahlgren, S., *Svensk farm. tidskr.* 1962, *66*, 585-91.

VARIABLES:

One temperature: 60°C

PREPARED BY:

R. Piekos

EXPERIMENTAL VALUES:

Solubility of sulfanilamide in poly(ethylene glycol) 3000 at 60°C is 17% by weight (0.99 mol kg^{-1}, compiler).

AUXILIARY INFORMATION

METHOD/APPARATUS/PROCEDURE:

The soly detns were made in 100-cm^3 Erlenmeyer flasks immersed in a const-temp bath. The suspension was stirred with an electrically driven propeller stirrer for at least 4 h.

SOURCE AND PURITY OF MATERIALS:

The source and purity of sulfanilamide was not specified. PEG 3000: mp 56°C, pH 6.4 (1.00 g in 20.0 g of water); ash content 0.025%; free acid: 0.05 cm^3 of 0.1 N NaOH was required to neutralize free acids in 5.0 g of PEG dissolved in 20 cm^3 of EtOH against phenolphthalein; average mol wt 3000; water content 0.4%.

ESTIMATED ERROR:

Temp: ±0.5°C (author).

Soly: duplicate tests were made of concns on both sides of the borderline value (author).

REFERENCES:

COMPONENTS:

(1) Benzenesulfonamide, 4-amino- (sulfanilamide); $C_6H_8N_2O_2S$; [63-74-1]

(2) Ethanol, 2,2'-oxybis- (diethylene glycol; $C_4H_{10}O_3$; [111-46-6]

ORIGINAL MEASUREMENTS:

Mingoia, Q.

Ann. Chim. Farm. (Suppl. to Farm. Ital.) *Apr.*, 1939, 48-58.

VARIABLES:

One temperature: 20°C

PREPARED BY:

R. Piekos

EXPERIMENTAL VALUES:

Solubility of sulfanilamide in diethylene glycol at 20°C is 39.66 wt.% (2.303 mol kg^{-1} solution, compiler).

AUXILIARY INFORMATION

METHOD/APPARATUS/PROCEDURE:

Five cm^3 of a sulfanilamide soln in diethylene glycol were diluted with distilled water to 50 cm^3 and sulfanilamide was assayed in this solution by known methods (probably colorimetric).

SOURCE AND PURITY OF MATERIALS:

Source and purity of sulfanilamide were not specified.

Diethylene glycol was from Carbide and Carbon Co. (purity not specified).

ESTIMATED ERROR:

Nothing specified.

REFERENCES:

COMPONENTS:

(1) Benzenesulfonamide, 4-amino- (sulfanilamide); $C_6H_8N_2O_2S$; [63-74-1]

(2) Benzene; C_6H_6; [71-43-2]

ORIGINAL MEASUREMENTS:

Lin, H. O.; Ph.D. Dissertation; The University of Iowa; Iowa City, IA; 1971; p. 77.

VARIABLES:

Four temperatures; four crystalline forms

PREPARED BY:

J. K. Guillory

EXPERIMENTAL VALUES:

Solubilities of Sulfanilamide Polymorphic Forms: mole dm^{-3}

t/°C	α	β	γ	δ
25.0	2.70×10^{-4}	2.60×10^{-4}	2.54×10^{-4}	2.68×10^{-4}
30.0	3.12×10^{-4}	2.90×10^{-4}	2.88×10^{-4}	2.93×10^{-4}
33.0	3.54×10^{-4}	3.52×10^{-4}	3.43×10^{-4}	3.65×10^{-4}
37.0	3.97×10^{-4}	3.96×10^{-4}	3.97×10^{-4}	4.25×10^{-4}

AUXILIARY INFORMATION

METHOD/APPARATUS/PROCEDURE:

An excess of crystalline sulfanilamide was placed in 20-ml screw-capped vials with 20 ml of benzene. Vials were rotated end-over-end in a bath whose temperature was controlled to $\pm 0.1^{\circ}C$. Equilibrium was attained after 36 h or less. Supernatant was filtered through sintered glass, diluted 1:9 with 95% ethanol, and analyzed spectrophotometrically at 272 nm. Measurements were performed in duplicate, triplicate or quadruplicate. Residual crystals were analyzed by differential thermal analysis to detect change in crystal form. None was observed.

SOURCE AND PURITY OF MATERIALS:

Sulfanilamide (Mallinckrodt) was recrystd from 95% ethanol, methanol, n-pentanol, or n-butanol. Crystal forms were identified using densities, refractive indexes, x-ray diffraction measurements (1) and by infrared spectrophotometry and differential thermal analysis (2).

Benzene (Mallinckrodt).

ESTIMATED ERROR:

Uncertainty of temperature ($\pm 0.1^{\circ}C$).
Uncertainty of solubility measurements probably ($\pm$1-2%) based on agreement of replicate absorbance measurements.

REFERENCES:

1. Lin, H. O.; Baenziger, N. C.; and Guillory, J. K., *J. Pharm. Sci.* 1974, *63*, 145-6.

2. Lin, H. O.; Guillory, J. K., *J. Pharm. Sci.* 1970, *59*, 972-5.

COMPONENTS:

(1) Benzenesulfonamide, 4-amino- (sulfanilamide); $C_6H_8N_2O_2S$; [63-74-1]

(2) Methane, trichloro- (chloroform); $CHCl_3$; [67-66-3]

ORIGINAL MEASUREMENTS:

Yamazaki, M.; Aoki, M.; Kamada, A.; Yata, N. Yakuzaigaku 1967, *27(1)*, 37-40.

VARIABLES:

One temperature: 30°C

PREPARED BY:

R. Piekos

EXPERIMENTAL VALUES:

Solubility of sulfanilamide in chloroform at 30°C is 1.32 mmol/L (0.227 g dm^{-3}, compiler).

AUXILIARY INFORMATION

METHOD/APPARATUS/PROCEDURE:

Sulfanilamide (0.5 g) was placed in an L-shaped tube together with 20 ml of chloroform. The mixt was shaken in a thermostat until equilibrium was attained. The sulfanilamide was then assayed in the supernatant spectrophotometrically at 545 nm on a Beckmann DU spectrophotometer. The results were taken from a calibration graph.

SOURCE AND PURITY OF MATERIALS:

Nothing specified.

ESTIMATED ERROR:

Soly: not specified.

Temp: ± 1°C (authors).

REFERENCES:

COMPONENTS:	ORIGINAL MEASUREMENTS:
(1) Benzenesulfonamide, 4-amino- (sulfanilamide); $C_6H_8N_2O_2S$; [63-74-1] (2) Methane, trichloro-; $CHCl_3$; [67-66-3]	Kitao, K.; Kubo, K.; Morishita, T.; Yata, N.; Kamada, A. *Chem. Pharm. Bull.* 1973, *21*, 2417-26.
VARIABLES:	PREPARED BY:
One temperature: 37°C	R. Piekos

EXPERIMENTAL VALUES:

Solubility of sulfanilamide in $CHCl_3$ at 37°C is 1.40 mmol dm^{-3} solution.

AUXILIARY INFORMATION

METHOD/APPARATUS/PROCEDURE:

One ml of the $CHCl_3$ soln of sulfanilamide at equilibrium was taken into a test tube. After evapn of the solvent, the residue was dissolved in 1N NaOH, the soln was properly dild with deionized water, and the concn of sulfanilamide was detd by diazotization.

SOURCE AND PURITY OF MATERIALS:

Comm available sulfanilamide was used as supplied.

Purity of the $CHCl_3$ was not specified.

ESTIMATED ERROR:

Soly: not specified.

Temp: ± 1°C (authors).

REFERENCES:

COMPONENTS:	ORIGINAL MEASUREMENTS:
(1) Benzenesulfonamide, 4-amino- (sulfanilamide); $C_6H_8N_2O_2S$; [63-74-1] (2) 2-Propanone (acetone); C_3H_6O; [67-64-1]	Gutierrez, F. H. *Anales fis. quim. (Madrid)* 1945, *41*, 537-60.
VARIABLES:	PREPARED BY:
Temperature	R. Piekos

EXPERIMENTAL VALUES:

$t/^{\circ}C$	G^a	E^b	$X_g/1^c$	mol/1 acetone	mmol/mol acetone	$1{:}X_g{}^d$	$1 + X_{cc}{}^e$
0	25.620	20.395	208.701	1.212	86	3.90	4.79
5	27.299	21.445	220.997	1.283	92	3.66	4.53
10	27.597	21.628	222.104	1.289	93	3.62	4.52
15	28.318	22.069	225.723	1.311	95	3.53	4.43
20	29.965	23.056	237.083	1.377	101	3.34	4.22
25	31.132	23.741	244.445	1.419	105	3.21	4.09
30	32.516	24.537	253.397	1.471	110	3.08	3.95
35	34.509	25.656	266.858	1.549	116	2.90	3.75
40	37.067	27.043	284.452	1.652	125	2.70	3.52
45	39.914	28.578	303.905	1.765	135	2.51	3.29
50	45.110	31.087	340.806	1.979	152	2.22	2.93

[a] $G = \frac{p\ 100}{P - p}$, where p and P are the weights of solute and solution, resp.

[b] $E = \frac{G\ 100}{G + 100}$.

[c] g/1 acetone.

[d] g of acetone required to dissolve 1 g of solute.

[e] volume (cm^3) of acetone required to dissolve 1 g of solute.

AUXILIARY INFORMATION

METHOD/APPARATUS/PROCEDURE:

A special all-glass app was constructed enabling the prepn of satd solns, agitation by bubbling a stream of Me_2CO-satd N, filtration, and distn off the solvent without the contact with air. Two exchangeable dissoln vessels of 15 and 8 cm^3 working capacity were used depending on the soly of the solute. The app was immersed in a thermostat. The vols of acetone used were 15 or 5 cm^3 and the equilibration time was 2-2.5 h. The satd solns were filtered, weighed, the solvent was distd off, the residues were dried at 105°C, weighed, and examd for the presence of solvated Me_2CO.

SOURCE AND PURITY OF MATERIALS:

The source of the materials was not specified. Pure, anhyd Me_2CO was used. The absence of impurities and water was confirmed by procedures of the German Pharmacopeia VI and Spanish Pharmacopoeia VIII.

The purity of sulfanilamide was not specified.

ESTIMATED ERROR:

Soly: measurements were repeated until obtaining 2 values not differing in the second decimal (author).

Temp: $\pm 0.1^{\circ}C$ (author).

REFERENCES:

COMPONENTS:

(1) Benzenesulfonamide, 4-amino- (sulfanilamide); $C_6H_8N_2O_2S$; [63-74-1]

(2) Methylcyclohexanone; $C_7H_{12}O$; [1331-22-2]

ORIGINAL MEASUREMENTS:

Barber, H. J.; Wilkinson, J. H.

Quart. J. Pharm. Pharmacol. 1946, *19*, 248-55.

VARIABLES:

One temperature: 25°C

PREPARED BY:

R. Piekos

EXPERIMENTAL VALUES:

Approximate solubility of sulfanilamide in methylcyclohexanone at 25°C is 4.7 per cent w/v (0.27 mol dm^{-3} solution, compiler).

AUXILIARY INFORMATION

METHOD/APPARATUS/PROCEDURE:

Nothing specified.

SOURCE AND PURITY OF MATERIALS:

Nothing specified.

ESTIMATED ERROR:

Nothing specified.

REFERENCES:

COMPONENTS:	ORIGINAL MEASUREMENTS:
(1) Benzenesulfonamide, 4-amino- (sulfanilamide); $C_6H_8N_2O_2S$; [63-74-1] (2) Methylcyclohexanone; $C_7H_{12}O$; [1331-22-2]	Barber, H. J.; Wilkinson, J. H. *Pharm. J.* 1946, 105-6.
VARIABLES:	**PREPARED BY:**
One temperature: 25°C	R. Piekos

EXPERIMENTAL VALUES:

Approximate solubility of sulfanilamide in methylcyclohexanone at 25°C is 4.7 per cent w/v (0.27 mol dm^{-3} solution, compiler).

AUXILIARY INFORMATION

METHOD/APPARATUS/PROCEDURE:	SOURCE AND PURITY OF MATERIALS:
Nothing specified.	Nothing specified.
	ESTIMATED ERROR:
	Nothing specified.
	REFERENCES:

COMPONENTS:

(1) Benzenesulfonamide, 4-amino- (sulfanilamide); $C_6H_8N_2O_2S$; [63-74-1]

(2) Sorbitan monolaurate (Span 20); $C_{18}H_{34}O_6$; [1338-39-2]

ORIGINAL MEASUREMENTS:

Khawam, M. N.; Tawashi, R.; Czetsch-Lindenwald, H. v. *Sci. Pharm.* 1964, *32*, 271-9.

VARIABLES:

One temperature: 45°C

PREPARED BY:

R. Piekos

EXPERIMENTAL VALUES:

Solubility of sulfanilamide in Span 20 at 45°C is 0 g/100 g Span 20.

AUXILIARY INFORMATION

METHOD/APPARATUS/PROCEDURE:

Small weighed samples of sulfanilamide were added to Span 20 under stirring until dissoln occurred.

SOURCE AND PURITY OF MATERIALS:

Sulfanilamide conformed to the requirements of USP XVI.

Source and purity of Span 20 was not specified.

ESTIMATED ERROR:

Soly: measurements were made in duplicate (authors).

Temp: ± 1°C (authors).

REFERENCES:

COMPONENTS:	ORIGINAL MEASUREMENTS:
(1) Benzenesulfonamide, 4-amino- (sulfanilamide); $C_6H_8N_2O_2S$; [63-74-1] (2) Sorbitan monooleate, polyoxyethylene derivatives (Tween 80); [9005-65-6]	Khawam, M. N.; Tawashi, R.; Czetsch-Lindenwald, H. v. *Sci. Pharm.* 1964, *32*, 271-9.
VARIABLES:	PREPARED BY:
One temperature: 45°C	R. Piekos

EXPERIMENTAL VALUES:

Solubility of sulfanilamide in Tween 80 at 45°C is 30 g/100 g Tween 80 (1.7 mol kg^{-1}, compiler).

AUXILIARY INFORMATION

METHOD/APPARATUS/PROCEDURE:

Small weighed samples of sulfanilamide were added to Tween 80 under stirring until dissoln occurred.

SOURCE AND PURITY OF MATERIALS:

Sulfanilamide conformed to the requirements of USP XVI.

Tween 80 was a commercially available reagent with a HLB value of 15.

ESTIMATED ERROR:

Soly: measurements were made in duplicate (authors).

Temp: ±1°C (authors).

REFERENCES:

COMPONENTS:	ORIGINAL MEASUREMENTS:
(1) Benzenesulfonamide, 4-amino-, (sulfanilamide); $C_6H_8N_2O_2S$; [63-74-1] (2) Butane, 1,1' - oxybis-, (butyl ether); $C_8H_{18}O$; [142-96-1]	Lin, H. O.; Ph.D. Dissertation; The University of Iowa; Iowa City, IA; 1971; p. 76.
VARIABLES:	**PREPARED BY:**
Four temperatures; four crystalline forms	J. K. Guillory

EXPERIMENTAL VALUES:

Solubilities of Sulfanilamide Polymorphic Forms: mol dm^{-3}

t/°C	α	β	γ	δ
30.0	2.55×10^{-4}	2.60×10^{-4}	2.92×10^{-4}	2.83×10^{-4}
33.0	2.76×10^{-4}	2.80×10^{-4}	3.05×10^{-4}	3.06×10^{-4}
37.0	2.95×10^{-4}	2.98×10^{-4}	3.26×10^{-4}	3.37×10^{-4}
45.0	3.56×10^{-4}	3.60×10^{-4}	3.84×10^{-4}	3.88×10^{-4}

AUXILIARY INFORMATION

METHOD/APPARATUS/PROCEDURE:

An excess of crystalline sulfanilamide was placed in 20-ml screw-capped vials with 20 ml of butyl ether. Vials were rotated end-over-end in a bath whose temperature was controlled to ±0.1°C. Equilibrium was attained after 36 h or less. Supernatant was filtered through sintered glass, diluted 2:23 with 95% ethanol, and analyzed spectrophotometrically at 262 nm. Measurements were performed in triplicate. Residual crystals were analyzed by differential thermal analysis to detect change in crystal form. None was observed.

SOURCE AND PURITY OF MATERIALS:

Sulfanilamide (Mallinckrodt) was recrystd from 95% ethanol, methanol, n-pentanol, or n-butanol. Crystal forms were identified using densities, refractive indexes, x-ray diffraction measurements (1) and by infrared spectrophotometry and differential thermal analysis (2).

Butyl ether (Matheson Coleman & Bell).

ESTIMATED ERROR:

Uncertainty of temperature (±0.1°C).
Uncertainty of solubility measurements probably (±1-2%) based on agreement of triplicate absorbance measurements.

REFERENCES:

1. Lin, H. O.; Baenziger, N. C.; and Guillory, J. K., *J. Pharm. Sci.* 1974, *63*, 145-6.
2. Lin, H. O.; Guillory, J. K., *J. Pharm. Sci.* 1970, *59*, 972-5.

COMPONENTS:

(1) Benzenesulfonamide, 4-amino-, monohydrate, (sulfanilamide monohydrate); $C_6H_8N_2O_2S \cdot H_2O$; [20203-81-0]

(2) Water; H_2O; [7732-18-5]

ORIGINAL MEASUREMENTS:

Burger, A.

Pharm. Ind. 1973, *35*, 626-33.

VARIABLES:

Temperature

PREPARED BY:

R. Piekos

EXPERIMENTAL VALUES:

$t/^{\circ}C$	Saturation solubility, C_s [a]	
	mg/100 ml solution	10^2 mol dm^{-3} [b]
4.4	182	0.957
10.2	251	1.32
15.0	337	1.77
20.0	465	2.44
25.0	600	3.15
30.0	820	4.31
35.0	1100	5.78
37.0	1270	6.68

[a] $C_s = [HA] + [A^-]$, where [HA] is the molar concentration of the dissolved, undissociated molecules of sulfanilamide and $[A^-]$ is the concentration of the dissolved anion of sulfanilamide.

[b] Calculated by compiler.

AUXILIARY INFORMATION

METHOD/APPARATUS/PROCEDURE:

Sulfanilamide was assayed spectrophotometrically at 258.5 nm using a Zeiss PMQ II spectrophotometer and a 1/15 M phosphate buffer of pH 7.00 ($E^{1\%}_{1\ cm} = 945$).

SOURCE AND PURITY OF MATERIALS:

Sulfanilamide monohydrate was obtained by vacuum evapn of a sulfanilamide (source and purity not specified) soln at 40°C and cooling the crystals to 20°C at normal pressure.

Purity of the water was not specified.

ESTIMATED ERROR:

Soly: not specified.

Temp: $\pm 0.1^{\circ}C$ (author).

REFERENCES:

COMPONENTS:

(1) Benzenesulfonamide, 4-amino-, monohydrate (sulfanilamide monohydrate); $C_6H_8N_2O_2S \cdot H_2O$; [20203-81-0]

(2) Water; H_2O; [7732-18-5]

ORIGINAL MEASUREMENTS:

Sekiguchi, K.; Tsuda, Y.; Kanke, M.

Chem. Pharm. Bull. 1975, *23*.

VARIABLES:

Temperature

PREPARED BY:

R. Piekos

EXPERIMENTAL VALUES:

t/°C	Solubility	
	g/100 g solution[a]	10^2 mol kg^{-1} water[a,b]
15	0.386	2.03
20	0.520	3.04
25	0.726	4.25
30	0.970	5.69
35	1.291[c]	7.80
40	1.709[c]	10.10
45	2.237	13.29
50	2.897[c]	17.33

[a] Based on anhydrous sulfanilamide.

[b] Calculated by compiler.

[c] Figure obtained by extrapolation of the experimental solubility above and below the transition temperature.

AUXILIARY INFORMATION

METHOD/APPARATUS/PROCEDURE:

A sufficient amt of sample powder was placed in 50 g of distd water in a dissoln measurement cell and stirred at 600 rpm. At appropriate time intervals samples were taken by glass syringes until the concn attained equilibrium. The sample solns were immediately filtered through a 0.45 μ membrane filter. The filtrate was weighed and dild for spectrophotometric assay at 225 nm on a Hitachi-139 UV spectrophotometer.

SOURCE AND PURITY OF MATERIALS:

The sulfanilamide monohydrate was isolated by cooling the warm satd aq soln of sulfanilamide rapidly and maintaining it below 15°C. The hydrate was characterized by instrumental methods.

Distilled water was used.

ESTIMATED ERROR:

Nothing specified.

REFERENCES:

COMPONENTS:

(1) Benzenesulfonamide, 4-amino-, monohydrate (sulfanilamide monohydrate); $C_6H_8N_2O_2S \cdot H_2O$ [20203-81-0]

(2) Hydrochloric acid; HCl; [7647-01-0]

(3) Water; H_2O; [7732-18-5]

ORIGINAL MEASUREMENTS:

Burger, A.

Pharm. Ind. 1973, *35*, 626-33.

VARIABLES:

One temperature; 20.0°C

PREPARED BY:

R. Piekos

EXPERIMENTAL VALUES:

Saturation solubility, C_s[a], of sulfanilamide monohydrate in 0.1 N hydrochloric acid (pH 1.79) at 20.0°C is 2006 mg/100 ml solution (117 mmol/liter based on anhydrous sulfanilamide).

[a] $C_s = [HA] + [A^-]$, where [HA] is the molar concentration of the dissolved, undissociated molecules of sulfanilamide and $[A^-]$ is the concentration of the dissolved anion of sulfanilamide.

AUXILIARY INFORMATION

METHOD/APPARATUS/PROCEDURE:

Sulfanilamide was assayed spectrophotometrically at 258.5 nm using a Zeiss PMQ II spectrophotometer and a 1/15 M phosphate buffer of pH 7.00 ($E^{1\%}_{1\ cm}$ = 945)

SOURCE AND PURITY OF MATERIALS:

Sulfanilamide monohydrate was obtained by vacuum evapn of a sulfanilamide (source and purity not specified) soln at 40°C and cooling the crystals to 20°C at normal pressure.

Source and purity of the remaining materials was not specified.

ESTIMATED ERROR:

Soly: not specified.

Temp: ±0.1°C (author).

REFERENCES:

COMPONENTS:

(1) Benzenesulfonamide, 4-amino-, monohydrate (sulfanilamide monohydrate) $C_6H_8N_2O_2S \cdot H_2O$ [20203-81-0]

(2) Sodium hydroxide; NaOH; [1310-73-2]

(3) Water; H_2O; [7732-18-5]

ORIGINAL MEASUREMENTS:

Burger, A.

Pharm. Ind. 1973, *35*, 626-33.

VARIABLES:

pH

PREPARED BY:

R. Piekos

EXPERIMENTAL VALUES:

Concn of NaOH soln	pH	Saturation solubility, C_s[a], at 20.0°C	
		mmol/liter[b]	mg/100 ml solution[b]
0.1 N	11.23	128.4	2212
0.1 N	11.23	129.0	2222
approx. 0.25 N	11.37	178.2	3069

[a] C_s = [HA] + [A^-] , where [HA] is the molar concentration of the dissolved, undissociated molecules of sulfanilamide and [A^-] is the concentration of the dissolved anion of sulfanilamide.

[b] Based on anhydrous sulfanilamide.

AUXILIARY INFORMATION

METHOD/APPARATUS/PROCEDURE:

Sulfanilamide was assayed spectrophotometrically at 258.5 nm using a Zeiss PMQ II spectrophotometer, in 1/15M phosphate buffer of pH 7.00 ($E_{1\ cm}^{1\%}$ = 945).

SOURCE AND PURITY OF MATERIALS:

Sulfanilamide monohydrate was obtained by vacuum evapn of a sulfanilamide (source and purity not specified) soln at 40°C and cooling the crystals to 20°C at normal pressure.

Source and purity of the remaining materials was not specified.

ESTIMATED ERROR:

Soly and pH: not specified.

Temp: ±0.1°C (author).

REFERENCES:

COMPONENTS:	ORIGINAL MEASUREMENTS:
(1) Benzenesulfonamide, 4-amino-, monohydrate (sulfanilamide monohydrate); $C_6H_8N_2O_2S\ H_2O$; [20203-81-0] (2) Phosphoric acid, disodium salt; Na_2HPO_4; [7558-94-4] (3) Phosphoric acid, monopotassium salt; KH_2PO_4; [7778-77-0] (4) Water; H_2O; [7732-18-5]	Burger, A. *Pharm. Ind.* 1973, *35*, 626-33.
VARIABLES:	**PREPARED BY:**
pH	R. Piekos

EXPERIMENTAL VALUES:

Concn of the phosphate buffer (1)	pH	Satn solubility[a] in phosphate buffer at 20.0°C	
		mol/liter soln[b]	mg/100 ml soln[b]
1/15 M	5.50	26.6	458
0.05 M	6.88	26.9	464

[a]Saturation solubility is the sum of the concentrations of the dissolved, undissociated molecules of sulfanilamide and of its dissolved anion, expressed in mol dm^{-3}.

[b]Based on anhydrous sulfanilamide.

AUXILIARY INFORMATION

METHOD/APPARATUS/PROCEDURE:

Sulfanilamide was assayed spectrophotometrically at 258.5 nm using a Zeiss PMQ II spectrophotometer, in 1/15M phosphate buffer of pH 7.00 ($E^{1\%}_{1\ cm}$ = 945).

SOURCE AND PURITY OF MATERIALS:

Sulfanilamide monohydrate was obtained by vacuum evapn of a sulfanilamide (source and purity not specified) soln at 40°C and cooling the crystals to 20°C at normal pressure.

Source and purity of the remaining materials was not specified.

ESTIMATED ERROR:

Soly and pH: not specified.

Temp: ±0.1°C (author).

REFERENCES:

1. Küster, F. W.; Thiel, A.; Fischbeck, K. *Logarithmische Rechentafeln, 100, Aufl., Walter de Gruyter, Berlin,* 1969.

COMPONENTS:

(1) Benzenesulfonamide, 4-amino-, monohydrate (sulfanilamide monohydrate); $C_6H_8N_2O_2S \cdot H_2O$; [20203-81-0]

(2) Boric acid, trisodium salt; Na_3BO_3; [14312-40-4]

(3) Hydrochloric acid; HCl; [7647-01-0]

(4) Water; H_2O; [7732-18-5]

ORIGINAL MEASUREMENTS:

Burger, A.

Pharm. Ind. 1973, *35*, 626-33.

VARIABLES:

One temperature: 20.0°C; one pH: 8.50

PREPARED BY:

R. Piekos

EXPERIMENTAL VALUES:

Saturation solubility, $C_s^{\,a}$, of sulfanilamide monohydrate in a borate buffer (1) of pH 8.50 at 20.0°C is 482 mg/100 ml solution or 28.0 mmol/liter based on anhydrous sulfanilamide.

[a] $C_s = [HA] + [A^-]$, where HA is the molar concentration of the dissolved, undissociated molecules of sulfanilamide and $[A^-]$ is the concentration of the dissolved anion of sulfanilamide.

AUXILIARY INFORMATION

METHOD/APPARATUS/PROCEDURE:

Sulfanilamide was assayed spectrophotometrically at 258.5 nm using a Zeiss PMQ II spectrophotometer.

SOURCE AND PURITY OF MATERIALS:

Sulfanilamide monohydrate was obtained by vacuum evapn of a sulfanilamide (source and purity not specified) soln at 40°C and cooling the crystals to 20°C at normal pressure.

Source and purity of the remaining materials was not specified.

ESTIMATED ERROR:

Soly: not specified.

pH: not specified.

Temp: ±0.1°C (author).

REFERENCES:

1. Küster, F. W.; Thiel, A.; Fischbeck, K. *Logarithmische Rechentafeln, 100. Aufl., Walter de Gruyter, Berlin* 1969.

COMPONENTS:

(1) Benzenesulfonamide, 4-amino-, monohydrate (sulfanilamide monohydrate); $C_6H_8N_2O_2S \cdot H_2O$; [20203-81-0]

(2) Hydrochloric acid; HCl; [7647-01-0]

(3) 1,2,3-Propanetricarboxylic acid, disodium salt; $C_6H_7Na_2O_7$; [144-33-2]

(4) Water; H_2O; [7732-18-5]

ORIGINAL MEASUREMENTS:

Burger, A.

Pharm. Ind. 1973, *35*, 626-33.

VARIABLES:

One temperature: 20.0°C

PREPARED BY:

R. Piekos

EXPERIMENTAL VALUES:

Saturation solubility, C_s[a], of sulfanilamide monohydrate in the buffer solution of pH 3.85 at 20.0°C is 485 mg/100 ml solution (28.2 mmol/liter based on anhydrous sulfanilamide).

[a] $C_s = [HA] + [A^-]$, where [HA] is the molar concentration of the dissolved, undissociated molecules of sulfanilamide and $[A^-]$ is the concentration of the dissolved anion of sulfanilamide.

AUXILIARY INFORMATION

METHOD/APPARATUS/PROCEDURE:

Sulfanilamide was assayed spectrophotometrically at 258.5 nm using a Zeiss PMQ II spectrophotometer.

SOURCE AND PURITY OF MATERIALS:

Sulfanilamide monohydrate was obtained by vacuum evapn of a sulfanilamide (source and purity not specified) soln at 40°C and cooling the crystals to 20°C at normal pressure.

Source and purity of the remaining materials was not specified.

ESTIMATED ERROR:

Soly and pH: not specified.

Temp: ±0.1°C (author).

REFERENCES:

1. Küster, F. W.; Thiel, A.; Fischbeck, K. *Logarithmische Rechentafeln, 100. Aufl., Walter de Gruyter, Berlin* 1969.

COMPONENTS:

(1) Benzenesulfonamide, 4-amino-, monohydrate (sulfanilamide monohydrate) $C_6H_8N_2O_2S \cdot H_2O$; [20203-81-0]

(2) Aminoacetic acid; $C_2H_5NO_2$; [56-40-6]

(3) Hydrochloric acid; HCl; [7646-01-0]

(4) Sodium chloride; NaCl; [7647-14-5]

(5) Water; H_2O; [7732-18-5]

ORIGINAL MEASUREMENTS:

Burger, A.

Pharm. Ind. 1973, *35*, 626-33.

VARIABLES:

pH

PREPARED BY:

R. Piekos

EXPERIMENTAL VALUES:

pH	Saturation solubility, C_s[a], in buffer solution (1) at 20°C	
	mmol liter^{-1}[b]	mg/100 ml solution[b]
2.00	82.9	1410
2.11	67.6	1164
2.22	59.6	1026
2.33	53.2	917
2.35	52.6	906
2.49	47.5	818
2.60	43.5	749
2.75	37.5	646
3.10	33.4	577

[a] $C_s = [HA] + [A^-]$, where [HA] is the molar concentration of the dissolved, undissociated molecules of sulfanilamide and $[A^-]$ is the concentration of the dissolved anion of sulfanilamide.

[b] Based on anhydrous sulfanilamide.

AUXILIARY INFORMATION

METHOD/APPARATUS/PROCEDURE:

Sulfanilamide was assayed spectrophotometrically at 258.5 nm using a Zeiss PMQ II spectrophotometer.

SOURCE AND PURITY OF MATERIALS:

Sulfanilamide monohydrate was obtained by vacuum evapn of a sulfanilamide (source and purity not specified) soln at 40°C and cooling the crystals to 20°C at normal pressure.

Source and purity of the remaining materials was not specified.

ESTIMATED ERROR:

Soly and pH: not specified.

Temp: ±0.1°C (author).

REFERENCES:

1. Küster, F. W.; Thiel, A.; Fischbeck, K. *Logarithmische Rechentafeln, 100. Aufl., Walter de Gruyter, Berlin* 1969.

COMPONENTS:

(1) Benzenesulfonamide, 4-amino-, monohydrate (sulfanilamide monohydrate) $C_6H_8N_2O_2S \cdot H_2O$; [20203-81-0]

(2) Aminoacetic acid; $C_2H_5NO_2$; [56-40-6]

(3) Sodium chloride; NaCl; [7647-14-5]

(4) Sodium hydroxide; NaOH; [1310-73-2]

(5) Water; H_2O; [7732-18-5]

ORIGINAL MEASUREMENTS:

Burger, A.

Pharm. Ind. 1973, *35*, 626-33.

VARIABLES:

pH

PREPARED BY:

R. Piekos

EXPERIMENTAL VALUES:

pH of buffer soln	Saturation solubility, C_s[a], at 20.0°C	
	mmol/liter[b]	mg/100 ml solution[b]
9.20	28.0	482
9.95	32.5	560
9.95 (sic!)	34.0	586
10.04	35.6	613
10.29	37.8	651
10.33	40.4	696
10.42	43.8	755
10.47	46.1	794
10.75	58.5	1007
10.84	72.3	1245
10.95	83.0	1429
11.15	113.5	1955

[a] $C_s = [HA] + [A^-]$, where [HA] is the molar concentration of the dissolved, undissociated molecules of sulfanilamide and $[A^-]$ is the concentration of the dissolved anion of sulfanilamide.

[b] Based on anhydrous sulfanilamide.

AUXILIARY INFORMATION

METHOD/APPARATUS/PROCEDURE:

Sulfanilamide was assayed spectrophotometrically at 258.5 nm using a Zeiss PMQ II spectrophotometer.

SOURCE AND PURITY OF MATERIALS:

Sulfanilamide monohydrate was obtained by vacuum evapn of a sulfanilamide (source and purity not specified) soln at 40°C and cooling the crystals to 20°C at normal pressure.

Source and purity of the remaining materials was not specified.

ESTIMATED ERROR:

Soly and pH: not specified.

Temp: ±0.1°C (author).

REFERENCES:

1. Küster, F. W.; Thiel, A.; Fischbeck, K. *Logarithmische Rechentafeln, 100. Aufl., Walter de Gruyter, Berlin* 1969.

COMPONENTS:

(1) Benzenesulfonamide, 4-amino-, monohydrate (sulfanilamide monohydrate); $C_6H_8N_2O_2O_2S{\cdot}H_2O$; [20203-81-0]

(2) Ethanol; C_2H_6O; [64-17-5]

(3) Water; H_2O; [7732-18-5]

ORIGINAL MEASUREMENTS:

Burger, A.

Pharm. Ind. 1973, *35*, 626-33.

VARIABLES:

Concentration of ethanol

PREPARED BY:

R. Piekos

EXPERIMENTAL VALUES:

Concentration of ethanol Volume %	Saturation solubility, C_s[a], at 20.0°C	
	mg/100 ml soln	10^2 mol dm^{-3} [b]
35	1441	7.576
30	1224	6.435
20	906	4.763
10	689	3.622
0	469	2.466

[a] C_s = [HA] + [A^-], where [HA] is the molar concentration of the dissolved, undissociated molecules of sulfanilamide and [A^-] is the concentration of the dissolved anion of sulfanilamide.

[b] Calculated by compiler.

AUXILIARY INFORMATION

METHOD/APPARATUS/PROCEDURE:

Sulfanilamide was assayed spectrophotometrically at 258.5 nm using a Zeiss PMQ II spectrophotometer, in 1/15M phosphate buffer of pH 7.00 ($E^{1\%}_{1\ cm}$ = 945).

SOURCE AND PURITY OF MATERIALS:

Sulfanilamide monohydrate was obtained by vacuum evapn of a sulfanilamide (source and purity not specified) soln at 40°C and cooling the crystals to 20°C at normal pressure.

Source and purity of EtOH and water was not specified.

ESTIMATED ERROR:

Soly: not specified.

Temp: ±0.1°C (author).

REFERENCES:

COMPONENTS:

(1) Benzenesulfonamide, 4-amino-, monohydrochloride (sulfanilamide-HCl); $C_6H_8N_2O_2S{\cdot}HCl$; [6101-31-1]

(2) 2-Propanone (acetone); C_3H_6O; [67-64-1]

ORIGINAL MEASUREMENTS:

Gutierrez, F. H.

Anales fis. quim. (Madrid) 1945, *41*, 537-60.

VARIABLES:

Temperature

PREPARED BY:

R. Piekos

EXPERIMENTAL VALUES:

$t/^{\circ}C$	G^a	E^b	$X_g/1^c$	mol/1[d] acetone	mmol/mol acetone	$1{:}X_g$ [e]	$1 + X_{cc}$ [f]
15	0.163	0.163	1.299	6.11	0.46	615.5	769.8
20	0.164	0.164	1.298	6.10	0.46	609.7	770.4
25	0.164	0.164	1.298	6.10	0.46	609.7	770.4

[a] $G = \frac{p\ 100}{P - p}$, where p and P are the weights of solute and solution, resp.

[b] $E = \frac{G\ 100}{G + 100}$; [c] g/l acetone; [d] should be mmol/l acetone (compiler);

[e] g of acetone required to dissolve 1 g of solute;

[f] volume (cm^3) of acetone required to dissolve 1 g of solute.

AUXILIARY INFORMATION

METHOD/APPARATUS/PROCEDURE:

A special all-glass app was constructed enabling the prepn of satd solns, agitation by bubbling a stream of acetone-satd N, filtration, and distn off the solvent without contact with air. Two exchangeable dissoln vessels of 15 and 8 cm^3 working capacity were used depending on the soly of solute. The app was immersed in a thermostat. The vols of acetone used were 15 or 5 cm^3 and the equilibration time was 2-2.5 h. The satd solns were filtered, weighed, the solvent was distd off, the residues were dried at 105°C, weighed, and examd for the presence of solvated acetone.

SOURCE AND PURITY OF MATERIALS:

The source of the materials was not specified. Pure, anhyd acetone was used. The absence of impurities and water was confirmed by procedures of the German Pharmacopeia VI and Spanish Pharmacopeia VIII.

The purity of sulfanilamide-HCl was not specified.

ESTIMATED ERROR:

Soly: measurements were repeated until obtaining 2 values not differing in the second decimal (author).

Temp: ±0.1°C (author).

REFERENCES:

COMPONENTS:

(1) Benzenesulfonamide, 4-amino-, benzenesulfonamide, 4-amino-N-2-thiazolyl- (molecular compound); $C_6H_8N_2O_2S \cdot C_9H_9N_3O_2S_2$; [1704-78-3]

(2) Water; H_2O; [7732-18-5]

ORIGINAL MEASUREMENTS:

Ito, K.; Sekiguchi, K.

Chem. Pharm. Bull. 1967, *15(4)*, 420-6.

VARIABLES:

Temperature

PREPARED BY:

R. Piekos

EXPERIMENTAL VALUES:

$t/^{\circ}C$	Solubility	
	10^3 mol dm^{-3} solution	g dm^{-3} [a]
20	0.173	0.074
25	0.231	0.099
30	0.320	0.137
35	0.412	0.176

[a] Calculated by compiler.

AUXILIARY INFORMATION

METHOD/APPARATUS/PROCEDURE:

The earlier described method (1) was used: in a 200-ml egg-plant type flask immersed in a thermostat, an excess (0.6 and 0.8 g per 100 ml of water at expts at 20°C and 30°C, resp) of the mol compd was placed with 100 ml of distd water which was previously kept at appropriate temp. Immediately after addn of water the mixt was vigorously agitated with an elec stirrer. The soly equilibrium at 20°C and 30°C was attained after 15 and about 9 h, resp. Aliquots were withdrawn at certain time intervals with a pipet equipped with a filter and the concn of solute was detd spectrophotometrically in the uv region.

SOURCE AND PURITY OF MATERIALS:

The source and purity of the mol compd was not specified.

Distd water was used.

ESTIMATED ERROR:

Nothing specified.

REFERENCES:

1. Sekiguchi, K.; Ito, K. *Chem. Pharm. Bull.* 1965, *13(4)*, 405.

COMPONENTS:

(1) Benzenesulfonamide, 4-amino-, monosodium salt (Na sulfanilamide); $C_6H_7N_2NaO_2S$; [10103-15-8]

(2) 2-Propanone (acetone); C_3H_6O; [67-64-1]

ORIGINAL MEASUREMENTS:

Gutierrez, F. H.

Anales fis. quim. (Madrid) 1945, *41*, 537-60.

VARIABLES:

Temperature

PREPARED BY:

R. Piekos

EXPERIMENTAL VALUES:

$t/^{\circ}C$	G^a	E^b	$X_g/1^c$	$mol/1^d$ acetone	mmol/mol acetone	$1:X_g^{\ e}$	$1 + X_{cc}^{\ f}$
10	0.240	0.239	1.927	9.9	0.72	416.66	518.94
20	0.248	0.247	1.962	10.1	0.74	404.86	509.68
30	0.256	0.255	2.000	10.3	0.77	392.16	500.00

[a] $G = \frac{p\ 100}{P - p}$, where p and P are the weights of solute and solution, resp.

[b] $E = \frac{G\ 100}{G + 100}$; [c] g/1 acetone; [d] should be mmol/1 acetone (compiler).

[e] g of acetone required to dissolve 1 g of solute;

[f] volume (cm^3) of acetone required to dissolve 1 g of solute.

AUXILIARY INFORMATION

METHOD/APPARATUS/PROCEDURE:

A special all-glass app was constructed enabling the prepn of satd solns, agitation by bubbling a stream of acetone-satd N, filtration, and distn off the solvent without the contact with air. Two exchangeable dissoln vessels of 15 and 8 cm^3 working capacity were used depending on the soly of solute. The app was immersed in a thermostat. The vols of acetone used were 15 or 5 cm^3 and the equilibration time was 2-2.5 h. The satd solns were filtered, weighed, the solvent was distd off, the residues were dried at $105^{\circ}C$, weighed, and examd for the presence of solvated acetone.

SOURCE AND PURITY OF MATERIALS:

The source of the materials was not specified. Pure, anhyd acetone was used. The absence of impurities and water was confirmed by procedures of the German Pharmacopeia VI and Spanish Pharmacopeia VIII.

The purity of Na sulfanilamide was not specified.

ESTIMATED ERROR:

Soly: measurements were repeated until obtaining 2 values not differing in the second decimal (author).

Temp: $\pm 0.1^{\circ}C$ (author).

REFERENCES:

COMPONENTS:	ORIGINAL MEASUREMENTS:
(1) Benzenesulfonamide, 4-amino-, zinc salt (2:1) (Zn(II) sulfanilamide); $C_{12}H_{14}N_4O_4S_2Zn$; [78739-60-3] (2) Water; H_2O; [7732-18-5]	Fox, Ch. L., Jr.; Modak, S.; Stanford, J. W.; Fox, P. L. *Scand. J. Plast. Reconstr. Surg.* 1979, *13(1)*, 89-94.
VARIABLES:	PREPARED BY:
One temperature: 28-30°C	R. Piekos

EXPERIMENTAL VALUES:

Solubility of Zn(II) sulfanilamide in water at room temperature (28-30°C)[a] is 197.0 mg% (4.831 x 10^{-3} mol dm^{-3} solution, compiler).

[a] Values given by one of the authors (S. M.) in personal communication.

AUXILIARY INFORMATION

METHOD/APPARATUS/PROCEDURE:

A satd soln of Zn(II) sulfanilamide was prepd in water and after 24 h aliquots from the clear supernatant were assayed for sulfanilamide content using the colorimetric method of Bratton and Marshall (1). The soly value was then calcd from the molecular formula.

SOURCE AND PURITY OF MATERIALS:

The Zn(II) sulfanilamide was prepd by the authors as follows: an inorg Zn salt was reacted with a Na salt of sulfanilamide and the ppt was said to be analyzed and characterized. No analytical results or spectral evidence was provided, however.

Purity of the water was not specified.

ESTIMATED ERROR:

Nothing specified.

REFERENCES:

1. Bratton, A. C.; Marshall, E. K., *J. Biol. Chem.* 1939, *120*, 537.

COMPONENTS:	ORIGINAL MEASUREMENTS:
(1) Acetamide, N-[(4-aminosulfonyl)-phenyl]- (acetyl sulfanilamide); $C_8H_{10}N_2O_3S$; [121-61-9] (2) Water; H_2O; [7732-18-5]	Durel, M. P.; Allinne, M. *Bull. Soc. Med. Hop. Paris III* 1941, 251-9.
VARIABLES:	**PREPARED BY:**
One temperature: 37°C	R. Piekos

EXPERIMENTAL VALUES:

Solubility of acetyl sulfanilamide in water at 37°C is 14.8 g/liter (6.90×10^{-2} mol dm^{-3}, compiler).

AUXILIARY INFORMATION

METHOD/APPARATUS/PROCEDURE:

A mixt of acetyl sulfanilamide and water was agitated for 24 hours at 37°C.

SOURCE AND PURITY OF MATERIALS:

Source and purity of acetyl sulfanilamide was not specified.

Distilled water was used.

ESTIMATED ERROR:

Nothing specified.

REFERENCES:

COMPONENTS:

(1) Acetamide, N-[4-(aminosulfonyl) phenyl]-; $C_8H_{10}N_2O_3S$; [121-61-9]

(2) Water; H_2O [7732-18-5]

ORIGINAL MEASUREMENTS:

Sapozhnikova, N. V.; Postovskii, I. Ya.

Zh. Prikl. Khim. 1944, *17*, 427-34.

VARIABLES:

Temperature

PREPARED BY:

R. Piekos

EXPERIMENTAL VALUES:

$t/^{\circ}C$	Solubility	
	Weight %	10^2 mol kg^{-1} water[a]
20	0.133	0.622
37	0.289	1.35
50	0.529[b]	2.48
75	1.50	7.11
99	3.55	17.2

[a] calculated by compiler.

[b] calculated from the heat of dissolution (9240 cal mol^{-1}).

AUXILIARY INFORMATION

METHOD/APPARATUS/PROCEDURE:

The sulfonamide was dissolved in water to form a satd soln which was occasionally agitated in a glass vessel immersed in a thermostat. The equilibrium was usually attained after 1 h. Five to 100- cm^3 samples of the satd soln were placed in Pt crucibles or dishes and evapd to dryness at temps lower than 110-115°C. The residue was dried to const weights at 105-110°C and weighed.

SOURCE AND PURITY OF MATERIALS:

Pure, recrystd sulfonamide was used. Its mp conformed to that reported in the literature.

Purity of the water was not specified.

ESTIMATED ERROR:

Soly: quite reliable results were obtained over the temp range 20-75°C. At higher temps the accuracy was poor due to evapn of water during sampling (authors).

Temp: $\pm$0.05°C (authors).

REFERENCES:

COMPONENTS:

(1) Acetamide, N-[4-(aminosulfonyl)-phenyl]- (acetyl sulfanilamide); $C_8H_{10}N_2O_3S$; [121-61-9]

(2) Phosphoric acid, monopotassium salt; KH_2PO_4; [7778-77-0]

(3) Water; H_2O; [7732-18-5]

ORIGINAL MEASUREMENTS:

Krüger-Thiemer, E.

Arch. Dermatol. Syphilis 1942, *183*, 90-116.

VARIABLES:

One temperature: ca 20°C; one pH: 4.37

PREPARED BY:

R. Piekos

EXPERIMENTAL VALUES:

Solubility of acetyl sulfanilamide in a 0.735M (10%) KH_2PO_4 solution of pH 4.37 at room temperature (about 20°C) is 0.128 g% (5.97×10^{-3} mol dm^{-3} solution, compiler).

AUXILIARY INFORMATION

METHOD/APPARATUS/PROCEDURE:

Acetyl sulfanilamide (0.5 g) was dissolved in 10 cm^3 of the 0.735 M (10%) KH_2PO_4 soln of pH 4.37, shaken for 2 h a room temp (about 20°C), and filtered. The filtrate was treated with equal vol of 2 N HCl, and refluxed for 15 min. After proper diln, a 1-cm^3 aliquot was withdrawn, acidified, cooled, and the sulfonamide content was detd colorimetrically (as sulfanilamide) by the Marshall method modified by Kimmig (1) using an Authenreith colorimeter. The pH was detd on an ultraionograph using a glass electrode.

SOURCE AND PURITY OF MATERIALS:

Acetyl sulfanilamide (source not specified) gave no coloration upon diazotization of its satd soln, thus showing absence of sulfanilamide.

The source and purity of the remaining materials was not specified.

ESTIMATED ERROR:

Soly: precision ±5% (author).

Temp: not specified.

pH: ±0.05 pH unit (author).

REFERENCES:

1. Kimmig, J. *Arch. Dermatol.* 1938, *176*, 722; *Erg. Hyg.* 1941, *24*, 398.

COMPONENTS:	ORIGINAL MEASUREMENTS:
(1) Acetamide, N-[4-(aminosulfonyl)-phenyl]- (acetyl sulfanilamide); $C_8H_{10}N_2O_3S$; [121-61-9] (2) Phosphoric acid, disodium salt; Na_2HPO_4; [7558-94-4] (3) Water; H_2O; [7732-18-5]	Krüger-Thiemer, E. *Arch. Dermatol. Syphilis* 1942, *183*, 90-116.
VARIABLES:	**PREPARED BY:**
One temperature: ca 20 °C; one pH: 8.74	R. Piekos

EXPERIMENTAL VALUES:

Solubility of acetyl sulfanilamide in a 0.705 M (10%) Na_2HPO_4 solution of pH 8.74 at room temperature (about 20 °C) is 0.278 g% (1.111 x 10^{-2} mol dm^{-3} solution, compiler).

AUXILIARY INFORMATION

METHOD/APPARATUS/PROCEDURE:

Acetyl sulfanilamide (0.5 g) was dissolved in 10 cm^3 of the 0.705 M (10%) Na_2HPO_4 soln, shaken for 2 h at room temp (about 20 °C), and filtered. The filtrate was treated with equal vol of 2 N HCl, and refluxed for 15 min. After proper diln, a 1-cm^3 aliquot was withdrawn, acidified, cooled, and the sulfonamide content was detd colorimetrically (as sulfanilamide) by the Marshall method modified by Kimmig (1) using an Autenrieth colorimeter. The pH was detd on an ultraionograph using a glass electrode.

SOURCE AND PURITY OF MATERIALS:

Acetyl sulfanilamide (source not specified) gave no coloration upon diazotization of its satd soln, thus showing absence of sulfanilamide.

The source and purity of the remaining materials was not specified.

ESTIMATED ERROR:

Soly: precision ±5% (author).

Temp: not specified.

pH: ±0.05 pH unit (author).

REFERENCES:

1. Kimmig, J. *Arch. Dermatol.* 1938, *176*, 722; *Erg. Hyg.* 1941, *24*, 398.

COMPONENTS:

(1) Acetamide, N-[4-(aminosulfonyl)phenyl]-(acetyl sulfanilamide); $C_8H_{10}N_2O_3S$; [121-61-9]

(2) Phosphoric acid, disodium salt; Na_2HPO_4; [7558-94-4]

(3) Phosphoric acid, monopotassium salt; KH_2PO_4; [7778-77-0]

(4) Water; H_2O; [7732-18-5]

ORIGINAL MEASUREMENTS:

Krüger-Thiemer, E.

Arch. Dermatol. Syphilis 1942, *183*, 90-116.

VARIABLES:

Temperature, pH

PREPARED BY:

R. Piekos

EXPERIMENTAL VALUES:

Composition of 1/15M phosphate buffer solutions			pH	Solubility			
				Room temp (ca 20°C)		37°C	
Na_2HPO_4	KH_2PO_4	% Content		g%	10^3 mol dm^{-3} solution[a]	g%	10^2 mol dm^{-3} solution
1.0	99.0	0.91	4.944	0.144	6.72	-	-
10.0	90.0	0.91	5.906	0.144	6.72	0.287	1.34
61.1	38.9	0.93	7.005	0.144	6.72	0.292	1.36
9.5	0.5	0.733[b]	7.51	0.127	5.93	-	-
94.7	5.3	0.95	8.018	0.143	6.11	-	-

[a] Calculated by compiler.

[b] Molar content; 10% buffer solution.

AUXILIARY INFORMATION

METHOD/APPARATUS/PROCEDURE:

Acetyl sulfanilamide (0.5 g) was dissolved in 10 cm^3 of a buffer soln, shaken for 2 h at 20°C (or left for 48 h at 37°C), and filtered at respective temp. The filtrate was treated with equal vol of 2N HCl and refluxed for 15 min. After proper diln, a 1-cm^3 aliquot was withdrawn, acidified, cooled, and the sulfonamide content was detd colorimetrically (as sulfanilamide) by the Marshall method modified by Kimmig (1) using an Authenreith colorimeter. The pH was detd on an ultraionograph using a glass electrode.

SOURCE AND PURITY OF MATERIALS:

Acetyl sulfanilamide (source not specified) gave no coloration upon diazotization of its satd soln, thus showing absence of sulfanilamide. The source and purity of the remaining materials was not specified.

ESTIMATED ERROR:

Soly: precision ±5% (author).

Temp: not specified.

pH: ±0.05 pH unit (author).

REFERENCES:

1. Kimmig, J. *Arch. Dermatol.* 1938, *176*, 722; *Erg. Hyg.* 1941, *24*, 398.

COMPONENTS:

(1) Acetamide, N- (4-aminosulfonyl)-phenyl - (N^4-acetylsulfanilamide); $C_8H_{10}N_2O_3S$; [121-61-9]

(2) Urea; CH_4N_2O; [57-13-6]

(3) Water; H_2O; [7732-18-5]

ORIGINAL MEASUREMENTS:

Rohdewald, P.

Pharmazie 1975, *30(7)*, 460-3.

VARIABLES:

Concentration of urea.

PREPARED BY:

R. Piekos

EXPERIMENTAL VALUES:

Concentration of urea mol/l[a]	Solubility at 20°C g/100 ml	Solubility at 20°C 10^2 mol dm^{-3}[b]
0.300	0.600	2.80
0.600	0.678	3.16
0.900	0.692	3.23

[a] Numerical values given by the author in personal communication.

[b] Calculated by compiler.

AUXILIARY INFORMATION

METHOD/APPARATUS/PROCEDURE:

The previously employed method (1) was used whereby the solns (50 cm^3) were placed in 100-cm^3 flasks with glass stoppers and rotated in a thermostated bath for 2 h. They were then filtered through a G3 glass filter and the undissolved N^4-acetylsulfanilamide was dried at 90°C to const wt and weighed.

SOURCE AND PURITY OF MATERIALS:

The source and purity of N^4-acetylsulfanilamide was not specified.

Urea (Schuchardt) was recrystd from aq MeOH. Purity of the water was not specified.

ESTIMATED ERROR:

Soly: not specified.

Temp: $\pm$0.05°C (author).

REFERENCES:

1. Schulte, K. E.; Rohdewald, P.; Weinhold, P. *Pharmazie* 1968, *23(5)*, 252.

COMPONENTS:

(1) Acetamide, N-[(4-aminosulfonyl)-phenyl]- (N^4-acetylsulfanilamide); $C_8H_{10}N_2O_3S$; [121-61-9]

(2) Urea, methyl-; $C_2H_6N_2O$; [598-50-5]

(3) Water; H_2O; [7732-18-5]

ORIGINAL MEASUREMENTS:

Rohdewald, P.

Pharmazie 1975, *30(7)*, 460-3.

VARIABLES:

Concentration of methylurea

PREPARED BY:

R. Piekos

EXPERIMENTAL VALUES:

Concentration of methylurea mol/l[a]	Solubility at 20°C	
	g/100 ml	10^2 mol dm^{-3} [b]
0.300	0.676	3.15
0.600	0.780	3.64
0.900	0.838	3.91

[a] Numerical values given by the author in personal communication.

[b] Calculated by compiler.

AUXILIARY INFORMATION

METHOD/APPARATUS/PROCEDURE:

The previously employed method (1) was used whereby the solns (50 cm^3) were placed in 100-cm^3 flasks with glass stoppers and rotated in a thermostated bath for 2 h. They were then filtered through a G3 glass filter and the undissolved N^4-acetylsulfanilamide was dried at 90°C to const wt and weighed.

SOURCE AND PURITY OF MATERIALS:

The source and purity of N^4-acetylsulfanilamide was not specified.

Methylurea(Schuchardt) was recrystd from aq MeOH.

Purity of the water was not specified.

ESTIMATED ERROR:

Soly: not specified.

Temp: ±0.05°C (author).

REFERENCES:

1. Schulte, K. E.; Rohdewald, P.; Weinhold, P. *Pharmazie* 1968, *23(5)*, 252.

COMPONENTS:

(1) Acetamide, N-[(4-aminosulfonyl)-phenyl]- (N^4-acetylsulfanilamide); $C_8H_{10}N_2O_3S$; [121-61-9]

(2) Urea, N,N'-dimethyl-; $C_3H_8N_2O$; [96-31-1]

(3) Water; H_2O; [7732-18-5]

ORIGINAL MEASUREMENTS:

Rohdewald, P.

Pharmazie 1975, *30(7)*, 460-3.

VARIABLES:

Concentration of N,N'-dimethylurea

PREPARED BY:

R. Piekos

EXPERIMENTAL VALUES:

Concentration of N,N'-dimethylurea mol/l[a]	Solubility at 20°C g/100 ml	10^2 mol dm^{-3} [b]
0.250	0.702	3.28
0.500	0.844	3.94

[a] Numerical values given by the author in personal conversation.

[b] Calculated by compiler.

AUXILIARY INFORMATION

METHOD/APPARATUS/PROCEDURE:

The previously employed method (1) was used whereby the solns (50 cm^3) were placed in 100-cm^3 flasks with glass stoppers and rotated in a thermostated bath for 2 h. They were then filtered through a G3 glass filter and the undissolved N^4-acetylsulfanilamide was dried at 90°C to const wt and weighed.

SOURCE AND PURITY OF MATERIALS:

The source and purity of N^4-acetylsulfanilamide was not specified. N,N'-dimethylurea (Schuchardt) was recrystd from aq MeOH. Purity of the water was not specified.

ESTIMATED ERROR:

Soly: not specified.

Temp: ±0.05°C (author).

REFERENCES:

1. Schulte, K. E.; Rohdewald, P.; Weinhold, P. *Pharmazie* 1968, *23(5)*, 252.

COMPONENTS:

(1) Acetamide, N-[(4-aminosulfonyl)-phenyl]- (N^4-acetylsulfanilamide); $C_8H_{10}N_2O_3S$; [121-61-9]

(2) Urea, N,N-dimethyl-; $C_3H_8N_2O$; [598-94-7]

(3) Water; H_2O; [7732-18-5]

ORIGINAL MEASUREMENTS:

Rohdewald, P.

Pharmazie 1975, *30(7)*, 460-3.

VARIABLES:

Concentration of N,N-dimethylurea

PREPARED BY:

R. Piekos

EXPERIMENTAL VALUES:

Concentration of N,N-dimethylurea	Solubility at 20°C	
mol/l[a]	g/100 ml	10^2 mol dm^{-3}[b]
0.197	0.748	3.49
0.388	0.862	4.02
0.573	0.992	4.63
0.753	1.100	5.13
0.927	1.234	5.76

[a] Numerical values given by the author in personal communication.

[b] Calculated by compiler.

AUXILIARY INFORMATION

METHOD/APPARATUS/PROCEDURE:

The previously employed method (1) was used whereby the solns (50 cm^3) were placed in 100-cm^3 flasks with glass stoppers and rotated in a thermostated bath for 2 h. They were then filtered through a G3 glass filter and the undissolved N^4-acetylsulfanilamide was dried at 90°C to const wt and weighed.

SOURCE AND PURITY OF MATERIALS:

The source and purity of N^4-acetylsulfanilamide was not specified.

N,N-dimethylurea (Schuchardt) was recrystd from aq MeOH.

Purity of the water was not specified.

ESTIMATED ERROR:

Soly: not specified.

Temp: ±0.05°C (author).

REFERENCES:

1. Schulte, K. E.; Rohdewald, P.; Weinhold, P. *Pharmazie* 1968, *23(5)* 252.

COMPONENTS:

(1) Acetamide, N-[(4-aminosulfonyl)-phenyl]- (N^4-acetylsulfanilamide); $C_8H_{10}N_2O_3S$; [121-61-9]

(2) Urea, tetramethyl-; $C_5H_{12}N_2O$; [632-22-4]

(4) Water; H_2O; [7732-18-5]

ORIGINAL MEASUREMENTS:

Rohdewald, P.

Pharmazie 1974, *30(7)*, 460-3.

VARIABLES:

Concentration of tetramethylurea

PREPARED BY:

R. Piekos

EXPERIMENTAL VALUES:

Concentration of tetramethylurea	Solubility at 20°C	
mol/l[a]	g/100 ml	10^2 mol dm^{-3} [b]
0.300	0.912	4.26
0.600	1.338	6.25
0.900	1.896	8.85

[a] Numerical values given by the author in personal communication.

[b] Calculated by compiler.

AUXILIARY INFORMATION

METHOD/APPARATUS/PROCEDURE:

The previously employed method (1) was used whereby the solns (50 cm^3) were placed in 100-cm^3 flasks with glass stoppers and rotated in a thermostated bath for 2 h. They were then filtered through a G3 glass filter and the undissolved N^4-acetylsulfanilamide was dried at 90°C to const wt and weighed.

SOURCE AND PURITY OF MATERIALS:

The source and purity of N^4-acetylsulfanilamide was not specified.

Tetramethylurea (Schuchardt) was recrystd from aq MeOH.

Purity of the water was not specified.

ESTIMATED ERROR:

Soly: not specified.

Temp: ±0.05°C (author).

REFERENCES:

1. Schulte, K. E.; Rohdewald, P.; Weinhold, P. *Pharmazie* 1968, *23(5)*, 252.

COMPONENTS:

(1) Acetamide, N-[(4-aminosulfonyl)-phenyl]- (N^4-acetylsulfanilamide); $C_8H_{10}N_2O_3S$; [121-61-9]

(2) Thiourea; CH_4N_2S; [62-56-6]

(3) Water; H_2O; [7732-18-5]

ORIGINAL MEASUREMENTS:

Rohdewald, P.

Pharmazie 1975, *30(7)*, 460-3.

VARIABLES:

Concentration of thiourea

PREPARED BY:

R. Piekos

EXPERIMENTAL VALUES:

Concentration of thiourea	Solubility at 20°C	
mol/l[a]	g/100 ml	10^2 mol dm^{-3}[b]
0.300	0.720	3.36
0.600	0.854	3.99
0.900	0.972	4.54

[a] Numerical values given by the author in personal communication.

[b] Calculated by compiler.

AUXILIARY INFORMATION

METHOD/APPARATUS/PROCEDURE:

The previously employed method (1) was used whereby the solns (50 cm^3) were placed in 100-cm^3 flasks with glass stoppers and rotated in a thermostated bath for 2 h. They were then filtered through a G3 glass filter and the undissolved N^4-acetylsulfanilamide was dried at 90°C to const wt and weighed.

SOURCE AND PURITY OF MATERIALS:

The source and purity of N^4-acetylsulfanilamide was not specified.

Thiourea (Schuchardt) was recrystd from aq MeOH.

Purity of the water was not specified.

ESTIMATED ERROR:

Soly: not specified.

Temp: ±0.05°C (author).

REFERENCES:

1. Schulte, K. E.; Rohdewald, P.; Weinhold, P. *Pharmazie* 1968, *23(5)*, 252.

COMPONENTS:

(1) Acetamide, N-[(4-(aminosulfonyl)-phenyl]- (acetyl sulfanilamide); $C_8H_{10}N_2O_3S$; [121-61-9]

(2) 2-Propanone (acetone); C_3H_6O; [67-64-1]

ORIGINAL MEASUREMENTS:

Gutierrez, F. H.

Anales fis. quim. (Madrid) 1945, *41*, 537-60.

VARIABLES:

Temperature

PREPARED BY:

R. Piekos

EXPERIMENTAL VALUES:

t/°C	G[a]	E[b]	X_g/1[c]	mol/1[d] acetone	mmol/mol acetone	1:X_g[e]	1 + X_{cc}[f]
0	2.106	2.063	17.155	8.0	5.7	47.48	58.29
5	2.230	2.181	18.036	8.4	6.0	44.84	55.44
10	2.344	2.290	18.822	8.8	6.3	42.66	53.13
15	2.516	2.454	20.055	9.4	6.8	39.75	49.86
20	2.652	2.584	20.983	9.8	7.2	37.71	47.66
25	2.779	2.704	21.821	10.1	7.5	35.98	45.83
30	3.006	2.918	23.426	10.9	8.1	33.27	42.69
35	3.335	3.228	25.789	12.0	9.0	29.99	38.78
40	3.502	3.383	26.874	12.6	9.2	28.56	37.21
45	3.748	3.613	28.537	13.3	10.1	26.68	35.04
50	3.871	3.728	29.245	13.7	10.5	25.83	34.19

[a] $G = \frac{p\ 100}{P - p}$, where p and P are the weights of solute and solution, resp.

[b] $E = \frac{G\ 100}{G + 100}$; [c] g/l acetone; [d] should be mmol/l (compiler);

[e] g of acetone required to dissolve 1 g of solute; [f] volume (cm^3) of acetone required to dissolve 1 g of solute.

AUXILIARY INFORMATION

METHOD/APPARATUS/PROCEDURE:

A special all-glass app was constructed enabling the prepn of satd solns, agitation by bubbling a stream of acetone-satd N, filtration, and distn off the solvent without the contact with air. Two exchangeable dissoln vessels of 15 and 8 cm^3 working capacity were used depending on the soly of solute. The app was immersed in a thermostat. The vols of acetone used were 15 or 5 cm^3, and the equilibration time was 2-2.5 h. The satd solns were filtered, weighed, the solvent was dist off, the residues were dried at 105°C, weighed, and examd for the presence of solvated acetone.

SOURCE AND PURITY OF MATERIALS:

The source of the materials was not specified. Pure, anhyd acetone was used. The absence of impurities and water was confirmed by procedures of the German Pharmacopeia VI and Spanish Pharmacopeia VIII. The purity of acetyl sulfacetamide was not specified.

ESTIMATED ERROR:

Soly: measurements were repeated until 2 values not differing in the second decimal were obtained.

Temp: ±0.1°C (author).

REFERENCES:

COMPONENTS:

(1) Glycine, N-[(4-(aminosulfonyl)-phenyl]-; $C_8H_{10}N_2O_4S$; [6138-11-0]

(2) 2-Propanone (acetone); C_3H_6O; [67-64-1]

ORIGINAL MEASUREMENTS:

Gutierrez, F. H.

Anales fis. quim. (Madrid) 1945, *41*, 537-60.

VARIABLES:

Temperature

PREPARED BY:

R. Piekos

EXPERIMENTAL VALUES:

$t/^{o}C$	G^a	E^b	X_g/l^c	mol/l[d] acetone	mmol/mol acetone	$1{:}X_g$ [e]	$1 + X_{cc}$ [f]
0	3.733	3.599	30.409	132	9	26.80	32.84
5	3.952	3.831	31.934	139	10	25.30	31.29
10	4.146	3.981	33.292	144	10.4	24.12	30.04
15	4.301	4.124	34.283	149	10.8	23.48	29.17
20	4.801	4.581	37.986	165	12	20.83	26.32
25	5.210	4.952	40.909	178	13	19.17	24.44
30	5.919	5.589	46.127	200	15	16.90	21.68
35	6.600	6.191	51.038	222	17	15.15	19.57
40	7.599	7.062	58.315	253	19	13.16	17.15
45	8.799	8.087	66.996	290	22	11.37	14.93
50	10.011	9.100	75.633	328	25	9.99	13.22

[a] $G = \frac{p\ 100}{P - p}$, where p and P are the weights of solute and solution, resp.

[b] $E = \frac{G\ 100}{G + 100}$; [c] g/l acetone; [d] should be mmol/l (compiler);

[e] g of acetone to dissolve 1 g of solute; [f] volume (cm^3) of acetone required to dissolve 1 g of solute.

AUXILIARY INFORMATION

METHOD/APPARATUS/PROCEDURE:

A special all-glass app was constructed enabling the prepn of satd solns, agitation by bubbling a stream of acetone-satd N, filtration, and distn off the solvent without the contact with air. Two exchangeable dissoln vessels of 15 and 8 cm^3 working capacity were used depending on the soly of solute. The app was immersed in a thermostat. The vols of acetone used were 15 or 5 cm^3, and the equilibration time was 2-2.5 h. The satd solns were filtered, weighed, the solvent was distd off, the residues were dried at $105^{o}C$, weighed, and examd for the presence of solvated acetone.

SOURCE AND PURITY OF MATERIALS:

The source of the materials was not specified. Pure, anhyd acetone was used. The absence of impurities and water was confirmed by procedures of the German Pharmacopeia VI and Spanish Pharmacopeia VIII.

The purity of the solute was not specified.

ESTIMATED ERROR:

Soly: measurements were repeated until 2 values not differing in the second decimal were obtained (author).

Temp: $\pm 0.1^{o}C$ (author).

REFERENCES:

COMPONENTS:

(1) Glycine, N-[(4-(aminosulfonyl)-phenyl]-, monosodium salt; $C_8H_9N_2NaO_4S$; [60364-23-0]

(2) 2-Propanone (acetone); C_3H_6O; [67-64-1]

ORIGINAL MEASUREMENTS:

Gutierrez, F. H.

Anales fis. quim. (Madrid) 1945, *41*, 537-60.

VARIABLES:

Temperature

PREPARED BY:

R. Piekos

EXPERIMENTAL VALUES:

$t/^{\circ}C$	G^a	E^b	$X_g/1^c$	$mol/1^d$ acetone	mmol/mol acetone	$1:X_g{}^e$	$1 + X_{cc}{}^f$
10	0.049	0.049	0.3975	1.6	0.11	2044.89	2541.29
20	0.085	0.085	0.6725	2.7	0.19	1176.47	1486.99
30	0.135	0.135	1.0520	4.2	0.31	740.74	950.57

[a] $G = \frac{p\ 100}{P - p}$, where p and P are the weights of solute and solution, resp.

[b] $E = \frac{G\ 100}{G + 100}$; [c] g/l acetone; [d] should be mmol/l acetone (compiler);

[e] g of acetone required to dissolve 1 g of solute;

[f] volume (cm^3) of acetone required to dissolve 1 g of solute.

AUXILIARY INFORMATION

METHOD/APPARATUS/PROCEDURE:

A special all-glass app was constructed enabling the prepn of satd solns, agitation by bubbling a stream of acetone-satd N, filtration, and distn off the solvent without contact with air. Two exchangeable dissoln vessels of 15 and 8 cm^3 working capacity were used depending on the soly of solute. The app was immersed in a thermostat. The vols of acetone used were 15 or 5 cm^3, and the equilibration time was 2-2.5 h. The satd solns were filtered, weighed, the solvent was distd off, the residues were dried at 105°C, weighed, and examd for the presence of solvated acetone.

SOURCE AND PURITY OF MATERIALS:

The source of the materials was not specified. Pure, anhyd acetone was used. The absence of impurities and water in it was confirmed by procedures of the German Pharmacopeia VI and Spanish Pharmacopeia VIII.

The purity of the solute was not specified.

ESTIMATED ERROR:

Soly: measurements were repeated until 2 values not differing in the second decimal were obtained (author).

Temp: $\pm 0.1^{\circ}C$ (author).

REFERENCES:

COMPONENTS:	ORIGINAL MEASUREMENTS:
(1) [[4-(Aminosulfonyl)phenyl]amino - methanesulfonate, sodium; $C_7H_9N_2NaO_5S_2$; [138-43-2] (2) 2-Propanone (acetone); C_3H_6O; [67-64-1]	Gutierrez, F. H. *Anales fis. quim. (Madrid)* 1945, *41*, 537-60.
VARIABLES:	PREPARED BY:
Temperature	R. Piekos

EXPERIMENTAL VALUES:

$t/^{o}C$	G^{a}	E^{b}	$X_g/1^{c}$	mol/1[d] acetone	mmol/mol acetone	$1:X_g$ [e]	$1 + X_{cc}$ [f]
0	0.0025	0.0025	0.020	0.069	0.00503	40000	50000
5	0.0045	0.0045	0.036	0.125	0.00906	22222	27777
10	0.0050	0.0050	0.040	0.139	0.00107	20000	25000
15	0.0066	0.0066	0.053	0.183	0.01329	15151	18867
20	0.0082	0.0082	0.065	0.225	0.01651	12195	15384
25	0.010	0.0099	0.079	0.273	0.02014	10000	12658
30	0.014	0.0139	0.109	0.379	0.02820	7143	9174
35	0.013	0.0139	0.108	0.374	0.02820	7143	9259
40	0.022	0.0219	0.169	0.586	0.04472	4545	5911
45	0.031	0.0309	0.236	0.888	0.06249	3290	4238
50	0.038	0.0379	0.287	0.995	0.07655	2631	3487

[a] $G = \frac{p\ 100}{P - p}$, where p and P are the weights of solute and solution, resp.

[b] $E = \frac{G\ 100}{G + 100}$; [c] g/l acetone; [d] should be mmol/l (compiler);

[e] g of acetone required to dissolve 1 g of solute;

[f] volume (cm^3) of acetone required to dissolve 1 g of solute.

AUXILIARY INFORMATION

METHOD/APPARATUS/PROCEDURE:

A special all-glass app was constructed enabling the prepn of satd solns, agitation by bubbling a stream of acetone-satd N, filtration, and distn off the solvent without contact with air. Two exchangeable dissoln vessels of 15 and 8 cm^3 capacity were used depending on the soly of solute. The app was immersed in a thermostat. The vols of acetone used were 15 or 8 cm^3, and the equilibration time was 2-2.5 h. The satd solns were filtered, the residues were dried at 105°C, weighed, and examd for the presence of solvated acetone.

SOURCE AND PURITY OF MATERIALS:

The source of the materials was not specified. Pure, anhyd acetone was used. The absence of impurities and water in it was confirmed by procedures of the German Pharmacopeia VI and Spanish Pharmacopeia VIII.

The purity of the solute was not specified.

ESTIMATED ERROR:

Soly: measurements were repeated until 2 values not differing in the second decimal were obtained (author).

Temp: $\pm 0.1^{o}C$ (author).

REFERENCES:

COMPONENTS:

(1) Benzenesulfonamide, 4-(galactosylamino)-; $C_{12}H_{18}N_2O_7S$; [77400-75-0]

(2) 2-Propanone (acetone); C_3H_6O; [67-64-1]

ORIGINAL MEASUREMENTS:

Gutierrez, F. H.

Anales fis. quim. (Madrid) 1945, *41*, 537-60.

VARIABLES:

Temperatue

PREPARED BY:

R. Piekos

EXPERIMENTAL VALUES:

$t/^{\circ}C$	G^a	E^b	$X_g/1^c$	mol/1[d] acetone	mmol/mol acetone	$1:X_g$[e]	$1 + X_{cc}$[f]
0	0.325	0.324	2.647	7.9	0.56	307.70	377.80
5	0.349	0.348	2.823	8.5	0.60	286.53	354.23
10	0.371	0.369	2.979	8.8	0.64	269.54	335.68
15	0.499	0.496	3.978	11.9	0.87	200.40	251.38
20	0.606	0.602	4.795	14.3	1.05	165.02	208.55
25	0.650	0.646	5.105	15.3	1.13	153.85	195.92
30	0.700	0.695	5.455	16.3	1.21	142.86	183.32
35	0.713	0.708	5.514	16.5	1.24	140.25	181.36
40	0.711	0.706	5.456	16.3	1.23	140.65	183.28
45	0.721	0.715	5.490	16.4	1.25	137.28	182.15
50	0.722	0.715	5.455	16.3	1.25	137.12	183.32

[a] $G = \frac{p\ 100}{P - p}$, where p and P are the weights of solute and solution, resp.

[b] $E = \frac{G\ 100}{G + 100}$; [c] g/1 acetone; [d] should be mmol/1 acetone (compiler);

[e] g of acetone required to dissolve 1 g of solute;

[f] volume (cm^3) of acetone required to dissolve 1 g of solute.

AUXILIARY INFORMATION

METHOD/APPARATUS/PROCEDURE:

A special all-glass app was constructed enabling the prepn of satd solns, agitation by bubbling a stream of acetone-satd N, filtration, and distn off the solvent without contact with air. Two exchangeable dissoln vessels of 15 and 8 cm^3 working capacity were used depending on the soly of solute. The app was immersed in a thermostat. The vols of acetone used were 15 or 8 cm^3, and the equilibration time was 2-2.5 h. The satd solns were filtered, weighed, the solvent was distd off, the residues were dried at 105°C, weighed, and examd for the presence of solvated acetone.

SOURCE AND PURITY OF MATERIALS:

The source of the materials was not specified. Pure, anhyd acetone was used. The absence of impurities and water in it was confirmed by procedures of the German Pharmacopeia VI and Spanish Pharmacopeia VIII.

The purity of the solute was not specified.

ESTIMATED ERROR:

Soly: measurements were repeated until 2 values not differing in the second decimal were obtained (author).

Temp: $\pm 0.1^{\circ}C$ (author).

REFERENCES:

COMPONENTS:	ORIGINAL MEASUREMENTS:
(1) Benzoic acid, 4-[[4-(aminosulfonyl)-phenyl]azo]-2-hydroxy-, monpotassium salt; $C_{13}H_{10}KN_3O_5S$; [77400-72-7] (2) 2-Propanone (acetone); C_3H_6O; [67-64-1]	Gutierrez, F. H. *Anales fis. quim. (Madrid)* 1945, *41*, 537-60.
VARIABLES:	**PREPARED BY:**
Temperature	R. Piekos

EXPERIMENTAL VALUES:

t/°C	G[a]	E[b]	X_g/1[c]	mol/1[d] acetone	mmol/mol acetone	1:X_g[e]	1 + X_{cc}[f]
0	0.265	0.264	2.159	6.07	0.38	377.36	462.96
5	0.272	0.271	2.200	6.12	0.39	367.65	454.55
10	0.286	0.285	2.297	6.39	0.42	349.63	434.83
15	0.301	0.300	2.399	6.67	0.44	332.22	416.66
20	0.315	0.314	2.492	6.93	0.46	317.46	405.29
25	0.344	0.343	2.701	7.51	0.51	290,69	370.23
30	0.365	0.364	2.842	7.91	0.53	273.97	351.87
35	0.376	0.374	2.908	8.74	0.55	265.95	343.88
40	0.421	0.419	3.231	8.98	0.62	237.22	309.50
45	0.444	0.442	3.381	9.41	0.65	225.22	295.77
50	0.479	0.477	3.619	10.04	0.70	208.75	276.32

[a] $G = \frac{p\ 100}{P - p}$, where p and P are the weights of solute and solution, resp.

[b] $E = \frac{G\ 100}{B + 100}$; [c] g/l acetone/ [d] Should be mmol/l acetone (compiler);

[e] g of acetone required to dissolve 1 g of solute;

[f] volume (cm^3) of acetone required to dissolve 1 g of solute.

AUXILIARY INFORMATION

METHOD/APPARATUS/PROCEDURE:

A special all-glass app was constructed enabling the prepn of satd solns, agitation by bubbling a stream of acetone-satd N, filtration, and distn off the solvent without contact with air. Two exchangeable dissoln vessels of 15 and 8 cm^3 working capacity were used depending on the soly of solute. The app was immersed in a thermostat. The vols of acetone used were 15 or 8 cm^3, and the equilibration time was 2-2.5 h. The satd solns were filtered, weighed, the solvent was dist off, the residues were dried at 105°C, weighed, and examd for the presence of solvated acetone.

SOURCE AND PURITY OF MATERIALS:

The source of the materials was not specified. Pure, anhyd acetone was used. The absence of impurities and water in it was confirmed by procedures of the German Pharmacopeia VI and Spanish Pharmacopeia VIII.

The purity of the solute was not specified.

ESTIMATED ERROR:

Soly: measurements were repeated until 2 values not differing in the second decimal were obtained (author).

Temp: $\pm$0.1°C (author).

REFERENCES:

COMPONENTS:	ORIGINAL MEASUREMENTS:
(1) Benzenesulfonamide, 4-[(2,4-diaminophenyl)azo]- (Prontosil); $C_{12}H_{13}N_5O_2S$; [103-12-8] (2) 2-Propanone (acetone); C_3H_6O; [67-64-1]	Gutierrez, F. H. *Anales fis. quim. (Madrid)* 1945, *41*, 537-60.
VARIABLES:	**PREPARED BY:**
Temperature	R. Piekos

EXPERIMENTAL VALUES:

$t/^oC$	G^a	E^b	X_g/l^c	mol/l[d] acetone	mmol/mol acetone	$1:X_g$ [e]	$1 + X_{cc}$ [f]
0	8.046	7.428	65.543	222	16.0	12.43	15.26
5	8.227	7.601	66.540	225	16.4	12.15	15.03
10	8.599	7.918	69.550	238	17.1	11.63	14.38
15	8.899	8.171	70.934	243	17.7	11.23	14.09
20	9.205	8.429	72.828	250	18.3	10.87	13.73
25	9.705	8.846	76.204	262	19.3	10.30	13.12
30	10.214	9.267	79.528	273	20.3	9.79	12.57
35	10.617	9.598	82.101	281	21.1	9.43	12.18
40	11.141	10.024	85.496	293	22.2	8.97	11.69
45	11.516	10.326	87.683	301	22.9	8.68	11.41
50	12.101	10.794	91.423	313	24.1	8.26	10.93

[a] $G = \frac{p\ 100}{P - p}$, where p and P are the weights of solute and solution, resp.

[b] $E = \frac{G\ 100}{G + 100}$; [c] g/l acetone; [d] should be mmol/l (compiler);

[e] g of acetone required to dissolve 1 g of solute;

[f] volume (cm^3) of acetone required to dissolve 1 g of solute.

AUXILIARY INFORMATION

METHOD/APPARATUS/PROCEDURE:

A special all-glass app was constructed enabling the prepn of satd solns, agitation by bubbling a stream of acetone-satd N, filtration, and distn off the solvent without contact with air. Two exchangeable dissoln vessels of 15 and 8 cm^3 working capacity were used depending on the soly of solute. The app was immersed in a thermostat. The vols of acetone used were 15 or 5 cm^3, and the equilibration time was 2-2.5 h. The satd solns were filtered, weighed, the solvent was distd off, the residues were dried at 105°C, weighed, and examd for the presence of solvated acetone.

SOURCE AND PURITY OF MATERIALS:

The source of the materials was not specified. Pure, anhyd acetone was used. The absence of impurities and water was confirmed by procedures of the German Pharmacopeia VI and Spanish Pharmacopeia VIII.

The purity of Prontosil was not specified.

ESTIMATED ERROR:

Soly: measurements were repeated until 2 values not differing in the second decimal were obtained (author).
Temp: ±0.1°C (author).

REFERENCES:

COMPONENTS:

(1) 2,7-Naphthalenedisulfonic acid, 6-(acetylamino)-3-[[4-(aminosulfonyl)-phenyl]azo]-4-hydroxy-, disodium salt (Prontosil S); $C_{18}H_{14}N_4Na_2O_{10}S_3$; [133-60-8]

(2) 2-Propanone (acetone); C_3H_6O; [67-64-1]

ORIGINAL MEASUREMENTS:

Gutierrez, F. H.

Anales fis. quim. (Madrid) 1945, *41*, 537-60.

VARIABLES:

One temperature: 20°C

PREPARED BY:

R. Piekos

EXPERIMENTAL VALUES:

$$G = \frac{p\ 100}{P - p} < 1 \times 10^{-3} \text{ at } 20^{\circ}\text{C},$$

where p is the weight of solute and P is the weight of solution.

AUXILIARY INFORMATION

METHOD/APPARATUS/PROCEDURE:

Acetone was agitated with Prontosil S for 2 h in a specially constructed all-glass app. A residue left after evapn of the solvent could not be detd gravimetrically. Prontosil S was thus said to be practically insol in acetone.

SOURCE AND PURITY OF MATERIALS:

The source of the materials was not specified. Pure, anhyd acetone was used. The absence of impurities and water was confirmed by procedures of the German Pharmacopeia VI and Spanish Pharmacopeia VIII. The purity of Prontosil S was not specified.

ESTIMATED ERROR:

Soly: not specified.

Temp: ±0.1°C (author).

REFERENCES:

COMPONENTS:

(1) Benzenesulfonamide, 4-amino-N-methyl; $C_7H_{10}N_2O_2S$; [1709-52-0]

(2) Water; H_2O; [7732-18-5]

ORIGINAL MEASUREMENTS:

Kitao, K.; Kubo, K.; Morishita, T.; Yata, N.; Kamada, A. *Chem. Pharm. Bull.* 1973, *21*, 2417-26.

VARIABLES:

One temperature: 37°C

PREPARED BY:

R. Piekos

EXPERIMENTAL VALUES:

Solubility of 4-amino-N-methylbenzenesulfonamide in water at 37°C is 94.5 mmol dm^{-3} solution.

AUXILIARY INFORMATION

METHOD/APPARATUS/PROCEDURE:

The sulfonamide was assayed by diazotization. No details were given.

SOURCE AND PURITY OF MATERIALS:

The sulfonamide was synthesized by the authors. Its purity was not specified.

Deionized water was used.

ESTIMATED ERROR:

Soly: not specified.

Temp: ± 1°C (authors).

REFERENCES:

COMPONENTS:	ORIGINAL MEASUREMENTS:
(1) Benzenesulfonamide, 4-amino-N-methyl-; $C_7H_{10}N_2O_2S$; [1709-52-0] (2) Methane, trichloro-; $CHCl_3$; [67-66-3]	Kitao, K.; Kubo, K.; Morishita, T.; Yata, N.; Kamada, A. *Chem. Pharm. Bull.* 1973, *21*, 2417-26.
VARIABLES:	**PREPARED BY:**
One temperature: 37°C	R. Piekos

EXPERIMENTAL VALUES:

Solubility of 4-amino-N-methylbenzenesulfonamide in $CHCl_3$ at 37°C is 70.0 mmol dm^{-3} solution.

AUXILIARY INFORMATION

METHOD/APPARATUS/PROCEDURE:

One ml of the sulfonamide soln in $CHCl_3$ at equilibrium was taken into a test tube. After evapn of the solvent, the residue was dissolved in 1N HCl, the soln was properly dild with deionized water, and the concn of the sulfonamide was detd by diazotization.

SOURCE AND PURITY OF MATERIALS:

The sulfonamide was synthesized by the authors. Its purity was not specified.

Neither source nor purity of the $CHCl_3$ was specified.

ESTIMATED ERROR:

Soly: not specified.

Temp: ± 1°C (authors).

REFERENCES:

COMPONENTS:	ORIGINAL MEASUREMENTS:
(1) Benzenesulfonamide, 4-amino-N,N-dimethyl-; $C_8H_{12}N_2O_2S$; [1709-39-7] (2) Water; H_2O; [7732-18-5]	Kitao, K.; Kubo, K.; Morishita, T.; Yata, N.; Kamada, A. *Chem. Pharm. Bull.* 1973, *21*, 2417-26.
VARIABLES:	**PREPARED BY:**
One temperature: 37°C	R. Piekos

EXPERIMENTAL VALUES:

Solubility of 4-amino-N,N-dimethylbenzenesulfonamide in water at 37°C is 3.13 mmol dm^{-3} solution.

AUXILIARY INFORMATION

METHOD/APPARATUS/PROCEDURE:

An aliquot of the soln at equilibrium (pH 6) was dild with EtOH and the concn of the sulfonamide was detd by diazotization.

SOURCE AND PURITY OF MATERIALS:

The sulfonamide was synthesized by the authors. Its purity was not specified.

Deionized water was used.

ESTIMATED ERROR:

Soly: not specified.

Temp: ±1°C (authors).

REFERENCES:

COMPONENTS:

(1) Benzenesulfonamide, 4-amino-N,N-dimethyl-; $C_8H_{12}N_2O_2S$; [1709-39-7]

(2) Methane, trichloro-; $CHCl_3$; [67-66-3]

ORIGINAL MEASUREMENTS:

Kitao, K.; Kubo, K.; Morishita, T.; Yata, N.; Kamada, A. *Chem. Pharm. Bull.* 1973, *21*, 2417-26.

VARIABLES:

One temperature: 37°C

PREPARED BY:

R. Piekos

EXPERIMENTAL VALUES:

Solubility of 4-amino-N,N-dimethylbenzenesulfonamide in $CHCl_3$ at 37°C is 95.9 mmol dm^{-3} solution.

AUXILIARY INFORMATION

METHOD/APPARATUS/PROCEDURE:

One ml of the sulfonamide soln in $CHCl_3$ at equilibrium was taken into a test tube. After evapn of the solvent, the residue was dissolved in EtOH, the soln was properly dild with deionized water and the concn of the sulfonamide was detd by diazotization.

SOURCE AND PURITY OF MATERIALS:

The sulfonamide was synthesized by the authors. Its purity was not specified.

Neither source nor the purity of the $CHCl_3$ was specified.

ESTIMATED ERROR:

Soly: not specified.

Temp: ±1°C (authors).

REFERENCES:

COMPONENTS:	EVALUATOR:
(1) Acetamide, N-[(4-aminophenyl)sulfonyl]- (sulfacetamide); $C_8H_{10}N_2O_3S$; [144-80-9] (2) Water	Anthony N. Paruta Department of Pharmaceutics University of Rhode Island Kingston, Rhode Island, USA and Ryszard Piekos Faculty of Pharmacy, University of Gdansk Gdansk, Poland 1986

CRITICAL EVALUATION:

Solubility values for sulfacetamide in water at different temperatures are shown in Table I. The value of Gusyakov, Likhol'ot and Kutna (7) refer to the temperature range of 294-298K and were not further considered. Two sets of concurring values existed for 293K, three values quite close to 3 x 10^{-2} mol dm^{-3}, and three values agreeing quite closely to 2.3 x 10^{-2} mol dm^{-3}. Sapozhnikova and Postovskii's (1) value of 2.94 x 10^{-2} mol kg^{-1} refers to an equilibrium time of one hour considered insufficient for saturation, and is larger than the equilibrium values (3,4,6). Shkadova (8), at equilibrium though the value given was about 25% higher than the others given at 293K. Rohdewald (9) used an equilibration time (with agitation) of 2 hours also considered insufficient saturation time. Additionally, the value is higher than those for appropriately equilibrated samples given by three workers (3,4,6). There is a possibility that Rohdewald's high values could be due to a metastable phase. Gusyakov et al. (3,4) and Likhol'ot (6), using 8-24 hours equilibration and satisfactory analytical methods over a seven year period (1960-1967) gave values within a rather narrow range. The recommended value for sulfacetamide can be stated as being 2.33 x 10^{-2} mol dm^{-3}. At 310K the value of Langecker (2) was determined using a boiled sample (≅ 373K), then allowing the system to stand at 310K for an unspecified length of time. The value 6.53 x 10^{-2} mol dm^{-3} is about 15% greater than the average of the other values, thus not considered further. The method used by Kaneniwa, Watari and Iijima (10) of 3-5 days saturation is sometimes required of these types of organic solutes. The value given by these workers is 5.6 x 10^{-2} mol dm^{-3}. The value of Sapozhnikova and Postovskii (1), 5.48 x 10^{-2} mol kg^{-1} was obtained after one hour of equilibration. Generally one hour is considered as being a pre-saturation solubility, the closeness of the value to that given by Kaneniwa et al. (10) allows the value to be conditionally acceptable. Therefore, the value of 5.54 x 10^{-2} mol dm^{-3} suggested here must necessarily be a tentative assignment.

Table I: Solubility of Sulfacetamide in water, 293K and 310K

	10^2 mol dm^{-3} (*indicates mol kg^{-1})		
Reference	293K	303K	310K
1	2.94*	-	5.48*
2	-	-	6.53
3	2.34*	-	-
4	2.34	-	-
5	-	4.14	-
6	2.3	-	-
7	2.334 (294-298K)	-	-
8	3.02*	-	-
9	2.93	-	-
10	-	-	5.60

REFERENCES:

(1) Sapozhnikova, N.V.; Postovskii, I.Ya. *Zh. Prikl. Khim.* 1944, *17*, 427-34.
(2) Langecker, H. *Arch. Exptl. Path. Pharmakol.* 1948, *205*, 291-301.
(3) Gusyakov, V.P.; Likhol'ot, N.M. *FArm. Zh. (Kiev)* 1960, *15(3)*, 21-4.
(4) Likhol'ot, N.M. *Farm. Zh. (Kiev)* 1965, *20(5)*, 44-6.
(5) Yamazaki, M.; Aoki, M.; Kamada, A.; Yata, N. *Yakuzaigaku* 1967, *27(1)*, 37-40.
(6) Gusyakov, V.P.; Likhol'ot, N.M; Kutna, I.M. *Farm. Zh. (Kiev)* 1967, *22(3)*, 34-9.
(7) Gusyakov, V.P.; Likhol'ot, N.M; Kutna, I.M. *Farm. Zh. (Kiev)* 1968, *23(6)*, 56-61.
(8) Shkadova, A.I. *Farm. Zh. (Kiev)* 1969, *24(3)*, 39-41.
(9) Rohdewald, P. *Pharm. Ztg.* 1971, *No. 38*, 1342-4.
(10) Kaneniwa, N.; Watari, N.; Iijima, H. *Chem. Pharm. Bull.* 1978, *26(9)*, 2603-14.

COMPONENTS:

(1) Acetamide, N-[(4-aminophenyl) sulfonyl]- (sulfacetamide); $C_8H_{10}N_2O_3S$; [144-80-9]

(2) Water; H_2O; [7732-18-5]

ORIGINAL MEASUREMENTS:

Sapozhnikova, N. V.; Postovskii, I. Ya.

Zh. Prikl. Khim 1944, *17*, 427-34.

VARIABLES:

Temperature

PREPARED BY:

R. Piekos

EXPERIMENTAL VALUES:

t/°C	Solubility Weight%	Solubility 10^2 mol kg^{-1} water[a]
20	0.627	2.94
37	1.16	5.48

[a] Calculated by compiler.

AUXILIARY INFORMATION

METHOD/APPARATUS/PROCEDURE:

Sulfacetamide was dissolved in water to form a satd soln which was occasionnally agitated in a glass vessel immersed in a thermostat. The equilibrium was usually attained after 1 h. Five- to 100-cm^3 samples of the satd soln were placed in Pt crucibles or dishes and evapd to dryness at temps lower than 110-115°C. The residue was dried to const wt at 105-110°C and weighed.

SOURCE AND PURITY OF MATERIALS:

Pure, recrystd sulfacetamide was used. Its mp conformed to that reported in the literature.

Purity of the water was not specified.

ESTIMATED ERROR:

Soly: quite reliable results were obtained over the temp range 20-75°C (authors).

Temp: ±0.05°C (authors).

REFERENCES:

COMPONENTS:

(1) Acetamide, N-[(4-aminophenyl) sulfonyl]- (sulfacetamide); $C_8H_{10}N_2O_3S$; [144-80-9]

(2) Water; H_2O; [7732-18-5]

ORIGINAL MEASUREMENTS:

Langecker, H.

Arch. Exptl. Path. Pharmakol. 1948, *205*, 291-301.

VARIABLES:

One temperature: 37°C

PREPARED BY:

R. Piekos

EXPERIMENTAL VALUES:

Solubility of sulfacetamide in water at 37°C is 1400 mg% (6.53×10^{-2} mol dm^{-3}, compiler).

AUXILIARY INFORMATION

METHOD/APPARATUS/PROCEDURE:

An excess of sulfacetamide in water was boiled for 1 h in a sealed ampul followed by keeping the soln at 37°C. Before assaying, the solute was treated with a 2.6N NaOH soln (1) to cleave the acetyl group and the sulfanilamide was detd colorimetrically by the method of Bratton and Marshall (2) using a Havemann colorimeter (3), as well as by microanal detn of the solid residue.

SOURCE AND PURITY OF MATERIALS:

Source and purity of sulfacetamide was not specified.

The water was free of oxidants.

ESTIMATED ERROR:

Nothing specified.

REFERENCES:

1. Scudi, J. V. *J. Lab. Clin. Med.* 1940, *25*, 404.
2. Bratton, A. G.; Marshall, E. K. *J. Biol. Chem.* 1939, *128*, 537.
3. Havemann R. *Klin. Wochenschr.* 1940, p. 503.

COMPONENTS:

(1) Acetamide, N-[(4-aminophenyl) sulfonyl]- (sulfacetamide); $C_8H_{10}N_2O_3S$; [144-80-9]

(2) Water; H_2O; [7732-18-5]

ORIGINAL MEASUREMENTS:

Gusyakov, V. P.; Likhol'ot, N. M.

Farm. Zh. (Kiev) 1960, *15(3)*, 21-4.

VARIABLES:

One temperature: 20°C

PREPARED BY:

R. Piekos

EXPERIMENTAL VALUES:

Solubility of sulfacetamide in water at 20°C is 0.501 g/100 g water (2.34×10^{-2} mol kg^{-1}, compiler).

AUXILIARY INFORMATION

METHOD/APPARATUS/PROCEDURE:

A small excess of sulfacetamide was equilibrated for 8 h in a 50-ml test tube with 20 ml of water. Aliquots were taken with a pipet fitted with a filter. Sulfacetamide was detd at 295 nm using a SF-4 spectrophotometer.

SOURCE AND PURITY OF MATERIALS:

Sulfacetamide was prepd from a purified Na salt by neutralizing it with equivalent quantity of aq HCl. The product was repeatedly washed with water and its purity conformed to the requirements of the State Pharmacopeia VIII.

Purity of the water was not specified.

ESTIMATED ERROR:

Soly: the accuracy corresponded to that of the colorimetric detns (authors).

Temp: not specified.

REFERENCES:

COMPONENTS:

(1) Acetamide, N-[(4-aminophenyl) sulfonyl]- (sulfacetamide); $C_8H_{10}N_2O_3S$; [144-80-9]

(2) Water; H_2O; [7732-18-5]

ORIGINAL MEASUREMENTS:

Likhol'ot, N. M. *Farm. Zh. (Kiev)* 1965, *20(5)*, 44-6.

VARIABLES:

One temperature: 20°C

PREPARED BY:

R. Piekos

EXPERIMENTAL VALUES:

Solubility of sulfacetamide in water at 20°C is 0.501 g/100 ml (2.34×10^{-2} mol dm^{-3}, compiler).

AUXILIARY INFORMATION

METHOD/APPARATUS/PROCEDURE:

An earlier described method was employed (1) whereby a small excess of sulfacetamide was equilibrated with 20 ml of water for 8 h in a 50-ml test tube. Aliquots were withdrawn through a filter and sulfacetamide was assayed bromatometrically.

SOURCE AND PURITY OF MATERIALS:

Nothing specified.

ESTIMATED ERROR:

Soly: not specified.

Temp: ± 0.1°C (authors).

REFERENCES:

1. Gusyakov, V. P.; Likhol'ot, N. M. *Farm. Zh. (Kiev)* 1960, *15(8)*, 21.

COMPONENTS:	ORIGINAL MEASUREMENTS:
(1) Acetamide, N-[(4-aminophenyl) sulfonyl]- (sulfacetamide); $C_8H_{10}N_2O_3S$; [144-80-9] (2) Water; H_2O; [7732-18-5]	Yamazaki, M.; Aoki, M.; Kamada, A.; Yata, N. *Yakuzaigaku* 1967, *27(1)*, 37-40.
VARIABLES:	PREPARED BY:
One temperature: 30°C	R. Piekos

EXPERIMENTAL VALUES:

Solubility of sulfacetamide in water at 30°C is 41.4 mmol/L (8.87 g dm^{-3}, compiler).

AUXILIARY INFORMATION

METHOD/APPARATUS/PROCEDURE:	SOURCE AND PURITY OF MATERIALS:
Sulfacetamide (0.5 g) was placed in an L-shaped tube together with 20 ml of water. The mixt was shaken in a thermostat until equilibrium was attained. The sulfacetamide was then assayed in the supernatant spectrophotometrically at 545 nm on a Beckmann DU spectrophotometer. The results were taken from a calibration graph.	Nothing specified. ESTIMATED ERROR: Soly: not specified. Temp: ± 1°C (authors). REFERENCES:

COMPONENTS:

(1) Acetamide, N-[(4-aminophenyl)-sulfonyl]- (sulfacetamide); $C_8H_{10}N_2O_3S$; [144-80-9]

(2) Water; H_2O [7732-18-5]

ORIGINAL MEASUREMENTS:

Gusyakov, V. P.; Likhol'ot, N. M.; Kutna, I. M. *Farm. Zh. (Kiev)* 1967, *22(3)*, 34-9.

VARIABLES:

One temperature: 20°C

PREPARED BY:

R. Piekos

EXPERIMENTAL VALUES:

Solubility of sulfacetamide in water at 20°C is 0.50 g/100 ml (2.3×10^{-2} mol dm^{-3}, compiler).

AUXILIARY INFORMATION

METHOD/APPARATUS/PROCEDURE:

An excess of sulfacetamide in water was equilibrated for 24 h in an ampul immersed in a water thermostat. Aliquots of the satd soln were withdrawn through a filter and the sulfacetamide content was assayed in the filtrate photometrically.

SOURCE AND PURITY OF MATERIALS:

Purity of the sulfacetamide conformed to the requirements of the State Pharmacopeia IX.

Purity of the water was not specified.

ESTIMATED ERROR:

Soly: not specified.

Temp: $\pm$0.1°C (authors).

REFERENCES:

COMPONENTS:	ORIGINAL MEASUREMENTS:
(1) Acetamide, N-[(4-aminophenyl) sulfonyl]- (sulfacetamide); $C_8H_{10}N_2O_3S$; [144-80-9] (2) Water; H_2O; [7732-18-5]	Gusyakov, V. P.; Likhol'ot, N. M.; Kutna, I. M. *Farm. Zh. (Kiev)* 1968, *23(6)*, 56-61.
VARIABLES:	PREPARED BY:
One temperature: 21-25°C	R. Piekos

EXPERIMENTAL VALUES:

Solubility of sulfacetamide in water at room temperature (21-25°C) is 0.501 g/100 ml (2.334 x 10^{-2} mol dm^{-3}, compiler).

AUXILIARY INFORMATION

METHOD/APPARATUS/PROCEDURE:	SOURCE AND PURITY OF MATERIALS:
Small quantities (2-4 mg) of sulfacetamide were added to a known quantity of water under stirring until satn was attained.	Sulfacetamide: neither source nor purity was specified. Purity of the water was not specified.
	ESTIMATED ERROR:
	Nothing specified.
	REFERENCES:

COMPONENTS:

(1) Acetamide, N-[(4-aminophenyl) sulfonyl]- (sulfacetamide); $C_8H_{10}N_2O_3S$; [144-80-9]

(2) Water; H_2O; [7732-18-5]

ORIGINAL MEASUREMENTS:

Shkadova, A. I.

Farm. Zh. (Kiev) 1969, *24(3)*, 39-41.

VARIABLES:

One temperature: 20°C

PREPARED BY:

R. Piekos

EXPERIMENTAL VALUES:

Solubility of sulfacetamide in water at 20°C is 3.02×10^{-2} mol/kg (0.647 g/100 g, compiler).

AUXILIARY INFORMATION

METHOD/APPARATUS/PROCEDURE:

A satd aqueous soln of sulfacetamide was equilibrated in a water thermostat at 20 ± 0.1°C. The concn of sulfacetamide was detd by alkalimetric titrn.

SOURCE AND PURITY OF MATERIALS:

Sulfacetamide was prepd from its Na salt by addn of equivalent quantity of 0.1N HCl. The product was washed with water and dried.

Distd. water was used.

ESTIMATED ERROR:

Soly: not specified.

Temp: ± 0.1°C (author).

REFERENCES:

COMPONENTS:

(1) Acetamide, N-[(4-aminophenyl) sulfonyl]- (sulfacetamide); $C_8H_{10}N_2O_3S$; [144-80-9]

(2) Water; H_2O; [7732-18-5]

ORIGINAL MEASUREMENTS:

Rohdewald, P. *Pharm. Ztg.* 1971, *No. 38* 1342-4.

VARIABLES:

One temperature: 20°C

PREPARED BY:

R. Piekos

EXPERIMENTAL VALUES:

Solubility of sulfacetamide in water at 20°C is 0.318_8 b/50 ml ($2.975_8 \times 10^{-2}$ mol dm^{-3}, compiler) and 0.315 g/50 ml (2.93×10^{-2} mol dm^{-3}, compiler)[a].

[a] Two values were given without comment (compiler).

AUXILIARY INFORMATION

METHOD/APPARATUS/PROCEDURE:

The soln was equilibrated by agitation for 2 h at 20°C and the sulfacetamide was assayed by differential gravimetric analysis. No details were given.

SOURCE AND PURITY OF MATERIALS:

The source and purity of the materials was not specified.

ESTIMATED ERROR:

Soly: mean std deviation (68.3% of results deviating by S g), S = 0.019; relative std deviation 6.09%; no of measurements 34 (author). Temp: ±0.05°C (author).

REFERENCES:

COMPONENTS:

(1) Acetamide, N-[(4-aminophenyl) sulfonyl]- (sulfacetamide); $C_8H_{10}N_2O_3S$; [144-80-9]

(2) Water; H_2O; [7732-18-5]

ORIGINAL MEASUREMENTS:

Kaneniwa, N.; Watari, N.; Iijima, H.
Chem. Pharm. Bull. 1978, *26(9)*, 2603-14.

VARIABLES:

One temperature: 37°C

PREPARED BY:

R. Piekos

EXPERIMENTAL VALUES:

Solubility of sulfacetamide in water at 37°C is 12.0 mg/ml solution (5.60×10^{-2} mol dm^{-3}, compiler).

AUXILIARY INFORMATION

METHOD/APPARATUS/PROCEDURE:

An excess of sulfacetamide was placed in a flask contg 25 ml of water. The flask was shaken (2 strokes/s at the amplitude of 3 cm) in a thermostatically controlled water bath at 37°C. One-ml sample was withdrawn every 6 h (total equilibration period was 3-5 days) using a warmed Millipore filter syringe with a filter pore size of 0.45μ (Millipore HAWP 01300) and the filtrate was dild with water and assayed spectrophotometrically (1).

SOURCE AND PURITY OF MATERIALS:

Comm sulfacetamide of the Japanese Pharmacopeia grade and distd water were used.

ESTIMATED ERROR:

Soly: not specified.

Temp: ±0.05°C (authors).

REFERENCES:

1. Kanenisa, N.; Watari, N.
Chem. Pharm. Bull. 1974, *22*, 1969.

COMPONENTS:

(1) Acetamide, N-[(4-aminophenyl)sulfonyl]- (sulfacetamide); $C_8H_{10}N_2O_3S$; [144-80-9]

(2) Lithium chloride; LiCl; [7447-41-8]

(3) Water; H_2O; [7732-18-5]

ORIGINAL MEASUREMENTS:

Likholet, N. M., Gusyakov, V. P.
Med. Prom. SSSR 1963, *17(5)*, 28-31.

VARIABLES:

Concentration of LiCl

PREPARED BY:

R. Piekos

EXPERIMENTAL VALUES:

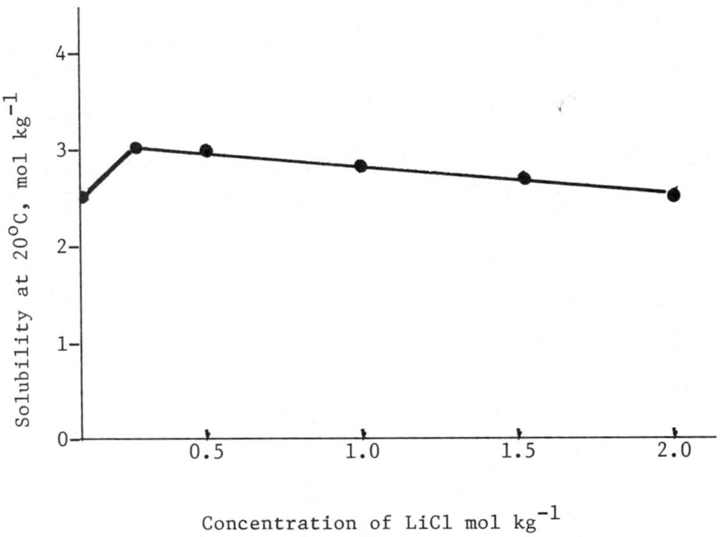

AUXILIARY INFORMATION

METHOD/APPARATUS/PROCEDURE:

Satd solns of sulfacetamide were prepd in 50-ml tightly closed ampuls in which 20 ml of a LiCl soln was placed and a small excess of sulfacetamide. The mixts were equilibrated at 20°C for 18 h. Aliquots were pipetted out through a filter, dild, and assayed spectrophotometrically at 285 nm on a SF-IV spectrophotometer.

SOURCE AND PURITY OF MATERIALS:

Sulfacetamide was purified by crystn.
LiCl was purified by a recommended procedure (1).
The source and purity of the materials were not specified.

ESTIMATED ERROR:

Soly: measurements were repeated several times (authors).

Temp: ±0.1°C (authors).

REFERENCES:

1. Karyakin, Ya. V.; Angelov, I. I. *Chistye khimicheskye reaktivy*, Moscow, 1955.

COMPONENTS:

(1) Acetamide, N-[(4-aminophenyl) sulfonyl]- (sulfacetamide); $C_8H_{10}N_2O_3S$; [133-80-9]

(2) Sodium chloride; NaCl; [7647-14-5]

(3) Water; H_2O; [7732-18-5]

ORIGINAL MEASUREMENTS:

Langecker, H.

Arch. Exptl. Path. Pharmakol. 1948, *205*, 291-301.

VARIABLES:

One temperature: 37°C

PREPARED BY:

R. Piekos

EXPERIMENTAL VALUES:

Solubility of sulfacetamide in physiological saline (0.9% w/w NaCl solution) at 37°C is 983 mg% (4.59×10^{-2} mol dm^{-3}, compiler).

AUXILIARY INFORMATION

METHOD/APPARATUS/PROCEDURE:

An excess of sulfacetamide was added to physiological saline and boiled for 1 h in a sealed ampul followed by keeping the ampul at 37°C. Before assaying, the solute was treated with a 2.6 N NaOH soln (1) to cleave the acetyl group and the sulfanilamide was detd colorimetrically by the method of Bratton and Marshall (2) using a Havemann colorimeter (3), as well as by microanal detn of the solid residue.

SOURCE AND PURITY OF MATERIALS:

Source and purity of the materials was not specified.

ESTIMATED ERROR:

Nothing specified.

REFERENCES:

1. Scudi, J. V. *J. Lab. Clin. Med.* 1940, *25*, 404.
2. Bratton, A. G.; Marshall, E. K. *J. Biol. Chem.* 1939, *128*, 537.
3. Havemann, R. *Klin. Wochenschr.* 1940, p. 503.

COMPONENTS:

(1) Acetamide, N - [(4-aminophenyl)sulfonyl- (sulfacetamide); $C_8H_{10}N_2O_3S$; [144-80-9]

(2) Sodium chloride; NaCl; [7647-14-5]

(3) Water; H_2O; [7732-18-5]

ORIGINAL MEASUREMENTS:

Likholet, N. M.; Gusyakov, V. P.

Med. Prom. SSSR 1963, *17(5)*, 28-31.

VARIABLES:

Concentration of NaCl

PREPARED BY:

R. Piekos

EXPERIMENTAL VALUES:

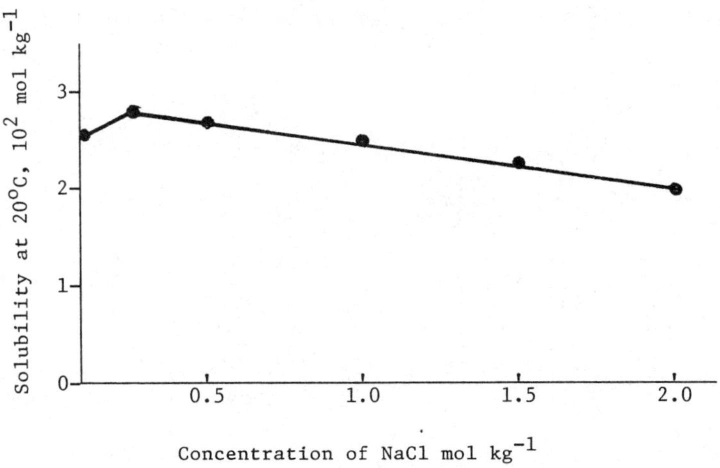

AUXILIARY INFORMATION

METHOD/APPARATUS/PROCEDURE:

Satd solns of sulfacetamide were prepd in 50-ml tightly closed ampuls in which 20 ml of a NaCl soln was placed and a small excess of sulfacetamide. The mixts were equilibrated for 18 h at 20°C. Aliquots were pipetted out through a filter, dild, and assayed spectrophotometrically at 285 nm on a SF-IV spectrophotometer.

SOURCE AND PURITY OF MATERIALS:

Sulfacetamide was purified by crystn. NaCl was purified by a recommended procedure (1). The source and purity of the materials were not specified.

ESTIMATED ERROR:

Soly: measurements were repeated several times (authors).

Temp: ±0.1°C (authors).

REFERENCES:

1. Karyakin, Ya. V. ; Angelov, I. I. *Chistye khimicheskye reaktivy*, Moscow, 1955.

COMPONENTS:

(1) Acetamide, N-[(4-aminophenyl) sulfonyl]- (sulfacetamide); $C_8H_{10}N_2O_3S$; [144-80-9]

(2) Phosphoric acid, disodium salt; Na_2HPO_4; [7558-94-4]

(3) Water; H_2O; [7732-18-5]

ORIGINAL MEASUREMENTS:

Krüger-Thiemer, E.

Arch. Dermatol. Syphilis 1942, *183*, 90-116.

VARIABLES:

One temperature: ea 20°C; one pH: 8.74

PREPARED BY:

R. Piekos

EXPERIMENTAL VALUES:

Solubility of sulfacetamide in a 0.705 M (10%) Na_2HPO_4 solution of pH 8.74 at room temperature (about 20°C) is 5.230 g% (2.441×10^{-1} mol dm^{-3} solution, compiler).

AUXILIARY INFORMATION

METHOD/APPARATUS/PROCEDURE:

Sulfacetamide (0.5 g) was dissolved in the 0.705 M (10%) Na_2HPO_4 soln (pH 8.74) at room temp (about 20°C), shaken for 2 h at 20°C, and filtered. A 1 cm^3 aliquot was then withdrawn, cooled, 1 cm^3 of 2N HCl was added, and the sulfacetamide content was detd colorimetrically by the Marshall method modified by Kimmig (1) using an Authenreith colorimeter. The pH was detd on an ultraionograph using a glass electrode.

SOURCE AND PURITY OF MATERIALS:

Sulfacetamide was the product manufd by Schering AG under the name Albucid. The source and purity of the remaining materials was not specified.

ESTIMATED ERROR:

Soly: precision ±5% (author).

Temp: not specified.

pH: ±0.05 pH unit (author).

REFERENCES:

1. Kimmig, J. *Arch. Dermatol.* 1938, *176*, 722; *Erg. Hyg.* 1941, *24*, 398.

COMPONENTS:

(1) Acetamide, N-[(4-aminophenyl) sulfonyl]- (sulfacetamide); $C_8H_{10}N_2O_3S$; [144-80-9]

(2) Potassium chloride; KCl; [7447-40-7]

(3) Water; H_2O; [7732-18-5]

ORIGINAL MEASUREMENTS:

Gusyakov, V. P.; Likhol'ot, N. M.

Farm. Zh. (Kiev) 1960, *15(3)*, 21-4.

VARIABLES:

Concentration of KCl

PREPARED BY:

R. Piekos

EXPERIMENTAL VALUES:

Concentration of KCl	Solubility at 20°C	
Weight %	g/100 g water	10^2 mol kg^{-1} [a]
0.74	0.493	2.30
1.82	0.551	2.57
3.59	0.566	2.64
6.93	0.541	2.53
12.97	0.454	2.12
15.70	0.394	1.84

[a] Calculated by compiler.

AUXILIARY INFORMATION

METHOD/APPARATUS/PROCEDURE:

A small excess of sulfacetamide was equilibrated for 8 h in a 50-ml test tube with 20 ml of aqueous KCl soln. Aliquots were taken with a pipet fitted with a filter. Sulfacetamide was detd in the filtrate at 285 nm using a SF-4 spectrophotometer.

SOURCE AND PURITY OF MATERIALS:

Sulfacetamide was prepd from a purified Na salt by neutralizing it with equivalent quantity of aq. HCl. The product was repeatedly washed with water and conformed to the requirements of the State Pharmacopeia VIII. KCl was doubly crystd. Purity of the water was not specified.

ESTIMATED ERROR:

Soly: the accuracy corresponded to that of colorimetric detns (authors).

Temp: not specified.

REFERENCES:

COMPONENTS:	ORIGINAL MEASUREMENTS:
(1) Acetamide, N-[(4-aminophenyl-)sulfonyl]- (sulfacetamide); $C_8H_{10}N_2O_3S$; [144-80-9] (2) Potassium chloride; KCl; [7447-40-7] (3) Water; H_2O; [7732-18-5]	Likholet, N. M.; Gusyakov, V. P. *Med. Prom. SSSR* 1963, *17(5)*, 28-31.
VARIABLES:	PREPARED BY:
Concentration of KCl	R. Piekos

EXPERIMENTAL VALUES:

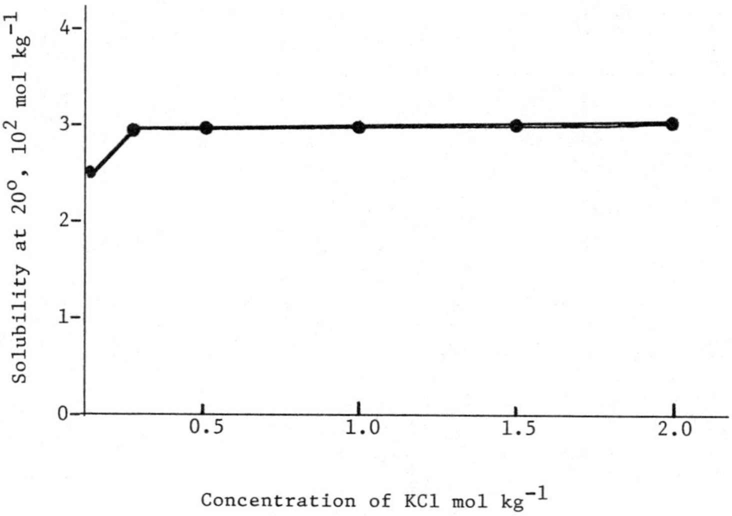

AUXILIARY INFORMATION

METHOD/APPARATUS/PROCEDURE:

Satd solns of sulfacetamide were prepd in 50-ml tightly closed ampuls in which 20 ml of a KCl soln was placed and a small excess of sulfacetamide. The mixts were equilibrated at 20°C for 18 h. Aliquots were pipetted out through a filter, dild, and assayed spectrophotometrically at 285 nm on a SF-IV spectrophotomer.

SOURCE AND PURITY OF MATERIALS:

Sulfacetamide was purified by crystn. KCl was purified by a recommended procedure (1). The source and purity of the materials were not specified.

ESTIMATED ERROR:

Soly: measurements were repeated several times (authors).

Temp: ±0.1°C (authors).

REFERENCES:

1. Karyakin, Ya. V.; Angelov, I. I. *Chistye khimicheskye reaktivy.* Moscow, 1955.

COMPONENTS:

(1) Acetamide, N-[(4-aminophenyl) sulfonyl]- (sulfacetamide); $C_8H_{10}N_2O_3S$; [144-80-9]

(2) Potassium bromide; KBr; [7758-02-3]

(3) Water; H_2O; [7732-18-5]

ORIGINAL MEASUREMENTS:

Gusyakov, V. P.; Likhol'ot, N. M.

Farm. Zh. (Kiev) 1960, *15(3)*, 21-4.

VARIABLES:

Concentration of KBr

PREPARED BY:

R. Piekos

EXPERIMENTAL VALUES:

Concentration of KBr	Solubility at 20°C	
Weight %	g/100 g water	10^2 mol kg^{-1} [a]
1.17	0.501	2.34
2.88	0.514	2.40
5.61	0.587	2.74
10.63	0.668	3.12
19.22	0.769	3.59

[a] Calculated by compiler.

AUXILIARY INFORMATION

METHOD/APPARATUS/PROCEDURE:

A small excess of sulfacetamide was equilibrated for 8 h in a 50-ml test tube with 20 ml of aqueous KBr soln. Aliquots were taken with a pipet fitted with a filter. Sulfacetamide was detd in the filtrate at 285 nm using a SF-4 spectrophotometer.

SOURCE AND PURITY OF MATERIALS:

Sulfacetamide was prepd from a purified Na salt by neutralizing it with equivalent quantity of aq HCl. The product was repeatedly washed with water and conformed to the requirements of the State Pharmacopeia VIII. KBr was doubly crystd. Purity of the water was not specified.

ESTIMATED ERROR:

Soly: the accuracy corresponded to that of colorimetric detns (authors).

Temp: not specified.

REFERENCES:

COMPONENTS:	ORIGINAL MEASUREMENTS:
(1) Acetamide, N-[(4-aminophenyl)sulfonyl] (sulfacetamide); $C_8H_{10}N_2O_3S$; [144-80-9] (2) Potassium bromide; KBr; [7758-02-3] (3) Water; H_2O; [7732-18-5]	Likholet, N. M.; Gusyakov, V. P. *Med. Prom. SSSR* 1963, *17(5)*, 28-31.
VARIABLES:	**PREPARED BY:**
Concentration of KBr	R. Piekos

EXPERIMENTAL VALUES:

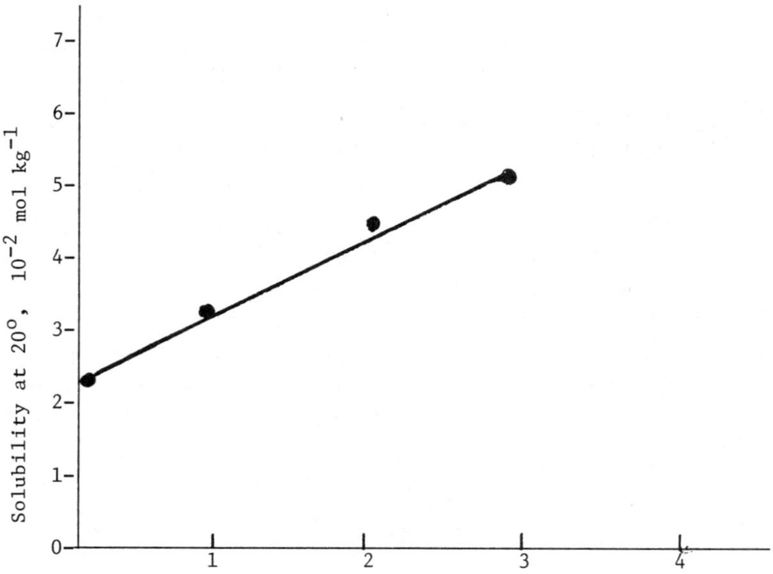

AUXILIARY INFORMATION

METHOD/APPARATUS/PROCEDURE:

Satd solns of sulfacetamide were prepd in 50-ml tightly closed ampuls in which 20 ml of a KBr soln was placed and a small excess of sulfacetamide. The mixts were equilibrated at 20°C for 18 h. Aliquots were pipetted out through a filter, dild, and assayed spectrophotometrically at 285 nm on a SF-IV spectrophotometer.

SOURCE AND PURITY OF MATERIALS:

Sulfacetamide was purified by crystn. KBr was purified by a recommended procedure (1). The source and purity of the reagents were not specified.

ESTIMATED ERROR:

Soly: measurements were repeated several times (authors).

Temp: ±0.1°C (authors).

REFERENCES:

1. Karyakin, Ya. V.; Angelov, I. I. *Chistye khimicheskye reaktivy*, Moscow, 1955.

COMPONENTS:	ORIGINAL MEASUREMENTS:
(1) Acetamide, N-[(4-aminophenyl) sulfonyl]- (sulfacetamide); $C_8H_{10}N_2O_3S$; [144-80-9] (2) Potassium iodide; KI; [7681-11-0] (3) Water; H_2O; [7732-18-5]	Likholet, N. M; Gusyakov, V. P. *Med. Prom. SSSR* 1963, *17(5)*, 28-31.

VARIABLES:	PREPARED BY:
Concentration of KI	R. Piekos

EXPERIMENTAL VALUES:

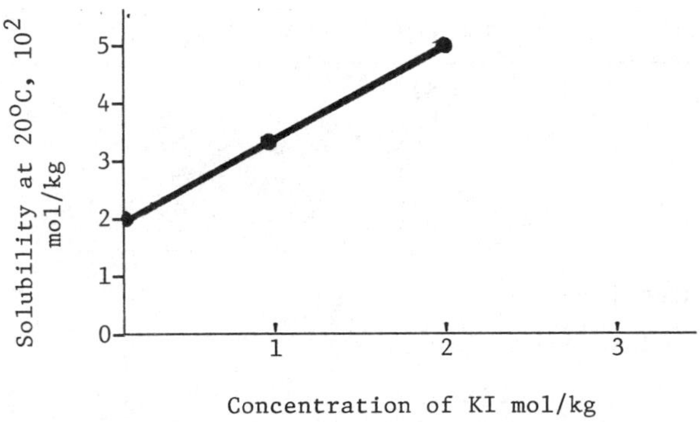

AUXILIARY INFORMATION

METHOD/APPARATUS/PROCEDURE:

Satd solns of sulfacetamide were prepd in 50-ml tightly closed ampuls in which 20 ml of a KI soln was placed and a small excess of sulfacetamide. The mixts were equilibrated at 20°C for 18 h. Aliquots were pipetted out through a filter, dild, and assayed spectrophotometrically at 285 nm on a SF-IV spectrophotometer.

SOURCE AND PURITY OF MATERIALS:

Sulfacetamide was purified by crystn.
KI was purified by a recommended procedure (1).
The source and purity of the materials were not specified.

ESTIMATED ERROR:

Soly: measurements were repeated several times (authors).

Temp: ±0.1°C (authors).

REFERENCES:

1. Karyakin, Ya. V.; Angelov, I. I. *Chistye khimicheskye reaktivy*, Moscow, 1955.

COMPONENTS:

(1) Acetamide, N-[(4-aminophenyl) sulfonyl]- (sulfacetamide); $C_8H_{10}N_2O_3S$; [144-80-9]

(2) Potassium iodide; KI; [7681-11-0]

(3) Water; H_2O [7732-18-5]

ORIGINAL MEASUREMENTS:

Gusyakov, V. P.; Likhol'ot, N. M.

Farm. Zh. (Kiev) 1960, *15(3)*, 21-4.

VARIABLES:

Concentration of KI

PREPARED BY:

R. Piekos

EXPERIMENTAL VALUES:

Concentration of KI	Solubility at 20°C	
Weight %	g/100 g water	10^2 mol kg^{-1} [a]
1.63	0.597	2.79
3.98	0.615	2.87
7.66	0.752	3.51
14.23	0.843	3.94

[a] Calculated by compiler.

AUXILIARY INFORMATION

METHOD/APPARATUS/PROCEDURE:

A small excess of sulfacetamide was equilibrated for 8 h in a 50-ml test tube with 20 ml of aqueous KI soln. Aliquots were taken with a pipet fitted with a filter. Sulfacetamide was detd in the filtrate at 285 nm using a SF-4 spectophotometer.

SOURCE AND PURITY OF MATERIALS:

Sulfacetamide was prepd from a purified Na salt by neutralizing it with equivalent quantity of aq HCl. The product was repeatedly washed with water and conformed to the requirements of the State Pharmacopeia VIII. KI was doubly crystd. Purity of the water was not specified.

ESTIMATED ERROR:

Soly: the accuracy corresponded to that of colorimetric detns (authors).

Temp: not specified.

REFERENCES:

COMPONENTS:

(1) Acetamide, N-[(4-aminophenyl) sulfonyl]- (sulfacetamide); $C_8H_{10}N_2O_3S$; [144-80-9]

(2) Potassium thiocyanate; KSCN; [333-20-0]

(3) Water; H_2O; [7732-18-5]

ORIGINAL MEASUREMENTS:

Gusyakov, V. P.; Likhol'ot, N. M.

Farm. Zh. (Kiev) 1960, *15(2)*, 21-4.

VARIABLES:

Concentration of KSCN

PREPARED BY:

R. Piekos

EXPERIMENTAL VALUES:

Concentration of KSCN	Solubility at 20°C	
Weight %	g/100 g water	10^2 mol kg^{-1} [a]
0.96	0.582	2.72
2.37	0.700	3.27
4.63	0.796	3.72
8.85	1.096	5.12

[a] Calculated by compiler.

AUXILIARY INFORMATION

METHOD/APPARATUS/PROCEDURE:

A small excess of sulfacetamide was equilibrated for 8 h in a 50-ml test tube with 20 ml of aqueous KSCN soln. Aliquots were taken with a pipet fitted with a filter. Sulfacetamide was detd in the filtrate at 285 nm using a SF-4 spectrophotometer.

SOURCE AND PURITY OF MATERIALS:

Sulfacetamide was prepd from a purified Na salt by neutralizing it with equivalent quantity of aq HCl. The product was repeatedly washed with water and conformed to the requirements of the State Pharmacopeia VIII. KSCN was doubly crystd.

Purity of the water was not specified.

ESTIMATED ERROR:

Soly: the accuracy corresponded to that of colorimetric detns (authors)

Temp: not specified.

REFERENCES:

COMPONENTS:

(1) Acetamide, N-[(4-aminophenyl)sulfonyl]-(sulfacetamide); $C_8H_{10}N_2O_3S$; [144-80-9]

(2) Potassium thiocyanate; KCNS; [333-20-0]

(3) Water; H_2O; [7732-18-5]

ORIGINAL MEASUREMENTS:

Likholet, N. M.; Gusyakov, V. P.

Med. Prom. SSSR 1963, *17(5)*, 28-31.

VARIABLES:

Concentration of KCNS

PREPARED BY:

R. Piekos

EXPERIMENTAL VALUES:

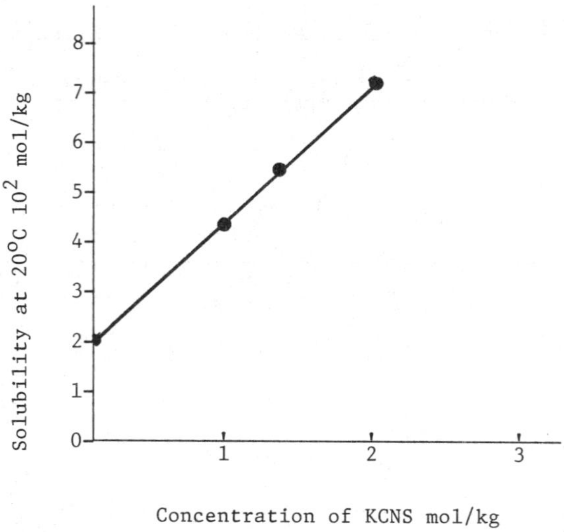

AUXILIARY INFORMATION

METHOD/APPARATUS/PROCEDURE:

Satd solns of sulfacetamide were prepd in 50-ml tightly closed ampuls in which 20 ml of a KCNS soln was placed and a small excess of sulfacetamide. The mixts were equilibrated at 20°C for 18 h. Aliquots were pipetted out through a filter, dild, and assayed spectrophotometrically at 285 nm on a SF-IV spectrophotometer.

SOURCE AND PURITY OF MATERIALS:

Sulfacetamide was purified by crystn. KCNS was purified by a recommended procedure (1). The source and purity of the materials were not specified.

ESTIMATED ERROR:

Soly: measurements were repeated several times (authors).

Temp: ±0.1°C (authors).

REFERENCES:

1. Karyakin, Ya. V.; Angelov, I. I. *Chistye khimicheskye reaktivy*, Moscow, 1955.

COMPONENTS:

(1) Acetamide, N-[(4-aminophenyl) sulfonyl]- (sulfacetamide); $C_8H_{10}N_2O_3S$; [144-80-9]

(2) Phosphoric acid, monopotassium salt; KH_2PO_4; [7778-77-0]

(3) Water; H_2O; [7732-18-5]

ORIGINAL MEASUREMENTS:

Krüger-Thiemer, E.

Arch. Dermatol. Syphilis 1942, *183*, 90-116.

VARIABLES:

One temperature: ca 20°C;
one pH: 4.37

PREPARED BY:

R. Piekos

EXPERIMENTAL VALUES:

Solubility of sulfacetamide in a 0.735M (10%) KH_2PO_4 solution of pH 4.37, at room temperature (about 20°C), is 0.632 g% (2.95 x 10^{-2} mol dm^{-3} solution, compiler).

AUXILIARY INFORMATION

METHOD/APPARATUS/PROCEDURE:

Sulfacetamide (0.5 g) was dissolved in the 0.735M (10%) KH_2PO_4 solution (pH 4.37) at room temp (about 20°C), shaken for 2 h at 20°C, and filtered. A 2 cm^3 aliquot was then withdrawn, cooled, 1 cm^3 of 2N HCl was added, and the sulfacetamide content was detd colorimetrically by the Marshall method modified by Kimmig (1) using an Autenrieth colorimeter. The pH was detd on an ultraionograph using a glass electrode.

SOURCE AND PURITY OF MATERIALS:

Sulfacetamide was the product manufd by Schering AG under the name Albucid. The source and purity of the remaining materials was not specified.

ESTIMATED ERROR:

Soly: precision ± 5% (author).

Temp: not specified.

pH: ±0.05 pH unit (author).

REFERENCES:

1. Kimmig, J. *Arch. Dermatol.* 1938, *176*, 722; *Erg. Hyg.* 1941, *24*, 398.

COMPONENTS:	ORIGINAL MEASUREMENTS:
(1) Acetamide, N-[(4-aminophenyl)sulfonyl]- (sulfacetamide); $C_8H_{10}N_2O_3S$; [144-80-9] (2) Ammonium chloride; NH_4Cl; [12125-02-9] (3) Water; H_2O; [7732-18-5]	Likholet, N. M.; Gusyakov, V. P. *Med. Prom. SSSR* 1963, *17(5)*, 28-31.
VARIABLES: Concentration of NH_4Cl	PREPARED BY: R. Piekos

EXPERIMENTAL VALUES:

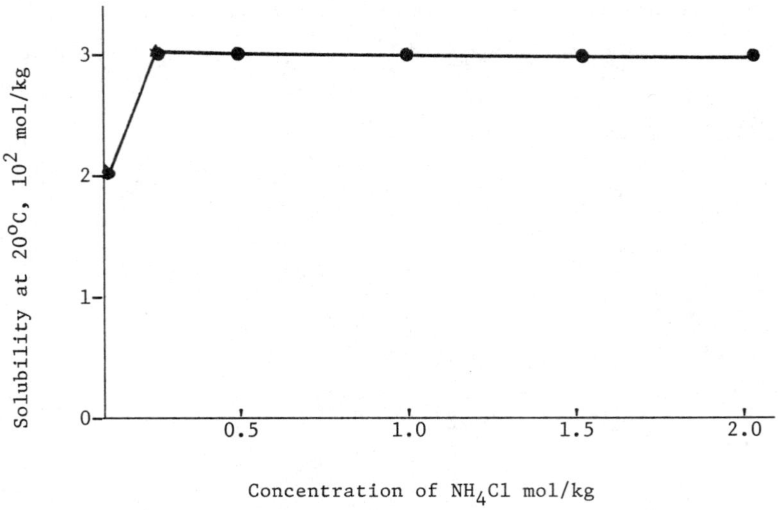

AUXILIARY INFORMATION

METHOD/APPARATUS/PROCEDURE:	SOURCE AND PURITY OF MATERIALS:
Satd solns of sulfacetamide were prepd in 50-ml tightly closed ampuls in which 20 ml of a NH_4Cl soln was placed and a small excess of sulfacetamide. The mixts were equilibrated for 18 h at 20°C. Aliquots were pipetted out through a filter, dild, and assayed spectrophotometrically at 285 nm on a SF-IV spectrophotometer.	Sulfacetamide was purified by crystn. NH_4Cl was purified by a recommended procedure (1). The source and purity of the materials were not specified.
	ESTIMATED ERROR: Soly: measurements were repeated several times (authors). Temp: ±0.1°C (authors).
	REFERENCES: 1. Karyakin, Ya. V.; Angelov, I. I. *Chistye khimicheskye reaktivy*, Moscow, 1955.

COMPONENTS:

(1) Acetamide, N-[(4-aminophenyl)sulfonyl]- (sulfacetamide); $C_8H_{10}N_2O_3S$; [144-80-9]

(2) Magnesium chloride; $MgCl_2$; [7786-30-3]

(3) Water; H_2O; [7732-18-5]

ORIGINAL MEASUREMENTS:

Likholet, N. M.; Gusyakov, V. P.

Med. Prom. SSSR 1963, *17(5)*, 28-31.

VARIABLES:

Concentration of $MgCl_2$

PREPARED BY:

R. Piekos

EXPERIMENTAL VALUES:

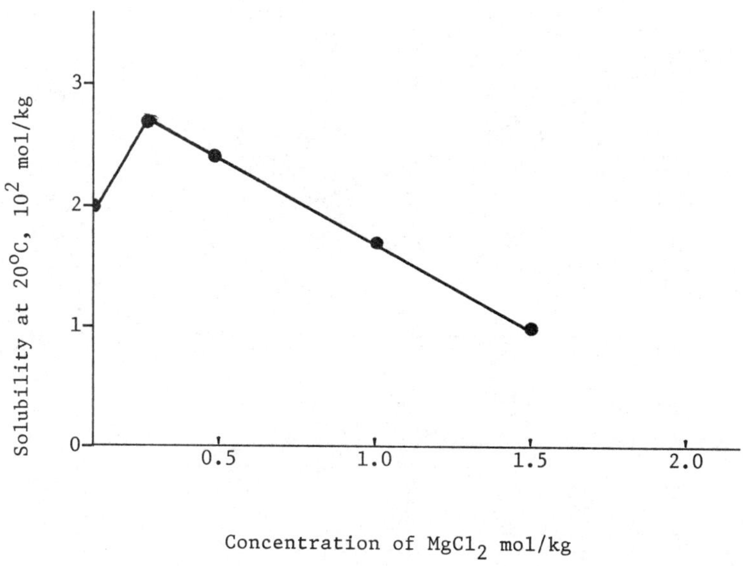

AUXILIARY INFORMATION

METHOD/APPARATUS/PROCEDURE:

Satd solns of sulfacetamide were prepd in 50-ml tightly closed ampuls in which 20 ml of a $MgCl_2$ soln was placed and a small excess of sulfacetamide. The mixts were equilibrated for 18 h at 20°C. Aliquots were pipetted out through a filter, dild, and assayed spectrophotometrically at 285 nm on a SF-IV spectrophotometer.

SOURCE AND PURITY OF MATERIALS:

Sulfacetamide was purified by crystn. $MgCl_2$ was purified by a recommended procedure (1). The source and purity of the materials were not specified.

ESTIMATED ERROR:

Soly: measurements were repeated several times (authors).

Temp: ±0.1°C (authors).

REFERENCES:

1. Karyakin, Ya. V.; Angelov, I. I. *Chistye khimicheskye reaktivy*, Moscow, 1955.

COMPONENTS:

(1) Acetamide, N-[(4-aminophenyl) sulfonyl]- (sulfacetamide); $C_8H_{10}N_2O_3S$; [144-80-9]

(2) Calcium chloride; $CaCl_2$; [10043-52-4]

(3) Water; H_2O; [7732-18-5]

ORIGINAL MEASUREMENTS:

Likholet, N. M.; Gusyakov, V. P.

Med. Prom. SSSR 1963, *17(5)*, 28-31.

VARIABLES:

Concentration of $CaCl_2$

PREPARED BY:

R. Piekos

EXPERIMENTAL VALUES:

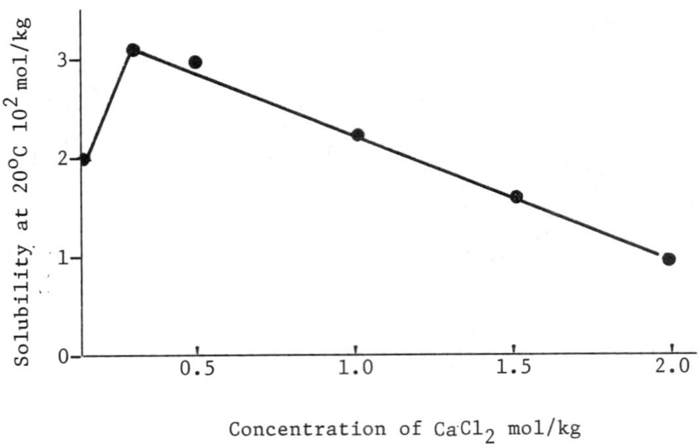

AUXILIARY INFORMATION

METHOD/APPARATUS/PROCEDURE:

Satd soln of sulfacetamide were prepd in 50-ml tightly closed ampuls in which 20 ml of a $CaCl_2$ soln was placed and a small excess of sulfacetamide. The mixts were equilibrated for 18 h at 20°C. Aliquots were pipetted out through a filter, dild, and assayed spectrophotometrically at 285 nm on a SF-IV spectrophotometer.

SOURCE AND PURITY OF MATERIALS:

Sulfacetamide was purified by crystn. $CaCl_2$ was purified by a recommended procedure (1). The source and purity of the materials were not specified.

ESTIMATED ERROR:

Soly: measurements were repeated several times (authors).

Temp: ±0.1°C (authors).

REFERENCES:

1. Karyakin, Ya. V.; Angelov, I. I. *Chistye Khimicheskye reaktivy*, Moscow, 1955.

COMPONENTS:	ORIGINAL MEASUREMENTS:
(1) Acetamide, N-[(4-aminophenyl) sulfonyl]- (sulfacetamide); $C_8H_{10}N_2O_3S$; [144-80-9] (2) Barium chloride; $BaCl_2$; [10361-37-2] (3) Water; H_2O; [7732-18-5]	Likholet, N. M.; Gusyakov, V. P. *Med. Prom. SSSR* 1963, *17(5)*, 28-31.
VARIABLES:	PREPARED BY:
Concentration of $BaCl_2$	R. Piekos

EXPERIMENTAL VALUES:

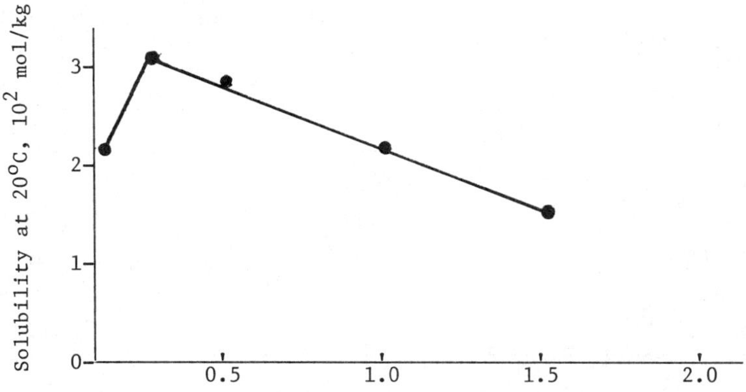

AUXILIARY INFORMATION

METHOD/APPARATUS/PROCEDURE:

Satd solns of sulfacetamide were prepd in 50-ml tightly closed ampuls in which 20 ml of a $CaCl_2$ soln was placed and a small excess of sulfacetamide. The mixts were equilibrated for 18 h at 20°C. Aliquots were pipetted out through a filter, dild, and assayed spectrophotometrically at 285 nm on a SF-IV spectrophotometer.

SOURCE AND PURITY OF MATERIALS:

Sulfacetamide was purified by crystn. $BaCl_2$ was purified by a recommended procedure (1). The source and purity of the materials were not specified.

ESTIMATED ERROR:

Soly: measurements were repeated several times (authors).

Temp: ±0.1°C (authors).

REFERENCES:

1. Karyakin, Ya. V.; Angelov, I. I. *Chistye khimcheskye reaktivy*, Moscow, 1955.

COMPONENTS:	ORIGINAL MEASUREMENTS:
(1) Acetamide, N-[(4-aminophenyl) sulfonyl]- (sulfacetamide); $C_8H_{10}N_2O_3S$; [144-80-9] (2) Phosphoric acid, monosodium salt; NaH_2PO_4; [7558-80-7] (3) Potassium chloride; KCl; [7447-40-7] (4) Water; H_2O; [7732-18-5]	Likhol'ot, N. M. *Farm. Zh. (Kiev)* 1965, *20(5)*, 44-6.

VARIABLES:	PREPARED BY:
Concentration of NaH_2PO_4 - KCl	R. Piekos

EXPERIMENTAL VALUES:

Concentration of NaH_2PO_4-KCl[a]	Solubility of sulfacetamide at 20 °C	
mol/l	g/100 ml	10^2 mol dm^{-3} [b]
0.088	0.602	2.05
0.112	0.597	2.03
0.139	0.597	2.03
0.165	0.597	2.03
0.182	0.596	2.03
0.185	0.596	2.03

[a] KCl was added in such amounts as to correct the ionic strength of solution.

[b] Calculated by compiler.

AUXILIARY INFORMATION

METHOD/APPARATUS/PROCEDURE:

An earlier described method was employed (1) whereby a small excess of sulfacetamide was equilibrated with 20 ml of a NaH_2PO_4-KCl soln for 8 h in a 50-ml test tube. Aliquots were withdrawn through a filter and sulfacetamide was assayed bromatometrically.

SOURCE AND PURITY OF MATERIALS:

Nothing specified.

ESTIMATED ERROR:

Soly: not specified.

Temp: $\pm 0.1\,^{\circ}C$ (authors).

REFERENCES:

1. Gusyakov, V. P.; Likhol'ot, N. M. *Farm. Zh. (Kiev)* 1960, *15(8)*, 21.

COMPONENTS:

(1) Acetamide, N-[(4-aminophenyl) sulfonyl]- (sulfacetamide); $C_8H_{10}N_2O_3S$; [144-80-9]

(2) Phosphoric acid, disodium salt; Na_2HPO_4; [7558-94-4]

(3) Phosphoric acid, monopotassium salt; KH_2PO_4; [7778-77-0]

(4) Water; H_2O; [7732-18-5]

ORIGINAL MEASUREMENTS:

Krüger-Thiemer, E.

Arch. Dermatol. Syphilis 1942, *183*, 90-116.

VARIABLES:

Temperature, pH

PREPARED BY:

R. Piekos

EXPERIMENTAL VALUES:

Composition of 1/15 M phosphate buffer solutions			pH	Solubility			
				Room temp (ca 20°C)		37°C	
Na_2HPO_4	KH_2PO_4	%Content		g%	10^2 mol dm^{-3} solution[a]	g%	10^2 mol dm^{-3} solution[a]
1.0	99.0	0.91	4.944	0.830	3.874	-	-
10.0	90.0	0.91	5.906	0.909	4.242	1.220	5.694
61.1	38.9	0.93	7.005	1.632	7.617	1.770	8.261
9.5	0.5	0.733[b]	7.51	4.710	21.98	-	-
94.7	5.3	0.95	0.018	2.232	10.42	-	-

[a] Calculated by compiler.

[b] Molar content; 10% buffer solution.

AUXILIARY INFORMATION

METHOD/APPARATUS/PROCEDURE:

Sulfacetamide (0.5 g) was dissolved in 10 cm^3 of a buffer soln, shaken for 2 h at 20°C (or left for 48 h at 37°C), and filtered at respective temp. A 1 cm^3 aliquot was then withdrawn, cooled, 1 cm^3 of 2 N HCl was added, and the sulfacetamide content was detd colorimetrically by the Marshall method modified by Kimmig (1) using an Authenreith colorimeter. The pH values were detd on an ultraionograph using a glass electrode.

SOURCE AND PURITY OF MATERIALS:

Sulfacetamide was the product manufd by Schering AG under the name Albucid. The source and purity of the remaining materials was not specified.

ESTIMATED ERROR:

Soly: precision ±5% (author).
Temp: not specified.
pH: ±0.05 pH unit (author).

REFERENCES:

1. Kimmig, J. *Arch. Dermatol.* 1938, *176*, 722; *Erg. Hyg.* 1941, *24*, 398.

COMPONENTS:

(1) Acetamide, N-[(4-aminophenyl)-sulfonyl]- (sulfacetamide); $C_8H_{10}N_2O_3S$; [144-80-9]

(2) Phosphoric acid, disodium salt; Na_2HPO_4; [7558-94-4]

(3) Phosphoric acid, monopotassium salt; KH_2PO_4; [7778-77-0]

(4) Water; H_2O; [7732-18-5]

ORIGINAL MEASUREMENTS:

Langecker, H.

Arch. Exptl. Path. Pharmakol. 1948, *205*, 291-301.

VARIABLES:

pH

PREPARED BY:

R. Piekos

EXPERIMENTAL VALUES:

pH of the 1/15M phosphate buffer	Solubility at 37°C	
	mg%	10^2 mol dm^{-3} [a]
4.9	978	4.56
5.9	974	4.56
6.9	1607	7.50
7.5	2241	10.46

[a] Calculated by compiler.

AUXILIARY INFORMATION

METHOD/APPARATUS/PROCEDURE:

An excess of sulfacetamide was added to the buffer soln and boiled for 1 h in a sealed ampul followed by keeping the ampul at 37°C. Before assaying the solute was treated with a 2.6N NaOH soln (1) to cleave the acetyl group and the sulfanilamide was detd colorimetrically by the method of Bratton and Marshall (2) using a Havemann colorimeter (3), as well as by microanal detn of the solid residue.

SOURCE AND PURITY OF MATERIALS:

Source and purity of the materials was not specified.

ESTIMATED ERROR:

Nothing specified.

REFERENCES:

1. Scudi, J. V. *J. Lab. Clin. Med.* 1940, *25*, 404.
2. Bratton, A. G.; Marshall, E. K. *J. Biol. Chem.* 1939, *128*, 537.
3. Havemann, R. *Klin. Wochenschr.* 1940, p. 503.

COMPONENTS:	ORIGINAL MEASUREMENTS:
(1) Acetamide, N-[(4-aminophenyl)-sulfonyl]- (sulfacetamide); $C_8H_{10}N_2O_3S$; [144-80-9] (2) Phosphoric acid, disodium salt; Na_2HPO_4; [7558-94-4] (3) Phosphoric acid, monopotassium salt; KH_2PO_4; [7778-77-0] (4) Water; H_2O; [7732-18-5]	Bandelin, F. J.; Malesh, W. *J. Am. Pharm. Assoc., Sci. Ed.* 1959, *48*, 177-81.
VARIABLES:	**PREPARED BY:**
pH	R. Piekos

EXPERIMENTAL VALUES:

Solubility of sulfacetamide in buffers of varying mixtures of $Na_2HPO_4 \cdot 7H_2O$ (71.6 g/l distilled water; 0.27 mol dm^{-3}, compiler) and KH_2PO_4 (36.3 g/l distilled water; 0.27 mol dm^{-3}, compiler) at 37°C

Initial pH	Solubility mg/100 ml	Solubility mol dm^{-3} [a]
5.0	1250	0.0583
5.5	1350	0.0629
6.0	2150	0.100
6.5	3020	0.141
7.0	4400	0.205
7.5	15,000	0.699
8.0	41,000	1.911

[a] Calculated by compiler.

AUXILIARY INFORMATION

METHOD/APPARATUS/PROCEDURE:

Solns were prepd by adding an excess of sulfacetamide to 10 ml of buffer soln at each pH level in 18 x 150-mm test tubes, stoppering the tubes, and placing them in a water bath at 37°C with gentle agitation for 24 h. The mixt was then filtered and a 1-ml aliquot was accurately pipetted into a volumetric flask for diln and analysis. The balance was retained for pH detn to ascertain any change in pH value. Sulfacetamide was assayed colorimetrically by the method of Bratton and Marshall as described in detail by Biamonte and Schneller (1). A standard curve was prepd using accurately prepd standard solutions.

SOURCE AND PURITY OF MATERIALS:

Neither source nor purity of the reagents were specified. Distilled water was used.

ESTIMATED ERROR:

Soly: av values of duplicate runs are reported (authors).

Temp and pH: not specified.

REFERENCES:

1. Biamonte, A. R.; Schneller, G. E. *J. Am. Pharm. Assoc., Sci. Ed.*, 1952, *41*, 341.

COMPONENTS:	ORIGINAL MEASUREMENTS:
(1) Acetamide, N-[(4-aminophenyl)-sulfonyl]- (sulfacetamide); $C_8H_{10}N_2O_3S$; [144-80-9] (2) Phosphoric acid, disodium salt; Na_2HPO_4; [7558-94-4] (3) Phosphoric acid, monopotassium salt; KH_2PO_4; [7778-77-0] (4) Water; H_2O; [7732-18-5]	Yamazaki, M.; Aoki, M.; Kamada, A.; Yata, N. *Yakuzaigaku* 1967, *27(1)*, 37-40.
VARIABLES:	PREPARED BY:
One temperature: 30°C; one pH: 7.4	R. Piekos

EXPERIMENTAL VALUES:

Solubility of sulfacetamide in a phosphate buffer solution of pH 7.4[a] (μ = 0.17) at 30°C is 91.0 mmol/L (19.50 g dm^{-3}, compiler).

[a] At the end of experiment the pH was 5.6.

AUXILIARY INFORMATION

METHOD/APPARATUS/PROCEDURE:

Sulfacetamide (0.5 g) was placed in an L-shaped tube together with 20 ml of a buffer soln. The mixt was shaken in a thermostat until equilibrium was attained. The sulfacetamide was then assayed in the supernatant spectrophotometrically at 545 nm on a Beckmann DU spectrophotometer. The results were taken from a calibration graph.

SOURCE AND PURITY OF MATERIALS:

Nothing specified.

ESTIMATED ERROR:

Soly and pH: not specified.

Temp: $\pm$1°C (authors).

REFERENCES:

COMPONENTS:

(1) Acetamide, N-[(4-aminophenyl)sulfonyl]- (sulfacetamide); $C_8H_{10}N_2O_3S$; [144-80-9]

(2) Phosphoric acid, disodium salt; Na_2HPO_4; [7558-94-4]

(3) 1,2,3-Propanetricarboxylic acid, 2-hydroxy- (citric acid); $C_6H_8O_7$; [77-92-91]

(4) Water; H_2O; [7732-18-5]

ORIGINAL MEASUREMENTS:

Likhol'ot, N. M. *Farm. Zh. (Kiev)* 1965, *20(5)*, 44-6.

VARIABLES:

pH

PREPARED BY:

R. Piekos

EXPERIMENTAL VALUES:

pH of McIlvaine's buffer solution	Solubility at 20°C	
	g/100 ml	10^2 mol dm^{-3} [a]
4.1	0.627	2.93
5.1	0.884	4.13
5.9	1.605	7.49
6.5	2.502	11.68
6.9	3.140	14.66
7.5	3.585	16.73

[a] Calculated by compiler.

AUXILIARY INFORMATION

METHOD/APPARATUS/PROCEDURE:

An earlier described method was employed (1) whereby a small excess of sulfacetamide was equilibrated with 20 ml of the McIlvaine's buffer soln for 8 hr in a 50-ml test tube. Aliquots were removed through a filter and sulfacetamide was assayed bromatometrically.

SOURCE AND PURITY OF MATERIALS:

Sulfacetamide: not specified.
McIlvaine's buffer solns were prepd from a 0.2M Na_2HPO_4 and a 0.1M citric acid solns. Source and purity of the buffer components were not specified.

ESTIMATED ERROR:

Soly: not specified.
Temp: $\pm$0.1°C (authors).
pH: not specified.

REFERENCES:

1. Gusyakov, V. P.; Likhol'ot, N. M. *Farm. Zh. (Kiev)* 1960, *15(8)*, 21.

COMPONENTS:

(1) Acetamide, N-[(4-aminophenyl)-sulfonyl]- (sulfacetamide) $C_8H_{10}N_2O_3S$; [144-80-9]
(2) Calcium chloride; $CaCl_2$; [10043-52-4]
(3) Magnesium chloride; $MgCl_2$; [7786-30-3]
(4) Phosphoric acid, monoammonium salt; $NH_4H_2PO_4$; [7722076-1]
(5) Potassium chloride; KCl; [7447-14-5]
(6) Sodium chloride; NaCl; [7647-14-5]
(7) Urea; CH_4N_2O; [57-13-6]
(8) Water; H_2O; [7732-18-5]

ORIGINAL MEASUREMENTS:

Bandelin, F. J.; Malesh, W.

J. Am. Pharm. Assoc., Sci. Ed. 1959, *48*, 177=81.

VARIABLES:

pH at 37°C

PREPARED BY:

R. Piekos

EXPERIMENTAL VALUES:

Solubility of sulfacetamide in a solution containing $CaCl_2$ 0.143, $MgCl_2$ 0.121, $NH_4H_2PO_4$ 0.300, KCl 1.660, NaCl 2.950 and urea 20 g/dm^3 (synthetic urine, Mosher Vehicle) at 37°C

Equilibrium pH	Solubility	
	mg/100 ml	mol/dm^3 [a]
4.5	1500	0.0699
5.0	1950	0.0909
5.5	3150	0.1468
5.8	6000	0.2797
6.2	15,000	0.6992
6.6	50,000	2.3307

[a] Calculated by compiler.

AUXILIARY INFORMATION

METHOD/APPARATUS/PROCEDURE:

Excess acetamide was added to aliquots of synthetic urine solns and 1% H_3PO_4 or 1% NaOH solns were used to adjust the pH to the required value. The solns were agitated for 24 h with addn of acid or base to keep them at the desired pH level until equilibrium in pH and concn was attained. Then the solns were filtered and in aliquots acetamide was assayed spectrophotometrically by the method described by Biamonte and Schneller (1). Before detn the soln was refluxed for 1 h with 5% H_2SO_4 to liberate the free amino compd.

SOURCE AND PURITY OF MATERIALS:

Nothing specified.

ESTIMATED ERROR:

Soly: average values of 2 detns were given.
Temp: not specified.
pH: not specified.

REFERENCES:

1. Biamonte, A. R.; Schneller, G. E., *J. Am. Pharm. Assoc., Sci. Ed.* 1952, *41*, 341.

COMPONENTS:	ORIGINAL MEASUREMENTS:
(1) Acetamide, N-[(4-aminophenyl)-sulfonyl]- (sulfacetamide); $C_8H_{10}N_2O_3S$; [144-80-9] (2) Ethanol; C_2H_6O; [64-17-5] (3) Water; H_2O; [7732-18-5]	Shkadova, A. I. *Farm. Zh. (Kiev)* 1969, *24(3)*, 39-41.
VARIABLES:	PREPARED BY:
Concentration of ethanol	R. Piekos

EXPERIMENTAL VALUES:

Concentration of ethanol		Solubility at 20°C	
mole %	weight %	10^2 mol kg^{-1}	g/100 g[a]
0	0	3.02	0.647
10	22.14	11.12	2.383
20	39.01	26.38	5.652
30	52.31	41.59	8.911
40	63.04	58.60	12.555
50	71.90	59.98	12.851
60	79.33	62.18	13.323
70	85.65	60.03	12.862
80	91.10	42.80	9.170
90	95.83	21.21	4.544

[a] Calculated by compiler.

AUXILIARY INFORMATION

METHOD/APPARATUS/PROCEDURE:	SOURCE AND PURITY OF MATERIALS:
Sulfacetamide was equilibrated with the solvent in a water thermostat at 20±0.1°C. The concn of sulfacetamide was detd by alkalimetric titration.	Sulfacetamide was prepd from its Na salt by addn of equivalent quantity of 0.1N HCl. The product was washed with water and dried. The EtOH - water mixts were prepd from abs EtOH (purity and source not specified) and distd water.
	ESTIMATED ERROR: Soly: not specified. Temp: ±0.1°C (author).
	REFERENCES:

COMPONENTS:

(1) Acetamide, N-[(4-aminophenyl)-sulfonyl]- (sulfacetamide); $C_8H_{10}N_2O_3S$; [144-80-9]

(2) Formamide; CH_3NO; [75-12-7]

(3) Water; H_2O; [7732-18-5]

ORIGINAL MEASUREMENTS:

Rohdewald, P. *Pharm. Ztg.* 1971, *No. 38*, 1342-4.

VARIABLES:

Concentration of formamide

PREPARED BY:

R. Piekos

EXPERIMENTAL VALUES:

$$-k_s = \log \frac{L_{H_2O}}{L_s\ c_s} = 0.12\ \text{l/mol},$$

where L_{H_2O} (0.318_8 g/50 ml = $2.975_8 \times 10^{-2}$ mol dm^{-3}, compiler) and L_s are solubilities of sulfacetamide in water and in aqueous formamide solutions, respectively, and c_s is the concentration of formamide.

L_s values were supplied by the author in personal communication and are shown below.

Concentration of formamide	L_s at 20°C	
mol/l	g/100 ml	10^2 mol dm^{-3} [a]
0.200	0.700	3.27
0.400	0.686	3.20
0.600	0.776	3.62

[a] Calculated by compiler.

AUXILIARY INFORMATION

METHOD/APPARATUS/PROCEDURE:

The solns were equilibrated by agitation for 2 h at 20°C and the sulfacetamide was assayed by differential gravimetric analysis. No details were given.

SOURCE AND PURITY OF MATERIALS:

The source and purity of sulfacetamide was not specified.
Anal reagent grade formamide (source not specified) dried over mol sieve was used.
Purity of the water was not specified.

ESTIMATED ERROR:

Soly: not specified.
Temp: ±0.05°C (author).

REFERENCES:

COMPONENTS:

(1) Acetamide, N-[(4-aminophenyl)-sulfonyl]- (sulfacetamide); $C_8H_{10}N_2O_3S$; [144-80-9]

(2) Acetamide; C_2H_5NO; [60-35-5]

(3) Water; H_2O; [7732-18-5]

ORIGINAL MEASUREMENTS:

Rohdewald, P. *Pharm. Ztg.* 1971, *No. 38*, 1342-4.

VARIABLES:

Concentration of acetamide

PREPARED BY:

R. Piekos

EXPERIMENTAL VALUES:

$$-k_s = \log \frac{L_{H_2O}}{L_s\, c_s} = 0.15 \text{ l/mol},$$

where L_{H_2O} (0.218_8 g/50 ml = $2.975_8 \times 10^{-2}$ mol dm^{-3}, compiler) and L_s are solubilities of sulfacetamide in water and in aqueous acetamide solutions, respectively, and c_s is the concentration of acetamide.

L_s values were supplied by the author in personal communication and are shown below.

Concentration of acetamide	L_s at 20°C	
mol/l	g/100 ml	10^2 mol dm^{-3} [a]
0.300	0.676	3.16
0.600	0.802	3.74
0.900	0.898	4.20
1.200	0.998	4.66

[a] Calculated by compiler.

AUXILIARY INFORMATION

METHOD/APPARATUS/PROCEDURE:

The solns were equilibrated by agitation for 2 h at 20°C and the sulfacetamide was assayed by differential gravimetric analysis. No details were given.

SOURCE AND PURITY OF MATERIALS:

The source and purity of sulfacetamide was not specified.

Anal reagent grade acetamide (source not specified) dried over mol sieve was used.

Purity of the water was not specified.

ESTIMATED ERROR:

Soly: not specified.

Temp: ±0.05°C (author).

REFERENCES:

COMPONENTS:

(1) Acetamide, N-[(4-aminophenyl)-sulfonyl]- (sulfacetamide); $C_8H_{10}N_2O_3S$; [144-80-9]

(2) Ethanethioamide; C_2H_5NS; [62-55-5]

(3) Water; H_2O; [7732-18-5]

ORIGINAL MEASUREMENTS:

Rohdewald, P. *Pharm. Ztg.* 1971, *No. 38*, 1342-4.

VARIABLES:

Concentration of ethanethioamide

PREPARED BY:

R. Piekos

EXPERIMENTAL VALUES:

$$-k_s = \log\frac{L_{H_2O}}{L_s\ c_s} = 0.34\ 1/\mathrm{mol},$$

where L_{H_2O} (0.318_8 g/50 ml $2.975_8 \times 10^{-2}$ mol dm^{-3}, compiler) and L_s are solubilities of sulfacetamide in water and in aqueous ethanethioamide solutions, respectively, and c_s is the concentration of ethanethioamide.

L_s values were supplied by the author in personal communication and are shown below.

Concentration of ethanethioamide mol/1	L_s at 20°C g/100 ml	10^2 mol dm^{-3} [a]
0.050	0.672	3.14
0.150	0.694	3.24
0.200	0.752	3.51
0.400	0.886	4.14
0.600	1.030	4.81

[a] Calculated by compiler

AUXILIARY INFORMATION

METHOD/APPARATUS/PROCEDURE:

The solns were equilibrated by agitation for 2 h at 20°C and the sulfacetamide was assayed bv differential gravimetric analysis. No details were given.

SOURCE AND PURITY OF MATERIALS:

The source and purity of sulfacetamide and water water was not specified. Anal reagent grade ethanethiomamide (source not specified) dried over mol sieve was used.

ESTIMATED ERROR:

Soly: Not specified.

Temp: ±0.05°C (author).

REFERENCES:

COMPONENTS:	ORIGINAL MEASUREMENTS:
(1) Acetamide, N-[(4-aminophenyl)-sulfonyl]- (sulfacetamide); $C_8H_{10}N_2O_3S$; [144-80-9] (2) Propanamide; C_3H_7NO; [79-05-0] (3) Water; H_2O; [7732-18-5]	Rohdewald, P. *Pharm. Ztg.* 1971, *No. 38*, 1342-4.
VARIABLES:	PREPARED BY:
Concentration of propanamide	R. Piekos

EXPERIMENTAL VALUES:

$$-k_s = \log\frac{L_{H_2O}}{L_s c_s} = 0.26 \text{ l/mol},$$

where L_{H_2O} (0.318_8 g/50 ml = $2.975_8 \times 10^{-2}$ mol dm^{-3}, compiler) and L_s are solubilities of sulfacetamide in water and in aqueous propanamide solutions, respectively, and c_s is the concentration of propanamide.

L_s values were supplied by the author in personal communication and are shown below.

Concentration of propanamide mol/l	L_s at 20°C	
	g/100 ml	10^2 mol dm^{-3} [a]
0.200	0.694	2.24
0.400	0.772	3.60
0.600	0.870	4.06
0.800	0.898	4.20
1.000	1.064	4.97

[a] Calculated by compiler.

AUXILIARY INFORMATION

METHOD/APPARATUS/PROCEDURE:	SOURCE AND PURITY OF MATERIALS:
The solns were equilibrated by agitation for 2 h at 20°C and the sulfacetamide was assayed by differential gravimetric analysis. No details were given.	The source and purity of sulfacetamide and water was not specified. Anal reagent grade propanamide (source not specified) dried over mol sieve was used.
	ESTIMATED ERROR:
	Soly: not specified. Temp: ±0.05°C (author).
	REFERENCES:

COMPONENTS:

(1) Acetamide, N-[(4-aminophenyl) sulfonyl]- (sulfacetamide); $C_8H_{10}N_2O_3S$; [144-80-9]

(2) Butanamide; C_4H_9NO; [541-35-5]

(3) Water; H_2O; [7732-18-5]

ORIGINAL MEASUREMENTS:

Rohdewald, P. *Pharm. Ztg.* 1971, *No. 38*, 1342-4.

VARIABLES:

Concentration of butanamide

PREPARED BY:

R. Piekos

EXPERIMENTAL VALUES:

$$-k_s = \log \frac{L_{H_2O}}{L_s \ c_s} = 0.30 \text{ l/mol},$$

where L_{H_2O} (0.318 g/50 ml = 2.975 x 10^{-2} mol dm^{-3}, compiler) and L_s are solubilities of sulfacetamide in water and in aqueous butanamide solutions, respectively, and c_s is the concentration of propanamide.

L_s values were supplied by the author in personal communication and are shown below.

Concentration of butanamide	L_s at 20^oC	
mol/l	g/100 ml	10^2 mol dm^{-3} [a]
0.200	0.710	3.31
0.400	0.798	3.72
0.600	0.934	3.36
0.800	1.080	5.04
1.000	1.204	5.62

[a] Calculated by compiler.

AUXILIARY INFORMATION

METHOD/APPARATUS/PROCEDURE:

The solns were equilibrated by agitation for 2 h at 20^oC and the sulfacetamide was assayed by differential gravimetric analysis. No details were given.

SOURCE AND PURITY OF MATERIALS:

The source and purity of sulfacetamide and water was not specified.

Anal reagent grade butanamide (source not specified) dried over mol sieve was used.

ESTIMATED ERROR:

Soly: not specified.

Temp: $\pm 0.05^oC$ (author).

REFERENCES:

COMPONENTS:	ORIGINAL MEASUREMENTS:
(1) Acetamide, N-[(4-aminophenyl) sulfonyl]- (sulfacetamide); $C_8H_{10}N_2O_3S$; [144-80-9] (2) Formamide, N,N-dimethyl-; C_3H_7NO; [68-12-2] (3) Water; H_2O; [7732-18-5]	Rohdewald, P. *Pharm. Ztg.* 1971, *No. 38*, 1342-4.
VARIABLES:	**PREPARED BY:**
Concentration of N,N-dimethylformamide	R. Piekos

EXPERIMENTAL VALUES:

$$-k_s = \log \frac{L_{H_2O}}{L_s c_s} = 0.32 \text{ l/mol},$$

where L_{H_2O} (0.318 g/50 ml = 2.975 x 10^{-2} mol dm^{-3}, compiler) and L_s are solubilities of sulfacetamide in water and in aqueous N,N-dimethylformamide solutions, respectively, and c_s is the concentration of N,N-dimethylformamide. L_s values were supplied by the author in personal communication and are shown below.

Concentration of N,N-dimethylformamide	L_s at 20°C	
mol/l	g/100 ml	10^2 mol dm^{-3} [a]
0.300	0.828	3.86
0.600	1.004	4.69
0.900	1.312	6.12

[a] Calculated by compiler.

AUXILIARY INFORMATION

METHOD/APPARATUS/PROCEDURE:	SOURCE AND PURITY OF MATERIALS:
The solns were equilibrated by agitation for 2 h at 20°C and the sulfacetamide was assayed by differential gravimetric analysis. No details were given.	The source and purity of sulfacetamide and water was not specified. Anal reagent grade N,N-dimethylformamide (source not specified) dried over mol sieve was used.
	ESTIMATED ERROR: Soly: not specified. Temp: $\pm$0.05°C (author).
	REFERENCES:

COMPONENTS:	ORIGINAL MEASUREMENTS:
(1) Acetamide, N-[(4-aminophenyl) sulfonyl]- (sulfacetamide); $C_8H_{10}N_2O_3S$; [144-80-9] (2) Adetamide, N-methyl-; C_3H_7NO; [79-16-3] (3) Water; H_2O; [7732-18-5]	Rohdewald, P. *Pharm. Ztg.* 1971, *No. 38*, 1342-4.
VARIABLES:	**PREPARED BY:**
Concentration of N-methylacetamide	R. Piekos

EXPERIMENTAL VALUES:

$$-k_s = \log \frac{L_{H_2O}}{L_s\, c_s} = 0.25 \text{ l/mol},$$

where L_{H_2O} (0.318 g/50 ml = 2.975 x 10^{-2} mol dm^{-3}, compiler) and L_s are solubilities of sulfacetamide in water and in aqueous N-methylacetamide solutions, respectively and c_s is the concentration of N-methylacetamide. L_s values were supplied by the author in personal communication and are shown below.

Concentration of N-methylacetamide	L_s at 20°C	
mol/l	g/100 ml	10^2 mol dm^{-3} [a]
0.300	0.836	3.90
0.600	0.944	4.41
0.900	1.088	5.08

[a] Calculated by compiler.

AUXILIARY INFORMATION

METHOD/APPARATUS/PROCEDURE:

The solns were equilibrated by agitation for 2 h at 20°C and the sulfacetamide was assayed by differential gravimetric analysis. No details were given.

SOURCE AND PURITY OF MATERIALS:

The source and purity of sulfacetamide and water was not specified. Anal reagent grade N-methylacetamide (source not specified) dried over mol sieve was used.

ESTIMATED ERROR:

Soly: not specified.
Temp: ±0.05°C (author).

REFERENCES:

COMPONENTS:	ORIGINAL MEASUREMENTS:
(1) Acetanide, N-[(4-aminophenyl) sulfonyl]- (sulfacetamide); $C_8H_{10}N_2O_3S$; [144-80-9] (2) Acetamide, N,N-dimethyl-; C_4H_9NO; [127-19-5] (3) Water; H_2O; [7732-18-5]	Rohdewald, P. *Pharm. Ztg.* 1971, *No. 38*, 1..2-4.
VARIABLES:	**PREPARED BY:**
Concentration of N,N-dimethylacetamide	R. Piekos

EXPERIMENTAL VALUES:

$$-k_s = \log \frac{L_{H_2O}}{L_s\ c_s} = 0.41\ 1/\text{mol},$$

where L_{H_2O} (0.318 g/50 ml = 2.975 x 10^{-2} mol dm^{-3}, compiler) and L_s are solubilities of sulfacetamide in water and in aqueous N,N-dimethylacetamide solutions, respectively, and c_s is the concentration of N,N-dimethylacetamide. L_s values were supplied by the author in personal communication and are shown below.

Concentration of N,N-dimethylacetamide	L_s at 20°C	
mol/1	g/100 ml	10^2 mol dm^{-3}[a]
0.400	0.926	4.32
0.500	0.760	3.55
0.600	1.080	5.04
0.800	1.254	5.85
1.000	1.500	7.00

[a] Calculated by compiler.

AUXILIARY INFORMATION

METHOD/APPARATUS/PROCEDURE:	SOURCE AND PURITY OF MATERIALS:
The solns were equilibrated by agitation for 2 h at 20°C and the sulfacetamide was assayed by differential gravimetric analysis. No details were given	The source and purity of sulfacetamide and water was not specified. Anal reagent grade N,N-dimethylacetamide (source not specified) dried over mol sieve was used.
	ESTIMATED ERROR: Soly: not specified. Temp: ±0.05°C (author).
	REFERENCES:

COMPONENTS:	ORIGINAL MEASUREMENTS:
(1) Acetamide, N-[(4-aminophenyl) sulfonyl]- (sulfacetamide); $C_8H_{10}N_2O_3S$; [144-80-9] (2) 3-Pyridinecarboxamide; $C_6H_6N_2O$; [98-92-0] (3) Water; H_2O; [7732-18-5]	Rohdewald, P. *Pharm. Ztg.* 1971, *No. 38*, 1342-4.
VARIABLES:	PREPARED BY:
Concentration of 3-pyridinecarboxamide	R. Piekos

EXPERIMENTAL VALUES:

$$-k_s = \log \frac{L_{H_2O}}{l_s\ c_s} = 0.88\ \text{l/mol},$$

where L_{H_2O} (0.318 g/50 ml = 2.975 x 10^{-2} mol dm^{-3}, compiler) and L_s are solubilities of sulfacetamide in water and in aqueous 3-pyridinecarboxamide solution, respectively, and c_s is the concentration of 3-pyridinecarboxamide. L_s values were supplied by the author in personal communication and are shown below.

Concentration of 3-pyridinecarboxamide	L_s at 20°C	
mol/l	g/100 ml	10^2 mol dm^{-3} [a]
0.100	0.854	3.99
0.150	0.948	4.42
0.200	1.000	4.67
0.400	1.424	6.65
0.600	1.860	8.68

[a] Calculated by compiler.

AUXILIARY INFORMATION

METHOD/APPARATUS/PROCEDURE:

The solns were equilibrated by agitation for 2 h at 20°C and the sulfacetamide was assayed by differential gravimetric analysis. No details were given.

SOURCE AND PURITY OF MATERIALS:

The source and purity of sulfacetamide and water was not specified.
Anal reagent grade 3-pyridinecarboxamide (source not specified) dried over mol sieve was used.

ESTIMATED ERROR:

Soly: not specified.
Temp: $\pm$0.05°C (author).

REFERENCES:

COMPONENTS:	ORIGINAL MEASUREMENTS:
(1) Acetamide, N-[(4-aminophenyl) sulfonyl]- (sulfacetamide); $C_8H_{10}N_2O_3S$; [144-80-9] (2) 3-Pyridinecarboxamide, N,N-diethyl- (nicetamide) $C_{10}H_{14}N_2O$; [59-26-7] (3) Water; H_2O; [7732-18-5]	Rohdewald, P. *Pharm. Ztg.* 1971, *No. 38*, 1342-4.
VARIABLES:	PREPARED BY:
Concentration of nicetamide	R. Piekos

EXPERIMENTAL VALUES:

$$-k_s = \log \frac{L_{H_2O}}{L_s\, c_s} = 0.94 \text{ l/mol},$$

where L_{H_2O} (0.318 g/50 ml = 2.975 x 10^{-2} mol dm^{-3}, compiler) and L_s are solubilities of sulfacetamide in water and in aqeuous nicetamide solutions, respectively, and c_s is the concentration of nicetamide.
L_s values were supplied by the author in personal communication and are shown below.

Concentration of nicetamide	L_s at 20°C	
mol/l	g/100 ml	10^2 mol dm^{-3} [a]
0.400	1.496	6.982
0.600	2.108	9.838
0.800	3.086	14.40
1.000	4.278	20.00

[a] Calculated by compiler.

AUXILIARY INFORMATION

METHOD/APPARATUS/PROCEDURE:

The solns were equilibrated by agitation for 2 h at 20°C and the sulfacetamide was assayed by differential gravimetric analysis. No details were given.

SOURCE AND PURITY OF MATERIALS:

The source and purity of sulfacetamide and water was not specified.
Anal reagent grade nicetamide (source not specified) dried over mol sieve was used.

ESTIMATED ERROR:

Soly: not specified.

Temp: ±0.05°C (author).

REFERENCES:

COMPONENTS:	ORIGINAL MEASUREMENTS:
(1) Acetamide, N-[(4-aminophenyl) sulfonyl]- (sulfacetamide); $C_8H_{10}N_2O_3S$; [144-80-9] (2) Poly(oxy-1,2-ethanediyl), α-hydro- -ω-hydroxy (PEG 400); $(C_2H_4O)_nH_2O$; [25322-68-3] 400 (3) Water; H_2O; [7732-18-5]	Khawam, M. N.; Yousef, R. T.; Czetach-Lindenwald, H. *Sci. Pharm.* 1966, *34*, 209-13.
VARIABLES:	**PREPARED BY:**
Concentration of PEG 400	R. Piekos

EXPERIMENTAL VALUES:

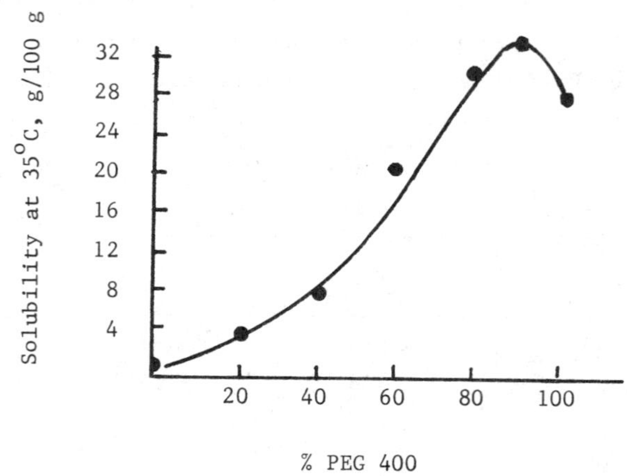

AUXILIARY INFORMATION

METHOD/APPARATUS/PROCEDURE:	SOURCE AND PURITY OF MATERIALS:
An earlier described method was employed (1) whereby a 100-ml conical flask contg PEG 400 soln was placed in a drying cabinet at 35°C and an excess of sulfa-cetamide was added under stirring for 1 h. After 12 h the soln was filtered and decanted and the solute was assayed in the filtrate spectrophotometrically using a Unicam SP 500 spectrophotometer and 1-ml quartz cuvets. Results were taken from a calibration graph.	Neither source nor purity of sulfacetamide and water was specified. PEG 400 was a product of Farbwerke Hoechst (purity not specified). **ESTIMATED ERROR:** Nothing specified. **REFERENCES:** 1. Khawam, M. N.; Tawashi, R.; Czetsch-Lindenwald, H. v. *Sci. Pharm.* 1965, *33*, 90.

COMPONENTS:

(1) Acetamide, N-[(4-aminophenyl) sulfonyl]- (Sulfacetamide); $C_8H_{10}N_2O_3S$; [144-80-9]

(2) Poly(oxy-1,2-ethanediyl), α-hydro-ω-hydroxy- (PEG 600); $(C_2H_4O)_nH_2O$; [25322-68-3] 600

(3) Water; H_2O; [7732-18-5]

ORIGINAL MEASUREMENTS:

Gusyakov, V. P.; Likhol'ot, N. M.; Kutna, I. M. *Farm. Zh. (Kiev)* 1968, *23(6)* 56-61.

VARIABLES:

One temperature: 21-25°C

PREPARED BY:

R. Piekos

EXPERIMENTAL VALUES:

Solubility of sulfacetamide in a 5% (by weight) aqueous α-hydro-ω-hydroxy-poly(oxy-1,2-ethanediyl) 600 at room temperature (21-25°C) is 0.852 g/100 ml (3.98×10^{-2} mol dm^{-3}, compiler).

AUXILIARY INFORMATION

METHOD/APPARATUS/PROCEDURE:

A small excess of sulfacetamide was added to a 5% (by wt) aq PEG 600 soln, the mixture was sealed in an ampul and agitated for 24 h (1). The concn of sulfacetamide was detd colorimetrically (2).

SOURCE AND PURITY OF MATERIALS:

Sulfacetamide: neither source nor purity was specified. PEG 600 was of the Austrian or West German origin. Its purity was not specified.

Purity of the water was not specified.

ESTIMATED ERROR:

Nothing specified.

REFERENCES:

1. Gusyakov, V. P.; Likhol'ot, N. M.; Kuta, I. M. *Farm. Zh. (Kiev)* 1967, *22(3)*, 34.
2. Predchetenskii, B. E.; Borovskaya, V. M.; Morgolina, L. T. *Laboratornye metody issledovaniya*, Medgiz, Moscow 1950, p. 371.

COMPONENTS:

(1) Acetamide, N-[(4-aminophenyl)-sulfonyl]- (sulfacetamide; $C_8H_{10}N_2O_3S$; [144-80-9]

(3) Poly(oxy-1,2-ethanediyl), α-hydro-ω-hydroxy- (PEG 4000); $C_2H_4O)_nH_2O$; [25322-68-3] 4000

(3) Water; H_2O; [7732-18-5]

ORIGINAL MEASUREMENTS:

Khawam, M. N.; Yousef, R. T.; Czetsch-Lindenwald, H. *Sci. Pharm.* 1966, *34*, 209-13.

VARIABLES:

Concentration of PEG 4000

PREPARED BY:

R. Piekos

EXPERIMENTAL VALUES:

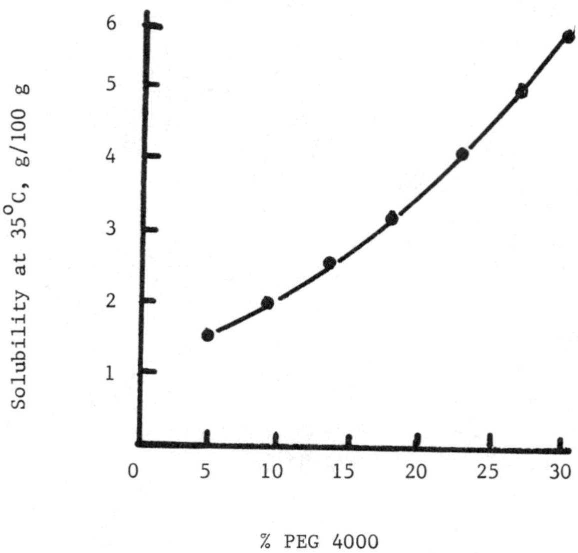

AUXILIARY INFORMATION

METHOD/APPARATUS/PROCEDURE:

An earlier described method was employed (1) whereby a 100-ml conical flask contg a PEG 4000 soln was placed in a drying cabinet at 35°C and an excess of sulfacetamide was added under stirring for 1 h. After 12 h the soln was filtered or decanted and the solute was assayed in the filtrate spectrophotometrically using a Unicam SP 500 spectrophotometer and 1-ml quartz cuvets. Results were taken from a calibration graph.

SOURCE AND PURITY OF MATERIALS:

Neither source nor purity of sulfacetamide and water were specified. PEG 4000 was a product of Farbwerke Hoechst (purity not specified).

ESTIMATED ERROR:

Nothing specified.

REFERENCES:

1. Khawam, M. N.; Tawashi, R.; Czetsch-Lindenwald, H. v. *Sci. Pharm.* 1965, *33*, 90.

COMPONENTS:	ORIGINAL MEASUREMENTS:
(1) Acetamide, N-[(4-aminophenyl) sulfonyl]- (sulfacetamide); $C_8H_{10}N_2O_3S$; [144-80-9] (2) Poly(oxy-1,2-ethanediyl), α-hydro-ω-hydroxy- (PEG 4000); $(C_2H_4O)_nH_2O$; [25322-68-3] 4000 (3) Water; H_2O; [7732-18-5]	Gusyakov, V. P.; Likhol'ot, N. M.; Kutna, I.M. *Farm. Zh. (Kiev)* 1968, *23(6)*, 56-61.
VARIABLES:	PREPARED BY:
One temperature: 21-25 °C	R. Piekos

EXPERIMENTAL VALUES:

Solubility of sulfacetamide in a 5% (by weight) aqueous α-hydro-ω-hydroxy-poly(oxy-1,2-ethanediyl) 4000 at room temperature (21-25 °C is 0.852 g/100 ml (3.98×10^{-2} dm^{-3}, compiler).

AUXILIARY INFORMATION

METHOD/APPARATUS/PROCEDURE:

A small excess of sulfacetamide was added to a 5% (by wt) aq PEG 4000 soln, the mixture was sealed in an ampul and agitated for 25 h (1).

The concn of sulfacetamide was detd colorimetrically (2).

SOURCE AND PURITY OF MATERIALS:

Sulfacetamide: neither source nor purity was specified.

PEG 4000 was of the Austrian or West German origin. Its purity was not specified.

Purity of the water was not specified.

ESTIMATED ERROR:

Nothing specified.

REFERENCES:

1. Gusyakov, V. P.; Likhol'ot, N. M.; Kutna, I. M. *Farm. Zh. (Kiev)* 1967, *22(3)*, 34.
2. Predchetenskii, B. E.; Borovskaya, V. M.; Morgolina, L. T. *Laboratornye metody issledovaniya, Medgiz, Moscow* 1950, *p.* 371.

COMPONENTS:	ORIGINAL MEASUREMENTS:
(1) Acetamide, N-[(4-aminophenyl) sulfonyl] · (sulfacetamide); $C_8H_{10}N_2O_3S$; [144-80-9] (2) Sorbitan monolaurate, polyoxyethylene derivatives (Tween 20); [9005-64-5] (3) Water; H_2O; [7732-18-5]	Khawam, M. N.; Yousef, R. T.; Czetsch-Lindenwald, H. *Sci. Pharm.* 1966, *34*, 209-13.
VARIABLES:	PREPARED BY:
Concentration of Tween 20	R. Piekos

EXPERIMENTAL VALUES:

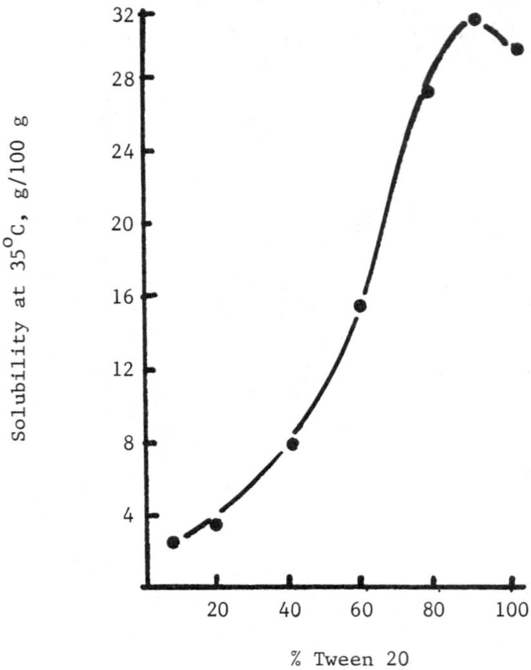

AUXILIARY INFORMATION

METHOD/APPARATUS/PROCEDURE:

An earlier described method was employed (1) whereby a 100-ml conical flask contg a Tween 20 soln was placed in a drying cabinet at 35°C and an excess of sulfacetamide was added under stirring for 1 h. After 12 h the soln was filtered or decanted and the solute was assayed in the filtrate spectrophotometrically using a Unicam SP 500 spectrophotometer and 1-ml quartz cuvets. Results were taken from a calibration graph.

SOURCE AND PURITY OF MATERIALS:

Neither source nor purity of sulfacetamide and water were reported.

Tween 20 was supplied by Atlas-Goldschmidt A. G., Essen (purity not specified).

ESTIMATED ERROR:

Nothing specified.

REFERENCES:

1. Khawam, M. N.; Tawashi, R.; Czetsch-Lindenwald, H. v. *Sci. Pharm.* 1965, *33*, 90.

COMPONENTS:	ORIGINAL MEASUREMENTS:
(1) Acetamide, N-[(4-aminophenyl)-sulfonyl]- (sulfacetamide); $C_8H_{10}N_2O_3S$; [144-80-9] (2) Sorbitan monolaurate, polyoxyethylene derivatives (Tween 20); [9005-64-5] (3) Water; H_2O; [7732-18-5]	Gusyakov, V. P.; Likhol'ot, N. M.; Kutna, I. M. *Farm. Zh. (Kiev)* 1967, *22(3)* 34-9.
VARIABLES:	**PREPARED BY:**
One temperature: 20°C	R. Piekos

EXPERIMENTAL VALUES:

$$S/S_o = 1.6 \text{ at } 20^{\circ}C$$

where S is the solubility of sulfacetamide in a 2% by weight aqueous Tween 20 solution, and

S_o is the solubility of sulfacetamide in water (0.50 g/100 ml).

Hence S = 0.80 g/100 ml (3.7 x 10^{-2} mol dm^{-3}) - compiler.

AUXILIARY INFORMATION

METHOD/APPARATUS/PROCEDURE:	SOURCE AND PURITY OF MATERIALS:
An excess of sulfacetamide in a 2% by wt aq Tween 20 soln was equilibrated for 24 h in an ampul immersed in a water thermostat. Aliquots of the satd soln were withdrawn through a filter and the sulfacetamide content was assayed in the filtrate photometrically.	Sulfacetamide conformed to the requirements of the State Pharmacopeia IX. Tween 20 was a product of Gee Lawson, England. Purity of the water was not specified.
	ESTIMATED ERROR: Soly: not specified. Temp: $\pm$0.1°C (authors).
	REFERENCES:

COMPONENTS:

(1) Acetamide, N-[(4-aminophenyl) sulfonyl]- (sulfacetamide); $C_8H_{10}N_2O_3S$; [144-80-9]

(2) Sorbitan monopalmitate, polyoxyethylene derivatives (Tween 40); [9005-66-7]

(3) Water; H_2O; [7732-18-5]

ORIGINAL MEASUREMENTS:

Gusyakov, V. P.; Likhol'ot, N. M.; Kutna, I. M. *Farm. Zh. (Kiev)* 1967, *22(3)*, 34-9.

VARIABLES:

One temperature: 20°C

PREPARED BY:

R. Piekos

EXPERIMENTAL VALUES:

$$S/S_o = 1.72 \text{ at } 20^{\circ}C$$

where S is the solubility of sulfacetamide in a 2% by weight aqueous Tween 40 solution, and

S_o is the solubility of sulfacetamide in water (0.50 g/100 ml).

Hence S = 0.86 g/100 ml (4.0×10^{-2} mol dm^{-3})- compiler.

AUXILIARY INFORMATION

METHOD/APPARATUS/PROCEDURE:

An excess of sulfacetamide in a 2% by wt aq Tween 40 soln was equilibrated for 24 h in an ampul immersed in a water thermostat. Aliquots of the satd soln were withdrawn through a filter and the sulfacetamide content was assayed in the filtrate photometrically.

SOURCE AND PURITY OF MATERIALS:

Sulfacetamide conformed to the requirements of the State Pharmacopeia IX. Tween 40 was a product of Gee Lawson, England. Purity of the water was not specified.

ESTIMATED ERROR:

Soly: not specified.

Temp: $\pm 0.1^{\circ}C$ (authors).

REFERENCES:

COMPONENTS:	ORIGINAL MEASUREMENTS:
(1) Acetamide, N-[(4-aminophenyl)-sulfonyl]- (sulfacetamide); $C_8H_{10}N_2O_3S$; [144-80-8] (2) Sorbitan monooleate, polyoxyethylene derivatives (Tween 80); [9005-65-6] (3) Water; H_2O; [7732-18-5]	Gusyakov, V. P.; Likhol'ot, N. M.; Kutna, I. M. *Farm. Zh. (Kiev)* 1967, *22(3)*, 34-9.
VARIABLES:	PREPARED BY:
One temperature: 20°C	R. Piekos

EXPERIMENTAL VALUES:

$$S/S_o = 1.7 \text{ at } 20^{\circ}C$$

where S is the solubility of sulfacetamide in a 2% by weight aqueous Tween 80 solution, and

S_o is the solubility of sulfacetamide in water (0.50 g/100 ml).

Hence S = 0.85 g/100 ml (4.0×10^{-2} mol dm^{-3}) - compiler.

AUXILIARY INFORMATION

METHOD/APPARATUS/PROCEDURE:	SOURCE AND PURITY OF MATERIALS:
An excess of sulfacetamide in a 2% by wt aq Tween 80 soln was equilibrated for 24 h in an ampul immersed in a water thermostat. Aliquots of the satd soln were withdrawn through a filter and the sulfacetamide content was assayed in the filtrate photometrically.	Sulfacetamide conformed to the requirements of the State Pharmacopeia IX. Tween 80 was a product of Gee Lawson, England. Purity of the water was not specified.
	ESTIMATED ERROR: Soly: not specified. Temp: ±0.1°C (authors).
	REFERENCES:

COMPONENTS:

(1) Acetamide, N-[(4-aminophenyl)-sulfonyl]- (sulfacetamide); $C_8H_{10}N_2O_3S$; [144-80-9]

(2) White petrolatum (liquid petrolatum)

ORIGINAL MEASUREMENTS:

Whitworth, C. W.; Becker, C. H.

J. Pharm. Sci. 1965, *54(4)*, 569-73.

VARIABLES:

One temperature: 37.5°C

PREPARED BY:

R. Piekos

EXPERIMENTAL VALUES:

Solubility of sulfacetamide in white petrolatum (liquid petrolatum) at 37.5°C is 0.089 mg% (4.1×10^{-6} mol dm^{-3} solution, compiler).

AUXILIARY INFORMATION

METHOD/APPARATUS/PROCEDURE:

A satd soln of sulfacetamide in liquid petrolatum was made and filtered carefully at a const temp to remove suspended particles. A portion of the soln was shaken for 4 h with 10 ml of EtOH. The alcoholic layer was centrifuged for 30 min. Aliquot portions of the alcoholic solns were allowed to evap to dryness, a trichloroacetic acid soln added, and subsequently the Marshall reagents. From the intensity of the color developed it was impossible to det the amt of the drug extd by the process utilized. A Klett-Summerson colorimeter with a No 54 filter was employed to det the color intensity, which was compared to that of standard solns.

SOURCE AND PURITY OF MATERIALS:

Sulfacetamide (N.F. grade) was from Ruger Chemical Co., Inc.

White petrolatum (liquid petrolatum) (U.S.P. grade) was from Fisher Scientific Co.

ESTIMATED ERROR:

Soly: not specified.

Temp: ±1°C (authors).

REFERENCES:

COMPONENTS:

(1) Acetamide, N-[(4-aminophenyl)-sulfonyl]- (sulfacetamide); $C_8H_{10}N_2O_3S$; [144-80-9]

(2) Sorbitan, (Z)-9-octadecenoate (2:3) (Arlacel 83); [8007-43-0]

(3) White petrolatum (liquid petrolatum)

ORIGINAL MEASUREMENTS:

Whitworth, C. W.; Becker, C. H.

J. Pharm. Sci. 1965, *54(4)*, 569-73

VARIABLES:

Concentration of Arlacel 83

PREPARED BY:

R. Piekos

EXPERIMENTAL VALUES:

Concentration of Arlacel 83 %	Solubility at 37.5°C	
	mg%	10^5 mol dm^{-3} soln[a]
1	0.150	0.700
5	0.906	4.22
10	1.761	8.21

[a] Calculated by compiler.

AUXILIARY INFORMATION

METHOD/APPARATUS/PROCEDURE:

A satd soln of sulfacetamide in the solvent was made and filtered carefully at a const temp to remove all suspended particles. A 5-ml portion of the soln was shaken for 4 h with 100 ml of EtOH. The alcoholic layer was centrifuged for 30 min. Aliquot portions of the alcoholic solns were allowed to evap to dryness, a trichloroacetic acid soln was added, and subsequently the Marshall reagents. From the intensity of the color developed it was possible to det the amt of the drug extd by the process utilized. A Klett-Summerson colorimeter with a No 54 filter was employed to det the color intensity, which was compared to that of standard solns.

SOURCE AND PURITY OF MATERIALS:

Sulfacetamide (N.F. grade) was from Ruger Chemical Co, Inc.

Arlacel 83 (Lot No 129) was from Atlas Powder Co. (Purity not specified).

White petrolatum (U.S.P. grade) was from Fisher Scientific Co.

ESTIMATED ERROR:

Soly: not specified.

Temp: ± 1°C (authors).

REFERENCES:

COMPONENTS:

(1) Acetamide, N-[(4-aminophenyl)-sulfonyl]- (sulfacetamide) $C_8H_{10}N_2O_3S$; [144-80-9]

(2) Cottonseed oil

ORIGINAL MEASUREMENTS:

Whitworth, C. W.; Becker, C. H.

J. Pharm. Sci. 1965, *54(4)*, 569-73.

VARIABLES:

One temperature: 37.5°C

PREPARED BY:

R. Piekos

EXPERIMENTAL VALUES:

Solubility of sulfacetamide in cottonseed oil at 37.5°C is 4.734 mg% (2.212 x 10^{-4} mol dm^{-3} solution, compiler).

AUXILIARY INFORMATION

METHOD/APPARATUS/PROCEDURE:

A satd soln of sulfacetamide in cottonseed oil was made and filtered carefully at a const temp to remove suspended particles. A portion of the soln was shaken for 4 h with 100 ml of EtOH. The alcoholic layer was centrifuged for 30 min. Aliquot portions of the alcoholic solns were allowed to evap to dryness, a trichloroacetic acid soln added, and subsequently the Marshall reagents. From the intensity of the color develped it was possible to det the amt of the drug extd by the process utilized. A Klett-Summerson colorimeter with a No 54 filter was employed to det the color intensity, which was compared to that of standard solns.

SOURCE AND PURITY OF MATERIALS:

Sulfacetamide (N.F. grade) was from Ruger Chemical Co., Inc.

Neither source nor purity of the cottonseed oil was specified.

ESTIMATED ERROR:

Soly: not specified.

Temp: ± 1°C (authors).

REFERENCES:

COMPONENTS:

(1) Acetamide, N-[(4-aminophenyl)-sulfonyl]- (sulfacetamide); $C_8H_{10}N_2O_3S$; [144-80-9]

(2) Cottonseed oil

(3) Sorbitan, (Z)-9-octadecenoate (2:3) (Arlacel 83); [8007-43-0]

ORIGINAL MEASUREMENTS:

Whitworth, C. W.; Becker, C. H.

J. Pharm. Sci. 1965, *54(4)*, 569-73.

VARIABLES:

Concentration of Arlacel 83

PREPARED BY:

R. Piekos

EXPERIMENTAL VALUES:

Concentration of Arlacel 83 %	Solubility at 37.5°C mg%	10^4 mol dm^{-3} soln[a]
1	5.675	2.648
5	6.950	3.244
10	8.45	3.94

[a] Calculated by compiler

AUXILIARY INFORMATION

METHOD/APPARATUS/PROCEDURE:

A satd soln of sulfacetamide in the solvent was made and filtered carefully at a const temp to remove all suspended matter. A 5-ml portion of the soln was shaken for 4 h with 100 ml of EtOH. The alcoholic layer was centrifuged for 30 min. Aliquot portions of the alcoholic solns were allowed to evap to dryness, a trichloroacetic acid soln was added, and subsequently the Marshall reagents. From the intensity of the color developed it was possible to det the amt of the drug extd by the process utilized. A Klett-Summerson colorimeter with a No 54 filter was employed to det the color intensity, which was compared to that of standard solns.

SOURCE AND PURITY OF MATERIALS:

Sulfacetamide (N.F. grade) was from Ruger Chemical Co., Inc.

Neither source nor purity of the cottonseed oil was specified.

Arlacel 83 (Lot No 129) was from Atlas Powder Co. (purity not specified).

ESTIMATED ERROR:

Soly: not specified.

Temp: $\pm 1^{\circ}C$ (authors).

REFERENCES:

COMPONENTS:	ORIGINAL MEASUREMENTS:
(1) Acetamide, N-[(4-aminophenyl)-sulfonyl]- (sulfacetamide); $C_8H_{10}N_2O_3S$; [144-80-9] (2) Methane, trichloro- (chloroform); $CHCl_3$; [67-66-3]	Yamazaki, M.; Aoki, M.; Kamada, A.; Yata, N. *Yakuzaigaku* 1967, *27(1)*, 37-40.
VARIABLES:	**PREPARED BY:**
One temperature: 30°C	R. Piekos

EXPERIMENTAL VALUES:

Solubility of sulfacetamide in chloroform at 30°C is 3.60 mmol/L (0.77 g dm^{-3}, compiler).

AUXILIARY INFORMATION

METHOD/APPARATUS/PROCEDURE:

Sulfacetamide (0.5 g) was placed in an L-shaped tube together with 20 ml of chloroform. The mixt was shaken in a thermostat until equilibrium was attained. The sulfacetamide was then assayed in the supernatant spectrophotometrically at 545 nm on a Beckmann DU spectrophotometer. The results were taken from a calibration graph.

SOURCE AND PURITY OF MATERIALS:

Nothing specified.

ESTIMATED ERROR:

Soly: not specified.

Temp: ± 1°C (authors).

REFERENCES:

COMPONENTS:

(1) Acetamide, N-[(4-aminophenyl)-sulfonyl]- (sulfacetamide); $C_8H_{10}N_2O_3S$; [144-80-9]

(2) 2-Propanone (acetone); C_3H_6O; [67-64-1]

ORIGINAL MEASUREMENTS:

Gutierrez, F. H.

Anales fis. quim. (Madrid) 1945, *41*, 537-60.

VARIABLES:

Temperature

PREPARED BY:

R. Piekos

EXPERIMENTAL VALUES:

$t/^{\circ}C$	G^a	E^b	X_g/l^c	mol/l^d acetone	mmol/mol acetone	$1{:}X_g{}^e$	$1 + X_{cc}{}^f$
0	13.858	12.085	112.887	527	37	7.22	8.86
5	15.005	13.047	121.360	566	41	6.66	8.02
10	16.450	14.126	132.094	617	45	6.08	7.57
15	18.408	14.546	146.730	685	50	5.43	6.82
20	20.948	17.328	165.741	774	57	4.77	6.03
25	23.904	19.336	187.694	876	65	4.19	5.33
30	27.751	21.723	216.265	1095	75	3.50	4.62
40	38.144	27.611	292.717	1361	103	2.62	3.42
45	45.913	31.465	349.582	1632	124	2.18	2.86
50	59.893	37.458	451.592	2127	163	1.67	2.21

[a] $G = \frac{p\ 100}{P - p}$, where p and P are the weights of solute and solution, resp.

[b] $E = \frac{G\ 100}{G + 100}$; [c] g/l acetone; [d] should be mmol/l acetone (compiler);

[e] g of acetone required to dissolve 1 g of solute;

[f] volume (cm^3) of acetone required to dissolve 1 g of solute.

AUXILIARY INFORMATION

METHOD/APPARATUS/PROCEDURE:

A special all-glass app was constructed enabling the prepn of satd solns, agitation by bubbling a stream of acetone-satd N, filtration, and distn off the solvent without contact with air. Two exchange-able dissoln vessels of 15 and 8 cm^3 working capacity were used depending on the soly of solute. The app was immersed in a thermostat. The vols of acetone used were 15 or 5 cm^3, and the equilibration time was 2-2.5 h. The satd solns were filtered, weighed, the solvent was distd off, the residues were dried at 105°C, weighed, and examd for the presence of solvated acetone.

SOURCE AND PURITY OF MATERIALS:

The source of the materials was not specified. Pure, anhydrous acetone was used. The absence of impurities and water was confirmed by procedures of the German Pharmacopeia VI and Spanish Pharmacopeia VIII.

The purity of sulfacetamide was not specified.

ESTIMATED ERROR:

Soly: measurements were repeated until 2 values not differing in the second decimal were obtained (author).

Temp: ±0.1°C (author).

REFERENCES:

COMPONENTS:

(1) Acetamide, N-[(4-aminophenyl)-sulfonyl]- (sulfacetamide); $C_8H_{10}N_2O_3S$; [144-80-9]

(2) Poly(oxy-1,2-ethanediyl), α-hydro-ω-hydroxy- (PEG 400); $(C_2H_4O)_nH_2O$; [25322-68-3]

ORIGINAL MEASUREMENTS:

Gusyakov, V. P.; Likhol'ot, N. M.; Kutna, I. M. *Farm. Zh. (Kiev)* 1968, *23(6)*, 56-61.

VARIABLES:

One temperature: 21-25°C

PREPARED BY:

R. Piekos

EXPERIMENTAL VALUES:

Solubility of sulfacetamide in α-hydro-ω-hydroxypoly(oxy-1,2-ethanediyl) 400 at room temperature (21-25°C) is 19.9% by weight (1.16 mol kg^{-1} PEG 400, compiler).

AUXILIARY INFORMATION

METHOD/APPARATUS/PROCEDURE:

Small quantities (2-4 mg) of sulfacetamide were added to a known quantity of PEG 400 under stirring until satn was attained.

SOURCE AND PURITY OF MATERIALS:

Sulfacetamide: neither source nor purity was specified.

PEG 400: source not specified; sp gr 1.127 g cm^{-3}; temp of solidification approx 6°C; refractive index 1.466 (temp not indicated).

ESTIMATED ERROR:

REFERENCES:

COMPONENTS:

(1) Acetamide, N-[[(4-acetylamino)-phenyl] sulfonyl] - (acetyl sulfacetamide); $C_{10}H_{12}N_2O_4S$; [5626-90-4]

(2) Phosphoric acid, disodium salt; Na_2HPO_4; [7558-94-4]

(3) Water; H_2O; [7732-18-5]

ORIGINAL MEASUREMENTS:

Krüger-Thiemer, E.

Arch. Dermatol. Syphilis 1942, *183*, 90-116.

VARIABLES:

One temperature: ca 20°C;

one pH: 8.74

PREPARED BY:

R. Piekos

EXPERIMENTAL VALUES:

Solubility of acetyl sulfacetamide in a 0.705M (10%) Na_2HPO_4 solution of pH 8.74 at room temperature (about 20°C) is 2.040 g% (7.959 x 10^{-2} mol dm^{-3} solution, compiler).

AUXILIARY INFORMATION

METHOD/APPARATUS/PROCEDURE:

Acetyl sulfacetamide (0.5 g) was dissolved in the 0.705M (10%) Na_2HPO_4 soln of pH 8.74, shaken for 2 h at room temp (about 20°C), and filtered. The filtrate was treated with equal vol of 2N HCl and refluxed for 15 min. After proper diln, a 1-cm^3 aliquot was withdrawn, acidified, cooled, and the sulfonamide content was detd colorimetrically by the Marshall method modified by Kimmig (1) using an Authenreith colorimeter. The pH value was detd on an ultraionograph using a glass electrode.

SOURCE AND PURITY OF MATERIALS:

Acetyl sulfacetamide (source not specified) gave no coloration upon diazotization of its satd soln, thus showing absence of sulfanilamide.

The source and purity of the remaining materials was not specified.

ESTIMATED ERROR:

Soly: precision ±5% (author).

Temp: not specified.

pH: ±0.05 pH unit (author).

REFERENCES:

1. Kimmig, J. *Arch. Dermatol.* 1938, *176*, 722; *Erg. Hyg.* 1941, *24*, 398.

COMPONENTS:

(1) Acetamide, N-[[(4-acetylamino) phenyl]sulfonyl] - (acetyl sulfacetamide); $C_{10}H_{12}N_2O_4S$; [5626-90-4]

(2) Phosphoric acid, monopotassium salt; KH_2PO_4; [7778-77-0]

(3) Water; H_2O; [7732-18-5]

ORIGINAL MEASUREMENTS:

Krüger-Thiemer, E.

Arch. Dermatol. Syphilis 1942, *183*, 90-116.

VARIABLES:

One temperature: ca 20°C;
one pH: 4.37

PREPARED BY:

R. Piekos

EXPERIMENTAL VALUES:

Solubility of acetyl sulfacetamide in a 0.735 M (10%) KH_2PO_4 solution of pH 4.37 at room temperature (about 20°C) is 0.028 g% (1.1×10^{-3} mol dm^{-3} solution, compiler).

AUXILIARY INFORMATION

METHOD/APPARATUS/PROCEDURE:

Acetyl sulfacetamide (0.5 g) was dissolved in the 0.735 M (10% KH_2PO_4 soln, shaken for 2 h at room temp (about 20°C), and filtered. The filtrate was treated with equal vol of 2N HCl, and refluxed for 15 min. After proper diln, a 1-cm^3 aliquot was withdrawn, acidified, cooled, and the sulfonamide content was detd colorimetrically by the Marshall method modified by Kimmig (1) using an Authenreith colorimeter. The pH value was detd on an ultraionograph using a glass electrode.

SOURCE AND PURITY OF MATERIALS:

Acetyl sulfacetamide (source not specified) gave no coloration upon diazotization of its satd soln, thus showing absence of sulfanilamide.

The source and purity of the remaining materials was not specified.

ESTIMATED ERROR:

Soly: precision ±5% (author).

Temp: not specified.

pH: ±0.05 pH unit (author).

REFERENCES:

1. Kimmig, J. *Arch. Dermatol.* 1938, *176*, 722; *Erg. Hyg.* 1941, *24*, 398.

COMPONENTS:

(1) Acetamide, N-[[(4-acetylamino)phenyl]-sulfonyl]- (acetyl sulfacetamide); $C_{10}H_{12}N_2O_4S$; [5626-90-4]

(2) Phosphoric acid, disodium salt; Na_2HPO_4; [7558-94-4]

(3) Phosphoric acid, monopotassium salt; KH_2PO_4; [7778-77-0]

(4) Water; H_2O; [7732-18-5]

ORIGINAL MEASUREMENTS:

Krüger-Thiemer, E.

Arch. Dermatol. Syphilis 1942, *183*, 90-116.

VARIABLES:

Temperature, pH

PREPARED BY:

R. Piekos

EXPERIMENTAL VALUES:

Composition of 1/15M phosphate buffer solutions			pH	Solubility			
				Room temp (ca 20°C		37°C	
Na_2HPO_4	KH_2PO_4	%Content		g%	10^3mol dm^{-3} solution[a]	g%	10^3 mol dm^{-3} solution[a]
1.0	99.0	0.91	4.944	0.043	1.7	-	-
10.0	90.0	0.91	5.906	0.087	3.4	0.122	4.76
61.1	38.9	0.93	7.005	0.638	24.9	0.699	27.3
9.5	0.5	0.733[b]	7.51	2.150	83.89	-	-
94.7	5.3	0.95	8.018	0.930	36.3	-	-

[a] Calculated by compiler.

[b] Molar content; 10% buffer solution

AUXILIARY INFORMATION

METHOD/APPARATUS/PROCEDURE:

Acetyl sulfacetamide (0.5 g) was dissolved in 10 cm^3 of a buffer soln, shaken for 2 h at 20°C (or left for 48 h at 37°C), and filtered at respective temp. The filtrate was treated with equal vol of 2N HCl and refluxed for 15 min. After proper diln, a 1-cm^3 aliquot was withdrawn, acidified, cooled, and the sulfonamide content was detd colorimetrically by the Marshall method modified by Kimmig (1) using an Authenreith colorimeter. The pH was detd on an ultraionograph using a glass electrode.

SOURCE AND PURITY OF MATERIALS:

Acetyl sulfacetamide (source not specified) gave no coloration upon diazotization of its satd soln, thus showing absence of sulfanilamide. The source and purity of the remaining materials was not specified.

ESTIMATED ERROR:

Soly: precision: ±5% (author).

Temp: not specified.

pH: ±0.05 pH unit

REFERENCES:

1. Kimmig, J. *Arch. Dermatol.* *176*, 722; *Erg. Hyg.* 1941, *24*, 398.

COMPONENTS:	ORIGINAL MEASUREMENTS:
(1) Acetamide, N-[[(4-acetylamino)phenyl]-sulfonyl]- (acetyl sulfacetamide); $C_{10}H_{12}N_2O_4S$; [5626-90-4] (2) Phosphoric acid, disodium salt; Na_2HPO_4; [7558-94-4] (3) Phosphoric acid, monopotassium salt; KH_2PO_4; [7778-77-0] (4) Water; H_2O; [7732-18-5]	Bandelin, F. J.; Malesh, W. *J. Am. Pharm. Assoc., Sci. Ed.* 1959, *48*, 177-81
VARIABLES: pH	PREPARED BY: R. Piekos

EXPERIMENTAL VALUES:

Solubility of acetyl sulfacetamide in buffers of varying mixtures of $Na_2HPO_4 \cdot 7H_2O$ (71.6 g/l distilled water; 0.27 mol dm^{-3}, compiler) and KH_2PO_4 (36.3 g/l distilled water; 0.27 mol dm^{-3}, compiler) at 37°C.

Equilibrium pH	Solubility (based on sulfacetamide) mg/100 ml	10^2 mol dm^{-3} [a]
4.5 (initial pH)	60	0.3
4.5	125	0.58
4.8	250	1.2
5.3	550	2.6
5.6	1150	5.37
5.9	2310	10.8
6.6	3900	18.2
7.0	3900	18.2

[a] Calculated by compiler.

AUXILIARY INFORMATION

METHOD/APPARATUS/PROCEDURE:

Solns were prepd by adding an excess of acetyl sulfacetamide to 10 ml of buffer soln at each pH level in 18 x 150-mm test tubes, stoppering the tubes, placing in water bath at 37°C with gentle agitation for 24 h. The solute was then hydrolyzed with 5% H_2SO_4 for 1 h to liberate sulfacetamide. One-ml aliquot of the hydrolyzate was accurately pipetted into a volumetric flask for diln and analysis. The sulfonamide was assayed colorimetrically by the method of Bratton and Marshall as described in detail by Biamonte and Schneller (1). A standard curve was prepd using accurately prepd standard solutions.

SOURCE AND PURITY OF MATERIALS:

Neither source nor purity of the reagents were specified. Distilled water was used.

ESTIMATED ERROR:

Soly: av values of duplicate runs are reported (authors).

Temp and pH: not specified.

REFERENCES:

1. Biamonte, A. R.; Schneller, G. E. *J. Am. Pharm. Assoc., Sci. Ed.*, 1952, *41*, 341.

COMPONENTS:

(1) Acetamide, N-[[(4-acetylamino)phenyl]-sulfonyl - (acetyl sulfacetamide); $C_{10}H_{12}N_2O_4S$; [5626-90-4]

(2) Calcium chloride; $CaCl_2$; [10043-52-4]

(3) Magnesium chloride; $MgCl_2$; [7786-30-3]

(4) Phosphoric acid, monoammonium salt; $NH_4H_2PO_4$; [7722-76-1]

(5) Potassium chloride; KCl; [7447-40-7]

(6) Sodium chloride; NaCl; [7647-14-5]

(7) Urea; CH_4N_2O; [57-13-6]

(8) Water; H_2O; [7732-18-5]

ORIGINAL MEASUREMENTS:

Bandelin, F. J.; Malesh, W.

J. Am. Pharm. Assoc., Sci. Ed. 1959, *48*, 177-81.

VARIABLES:

pH at 37°C

PREPARED BY:

R. Piekos

EXPERIMENTAL VALUES:

Solubility of acetyl sulfacetamide in a solution containing $CaCl_2$ 0.143, $MgCl_2$ 0.121, $NH_4H_2PO_4$ 0.300, KCl 1.660, NaCl 2.950 and urea 20 g/dm^3 (synthetic urine, Mosher Vehicle) at 37°C.

Equilibrium pH	Solubility (based on sulfacetamide)	
	mg/100 ml	mol/dm^3 [a]
4.5	240	0.94
5.0	310	1.2
5.5	505	2.0
6.0	1050	4.1
6.5	2520	9.8
7.0	5600	21.8

[a] Calculated by compiler.

AUXILIARY INFORMATION

METHOD/APPARATUS/PROCEDURE:

Excess acetyl sulfacetamide was added to aliquots of synthetic urine solns and 1% H_3PO_4 or 1% NaOH solns were used to adjust the pH to the required value. The solns were agitated for 24 h with addn of acid or base to keep them at the desired pH level until equilibrium was attained. Then the solns were filtered and in aliquots acetyl sulfacetamide was assayed spectrophotometrically by the method described by Biamonte and Schneller (1). Before detn the soln was refluxed with 5% H_2SO_4 for 1 h to liberate the free amino compd.

SOURCE AND PURITY OF MATERIALS:

Nothing specified.

ESTIMATED ERROR:

Soly: average values of 2 detns were given.
Temp: not specified.
pH: not specified.

REFERENCES:

1. Biamonte, A. R.; Schneller, G. E., *J. Am. Pharm. Assoc., Sci. Ed.* 1952, *41*, 341.

COMPONENTS:	ORIGINAL MEASUREMENTS:
(1) Acetamide, N-[[(4-acetylamino)phenyl]-sulfonyl]-; $C_{10}H_{12}N_2O_4S$; [5626-90-4] (2) Ethanol; C_2H_6O; [64-17-5] (3) Water; H_2O; [7732-18-5]	Sapozhnikova, N. V.; Postovskii, I. Ya. *Zh. Prikl. Khim.* 1944, *17*, 427-34.
VARIABLES:	PREPARED BY:
Concentration of ethanol	R. Piekos

EXPERIMENTAL VALUES:

Concentration of ethanol	Solubility at 75°C	
Weight%	Weight%	10^2 mol kg^{-1} solvent[a]
0	0.24	0.90
19.2	0.71	2.66
38.3	1.59	6.01
57.6	2.3	8.8
67.2	3.6	14
76.4	3.7	14
96	4.3	17

[a] Calculated by compiler.

AUXILIARY INFORMATION

METHOD/APPARATUS/PROCEDURE:

The acetylated sulfonamide was dissolved in EtOH - water mixts to form satd solns which were occasionally agitated in glass vessels immersed in a thermostat. The equilibrium was usually attained after 1 h. Five- to 100-cm^3 samples of the satd soln were placed in Pt crucibles or dishes and evapd to dryness at temps lower than 110-115°C. The residue was dried to const wt at 105-110°C and weighed.

SOURCE AND PURITY OF MATERIALS:

Pure recrystd acetylated sulfonamide was used. Its mp conformed to that reported in the literature. The purity of ethanol and water was not specified.

ESTIMATED ERROR:

Soly: quite reliable results were obtained (authors).

Temp: $\pm$0.05°C (authors).

REFERENCES:

COMPONENTS:

(1) Crotonamide, 3-methyl-N-[(4-aminophenyl)sulfonyl]- (irgamide); $C_{11}H_{14}N_2O_2S$; [115-68-14]
(2) Phosphoric acid, disodium salt; Na_2HPO_4; [7558-94-4]
(3) Phosphoric acid, monopotassium salt; KH_2PO_4; [7778-77-0]
(4) Water; H_2O [7732-18-5]

ORIGINAL MEASUREMENTS:

Pulver, R.; Suter, R.
Schweiz. Med. Wochenschr. 1943, *73(13)*, 403-8.

VARIABLES:

pH

PREPARED BY:

R. Piekos

EXPERIMENTAL VALUES:

Solubility of irgamide in M/15 phosphate buffers (according to Sørensen) at 20 °C

pH	mg%	10^2 mol dm^{-3} [a]
6.0	95	0.37
7.0	385	1.51
8.0	620	2.44

[a] calculated by compiler.

AUXILIARY INFORMATION

METHOD/APPARATUS/PROCEDURE:

Nothing specified.

SOURCE AND PURITY OF MATERIALS:

Nothing specified.

ESTIMATED ERROR:

Nothing specified.

REFERENCES:

COMPONENTS:

(1) 2-Butenamide, N-[(4-aminophenyl)-sulfonyl]-3-methyl- (sulfadicramide); $C_{11}H_{14}N_2O_3S$; [115-68-4]

(2) 2-Propanone (acetone) C_3H_6O; [67-64-1]

ORIGINAL MEASUREMENTS:

Gutierrez, F. H.

Anales fis. quim. (Madrid) 1945, *41*, 537-60.

VARIABLES:

Temperature

PREPARED BY:

R. Piekos

EXPERIMENTAL VALUES:

$t/^{o}C$	G^{a}	E^{b}	X_g/I^{c}	$mol/1^{d}$ acetone	mmol/mol acetone	$1:X_g^{\ e}$	$1 + X_{cc}^{\ f}$
0	17.498	14.893	142.539	561	39.9	5.71	7.01
5	18.500	15.611	149.628	588	42.5	5.40	6.69
10	19.591	16.348	157.316	619	44.7	5.10	6.36
15	20.699	17.149	164.992	648	47.3	4.83	6.06
20	22.261	18.208	176.129	692	50.9	4.49	5.67
25	25.099	18.762	181.373	713	52.7	4.33	5.51
30	24.922	19.950	194.217	764	56.8	4.01	5.15
35	26.799	21.135	207.237	815	61.2	3.73	4.82
40	29.400	22.720	225.616	887	67.1	3.40	4.43
45	33.029	24.828	251.483	988	75.4	3.03	3.94
50	39.451	28.290	298.052	1172	90.1	2.53	3.35

[a] $G = \frac{p\ 100}{P - p}$, where p and P are the weights of solute and solution, resp.

[b] $E = \frac{G\ 100}{G + 100}$; [c] g/l acetone; [d] should be mmol/l acetone (compiler);

[e] g of acetone required to dissolve 1 g of solute; [f] volume (cm^3) of acetone required to dissolve 1 g of solute.

AUXILIARY INFORMATION

METHOD/APPARATUS/PROCEDURE:

A special all-glass app was constructed enabling the prepn of satd solns, agitation by bubbling a stream of acetone-satd N, filtration, and distn off the solvent without contact with air. Two exchangeable dissoln vessels of 15 and 8 cm^3 working capacity were used depending on the soly of solute. The app was immersed in a thermostat. The vols of acetone used were 15 or 5 cm^3, and the equilibration time was 2-2.5 h. The satd solns were filtered, weighed, the solvent was distd off, the residues were dried at 105°C, weighed, and examd for the presence of solvated acetone.

SOURCE AND PURITY OF MATERIALS:

The source of the materials was not specified. Pure, anhyd acetone was used. The absence of impurities and water was confirmed by procedures of the German Pharmacopeia VI and Spanish Pharmacopeia VIII.

The purity of sulfadicramide was not specified.

ESTIMATED ERROR:

Soly: measurements were repeated until 2 values not differing in the second decimal were obtained (author).

Temp: ±0.1°C (author).

REFERENCES:

COMPONENTS:	ORIGINAL MEASUREMENTS:
(1) 2-Butenamide, N- (4-aminophenyl)-sulfonyl]-3-methyl-, monosodium salt (Na sulfadicramide); $C_{11}H_{13}N_2NaO_3S$; [78739-59-0] (2) 2-Propanone (acetone); C_3H_6O; [67-64-1]	Gutierrez, F. H. *Anales fis. quim. (Madrid)* 1945, *41*, 537-60.
VARIABLES:	PREPARED BY:
Temperature	R. Piekos

EXPERIMENTAL VALUES:

t/°C	G[a]	E[b]	X_g/l[c]	mol/l[d] acetone	mmol/mol acetone	$1:X_g$[e]	$1 + X_{cc}$[f]
0	0.087	0.087	0.709	2.57	0.18	1149.42	1410.44
5	0.092	0.092	0.744	2.69	0.19	1089.13	1344.09
10	0.098	0.098	0.787	2.87	0.20	1020.41	1270.65
15	0.107	0.107	0.853	3.08	0.22	934.58	1160.61
20	0.111	0.111	0.872	3.15	0.23	900.90	1146.67
25	0.112	0.112	0.879	3.18	0.24	892.86	1137.65
30	0.112	0.112	0.873	3.16	0.24	892.86	1145.30
35	0.112	0.112	0.866	3.13	0.24	892.86	1154.73
40	0.116	0.116	0.890	3.22	0.24	862.07	1123.60
45	0.127	0.127	0.967	3.49	0.26	787,40	1034.18
50	0.129	0.129	0.975	3.53	0.27	775.19	1025.64

[a] $G = \frac{p\ 100}{P - p}$, where p and P are the weights of solute and solution, resp.

[b] $E = \frac{G\ 100}{G + 100}$; [c] g/l acetone; [d] should be mmol/l (compiler);

[e] g of acetone required to dissolve 1 g of solute; [f] volume (cm^3) of acetone required to dissolve 1 g of solute.

AUXILIARY INFORMATION

METHOD/APPARATUS/PROCEDURE:	SOURCE AND PURITY OF MATERIALS:
A special all-glass app was constructed enabling the prepn of satd solns, agitation by bubbling a stream of acetone-satd N, filtration, and distn off the solvent without contact with air. Two exchangeable dissoln vessels of 15 and 8 cm^3 capacity were used depending on the soly of solute. The app was immersed in a thermostat. The volsof acetone used were 15 or 8 cm^3, and the equilibration time was 2-2.5 h. The satd solns were filtered, weighed, the solvent was distd off, the residue was dried at 105°C, weighed, and examd for the presence of solvated acetone.	The source of the materials was not specified. Pure, anhyd acetone was used. The absence of impurities and water in it was confirmed by procedures of the German Pharmacopeia VI and Spanish Pharmacopeia VIII. The purity of the solute was not specified. ESTIMATED ERROR: Soly: measurements were repeated until 2 values not differing in the second decimal were obtained (author). Temp: ±0.1°C (author). REFERENCES:

COMPONENTS:	ORIGINAL MEASUREMENTS:
(1) 2-Butenamide,-N-[[(4-acetylamino)phenyl] sulfonyl]-3-methyl-; (acetyl irgamide); $C_{13}H_{16}N_2O_4S$; [71119-41-0] (2) Phosphoric acid, disodium salt; Na_2HPO_4; [7558-94-4] (3) Phosphoric acid, monopotassium salt; KH_2PO_4; [7778-77-0] (4) Water; H_2O; [7732-18-5]	Pulver, R.; Suter, R. *Schweiz. Med. Wochenschr.* <u>1943</u>, *73(13)*, 403-8.
VARIABLES:	PREPARED BY:
pH	R. Piekos

EXPERIMENTAL VALUES:

pH	Solubility of acetyl irgamide in M/15 phosphate buffers (according to Sørensen) at 20°C	
	mg%	10^2 mol dm^{-3} [a]
6.0	159	0.54
7.0	673	2.27
8.0	880	2.97

[a] Calculated by compiler

AUXILIARY INFORMATION

METHOD/APPARATUS/PROCEDURE:

Nothing specified

SOURCE AND PURITY OF MATERIALS:

Nothing specified

ESTIMATED ERROR:

Nothing specified

REFERENCES:

COMPONENTS:	ORIGINAL MEASUREMENTS:
(1) Benzamide, N-[(4-aminophenyl)sulfonyl]- (sulfabenzamide); $C_{13}H_{12}N_2O_3S$; [127-71-9] (2) Water; H_2O; [7732-18-5]	Bhattacharyya, R.; Basu, U. P. *Indian Pharmacist* 1950, *6(3)*, 77-8, 86.
VARIABLES:	**PREPARED BY:**
One temperature;: 30°C	R. Piekos

EXPERIMENTAL VALUES:

Solubility of sulfabenzamide in water at 30°C is 207.0 mg per ml of solution[a] (0.774 mol dm^{-3} solution, compiler).

[a] The final pH was 3.6.

AUXILIARY INFORMATION

METHOD/APPARATUS/PROCEDURE:

A weighed sample of sulfabenzamide was placed in a clean reagent bottle and a known vol of water was added. The mixt was shaken in a mech shaker at 80-100 strokes/min. After at least 24 h the mixt was filtered through a clean, dried and weighed sintered-glass crucible. At the end of the filtration the crucible was washed with about 1 ml of water, dried at 105°C for 2-3 h, cooled, and weighed to const wt. The pH was detd with a Cambridge bench type pH meter using a glass electrode.

SOURCE AND PURITY OF MATERIALS:

Neither source nor purity of the sulfabenzamide was specified.

Doubly distd water was used.

ESTIMATED ERROR:

Soly: not specified.

Temp: ±0.2°C (authors).

pH: ±0.01 unit (authors).

REFERENCES:

COMPONENTS:

(1) Benzamide, N-[(4-aminophenyl)sulfonyl]- (sulfabenzamide); $C_{13}H_{12}N_2O_3S$; [127-71-9]

(2) Phosphoric acid, monopotassium salt; KH_2PO_4; [7778-77-0]

(3) Sodium hydroxide; NaOH; [1310-73-2]

(4) Water; H_2O; [7732-18-5]

ORIGINAL MEASUREMENTS:

Bhattacharyya, R.; Basu, U. P.

Indian Pharmacist 1950, *6(3)*, 77-8, 86.

VARIABLES:

pH

PREPARED BY:

R. Piekos

EXPERIMENTAL VALUES:

Initial pH	Solubility at 30°C in M/20 KH_2PO_4 solution of pH corrected with M/20 naOH solution		Final pH
	mg/ml solution	mol dm^{-3} [a]	
6.18	451.4	1.634	5.55
7.05	1153.8	4.176	5.9

[a] Calculated by compiler.

AUXILIARY INFORMATION

METHOD/APPARATUS/PROCEDURE:

A weighed sample of sulfabenzamide was placed in a clean reagent bottle and a known vol of the M/20 KH_2PO_4 soln was added, and the pH was adjusted to the desired value with M/20 NaOH soln. The mixt was shaken in a mech shaker at 80-100 strokes/min. After at least 24 h the mixt was filtered through a clean, dried and weighed sintered-glass crucible. At the end of the filtration the crucible was washed with about 1 ml of water, dried at 105°C for 2-3 h, cooled, and weighed to const wt. The pH was detd with a Cambridge bench type pH meter using a glass electrode.

SOURCE AND PURITY OF MATERIALS:

Neither source nor purity of the materials, with the exception was water, was specified.

The water was doubly distilled.

ESTIMATED ERROR:

Soly: not specified.

Temp: ± 0.2°C (authors).

pH: ± 0.01 unit (authors).

REFERENCES:

COMPONENTS:

(1) Benzamide, N-[(4-aminophenyl)sulfonyl]-3,4-dimethyl- (xyloylsulfamine); $C_{15}H_{16}N_2O_3S$; [120-34-3]
(2) Phosphoric acid, disodium salt; Na_2HPO_4; [7558-94-4]
(3) Phosphoric acid, monopotassium salt; KH_2PO_4; [7778-77-0]
(4) Water; H_2O; [7732-18-5]

ORIGINAL MEASUREMENTS:

Yamazaki, M.; Aoki, M.; Kamada, A.; Yata, N. *Yakuzaigaku* 1967, *27(1)*, 37-40.

VARIABLES:

One temperature: 30 °C; one pH: 7.4

PREPARED BY:

R. Piekos

EXPERIMENTAL VALUES:

Solubility of xyloylsulfamine in a phosphate buffer solution of pH 7.4[a] (μ = 0.17) at 30 °C is 8.80 mmol/L (2.68 g dm^{-3}, compiler).

[a]At the end of the experiment the pH was 7.2

AUXILIARY INFORMATION

METHOD/APPARATUS/PROCEDURE:

Xyloylsulfamine (0.5 g) was placed in an L-shaped tube together with 20 ml of the phosphate buffer soln. The mixt was shaken in a thermostat until equilibrium was attained. The xyloylsulfamine was assayed in the supernatant spectrophotometrically at 545 nm on a Beckmann DU spectrophotometer. The results were taken from a calibration graph.

SOURCE AND PURITY OF MATERIALS:

Nothing specified.

ESTIMATED ERROR:

Soly and pH: not specified.

Temp: ± 1 °C (authors).

REFERENCES:

COMPONENTS:	ORIGINAL MEASUREMENTS:
(1) Benzamide, N-[(4-aminophenyl)sulfonyl]-3,4-dimethyl- (xyloylsulfamine); $C_{15}H_{16}N_2O_3S$; [120-34-3] (2) Water; H_2O; [7732-18-5]	Yamazaki, M.; Aoki, M.; Kamada, A.; Yata, N. *Yakuzaigaku* 1967, *27(1)*, 37-40.
VARIABLES:	PREPARED BY:
One temperature: 30 °C	R. Piekos

EXPERIMENTAL VALUES:

Solubility of xyloylsulfamine in water at 30 °C is 0.20 mmol/L (6.09×10^{-2} g dm^{-3}, compiler).

AUXILIARY INFORMATION

METHOD/APPARATUS/PROCEDURE:	SOURCE AND PURITY OF MATERIALS:
Xyloylsulfamine (0.5 g) was placed in an L-shaped tube together with 20 ml of water. The mixt was shaken in a thermostat until equilibrium was attained. The xyloylsulfamine was assayed in the supernatant spectrophotometrically at 545 nm on a Beckmann DU spectrophotometer. The results were taken from a calibration graph.	Nothing specified. ESTIMATED ERROR: Soly: not specified. Temp: ± 1 °C (authors). REFERENCES:

COMPONENTS:	ORIGINAL MEASUREMENTS:
(1) Benzamide, N-[(4-aminophenyl)sulfonyl]-3,4-dimethyl- (xyloylsulfamine); $C_{15}H_{16}N_2O_3S$; [120-34-3] (2) Methane, trichloro- (chloroform); $CHCl_3$; [67-66-3]	Yamazaki, M.; Aoki, M.; Kamada, A.; Yata, N. *Yakuzaigaku* 1967, *27(1)*, 37-40.
VARIABLES:	PREPARED BY:
One temperature: 30°C	R. Piekos

EXPERIMENTAL VALUES:

Solubility of xyloylsulfamine in chloroform at 30°C is 5.42 mmol/L (1.65 g dm^{-3}, compiler).

AUXILIARY INFORMATION

METHOD/APPARATUS/PROCEDURE:

Xyloylsulfamine (0.5 g) was placed in an L-shaped tube together with 20 ml of chloroform. The mixt was shaken in a thermostat until equilibrium was attained. The xyloylsulfamine was assayed in the supernatant spectrophotometrically at 545 nm on a Beckmann DU spectrophotometer. The results were taken from a calibration graph.

SOURCE AND PURITY OF MATERIALS:

Nothing specified.

ESTIMATED ERROR:

Soly: not specified.

Temp: ± 1°C (authors).

REFERENCES:

COMPONENTS:

(1) Benzamide, N-[(4-aminophenyl)sulfonyl]-3,4-dimethyl- (irgafene); $C_{15}H_{16}N_2O_3S$; [120-34-3]

(2) Phosphoric acid, disodium salt; Na_2HPO_4; [7558-94-4]

(3) Phosphoric acid, monopotassium salt; KH_2PO_4; [7778-77-0]

(4) Water; H_2O; [7732-18-5]

ORIGINAL MEASUREMENTS:

Pulver, R.; Suter, R.

Schweiz. Med. Wochenschr. 1943, *73(13)*, 403-8.

VARIABLES:

pH

PREPARED BY:

R. Piekos

EXPERIMENTAL VALUES:

pH	Solubility of irgafene in M/15 phosphate buffers (according to Sørensen) at 20 °C	
	mg%	10^3 mol dm^{-3} [a]
6.0	33	1.1
7.0	184	6.1
8.0	370	12.2

[a] Calculated by compiler.

AUXILIARY INFORMATION

METHOD/APPARATUS/PROCEDURE:

Nothing specified.

SOURCE AND PURITY OF MATERIALS:

Nothing specified.

ESTIMATED ERROR:

Nothing specified.

REFERENCES:

COMPONENTS:	ORIGINAL MEASUREMENTS:
(1) Benzamide, N-[(4-aminophenyl)sulfonyl]-3,4-dimethyl- (xyloylsulfamine); $C_{15}H_{16}N_2O_3S$; [120-34-3] (2) 2-Propanone (acetone); C_3H_6O; [67-64-1]	Gutierrez, F. H. *Anales fis. quim. (Madrid)* 1945, *41*, 537-60.
VARIABLES:	PREPARED BY:
Temperature	R. Piekos

EXPERIMENTAL VALUES:

t/°C	G[a]	E[b]	X_g/1[c]	mol/1[d] acetone	mmol/mol acetone	1:X_g[e]	1 + X_{cc}[f]
0	7.651	7.107	62.325	205	14.6	13.07	16.06
5	8.097	7.491	65.489	215	15.7	12.35	15.27
10	9.005	8.261	72.310	237	17.2	11.15	13.83
15	9.771	8.901	76.085	249	18.6	10.24	13.14
20	10.496	9.499	83.044	273	20.0	9.53	12.04
25	11.111	10.000	87.244	287	21.2	9.00	11.46
30	12.332	10.978	96.103	316	23.5	8.18	10.41
35	13.207	11.664	102.130	335	25.2	7.57	9.79
40	13.997	12.278	107.413	353	26.7	7.15	9.35
45	15.609	13.501	118.847	390	29.8	6.41	8.41
50	17.211	14.684	130.009	427	32.6	5.81	7.69

[a] $G = \frac{p\ 100}{P - p}$, where p and P are the weights of solute and solution, resp.

[b] $E = \frac{G\ 100}{G + 100}$; [c] g/1 acetone; [d] should be mmol/1 acetone (compiler);

[e] g of acetone required to dissolve 1 g of solute;

[f] volume (cm^3) of acetone required to dissolve 1 g of solute.

AUXILIARY INFORMATION

METHOD/APPARATUS/PROCEDURE:

A special all-glass app was constructed enabling the prepn of satd solns, agitation by bubbling a stream of acetone-satd N, filtration, and distn off the solvent without contact with air. Two exchangeable dissoln vessels of 15 and 8 cm^3 working capacity were used depending on the soly of solute. The app was immersed in a thermostat. The vols of acetone used were 15 or 5 cm^3, and the equilibration time was 2-2.5 h. The satd solns were filtered, weighed, the solvent was distd off, the residues were dried at 105°C, weighed, and examd for the presence of solvated acetone.

SOURCE AND PURITY OF MATERIALS:

The source of the materials was not not specified. Pure, anhyd acetone was used. The absence of impurities and water was confirmed by procedures of the German Pharmacopeia VI and Spanish Pharmacopeia VIII.

The purity of xyloysulfamine was not specified.

ESTIMATED ERROR:

Soly: measurements were repeated until 2 values not differing in the second decimal were obtained (author).

Temp: $\pm$0.1°C (author).

REFERENCES:

COMPONENTS:

(1) Benzamide, N-[[(4-acetylamino)phenyl] sulfonyl]-3,4-dimethyl- (acetyl irgafene); $C_{17}H_{18}N_2O_4S$; [71119-40-9]
(2) Phosphoric acid, disodium salt; Na_2HPO_4; [7558-94-4]
(3) Phosphoric acid, monopotassium salt; KH_2PO_4; [7778-77-0]
(4) Water; H_2O; [7732-18-5]

ORIGINAL MEASUREMENTS:

Pulver, R.; Suter, R.
Schweiz. Med. Wochenschr. 1943, *73(13)*, 403-8.

VARIABLES:

pH

PREPARED BY:

R. Piekos

EXPERIMENTAL VALUES:

pH	Solubility of acetyl irgafene in M/15 phosphate buffer (according to Sørensen) at 20°C	
	mg%	10^3 mol dm^{-3} [a]
6.0	29	0.84
7.0	170	4.90
8.0	210	6.06

[a] Calculated by compiler.

AUXILIARY INFORMATION

METHOD/APPARATUS/PROCEDURE:

Nothing specified.

SOURCE AND PURITY OF MATERIALS:

Nothing specified.

ESTIMATED ERROR:

Nothing specified.

REFERENCES:

COMPONENTS:

(1) Benzenesulfonamide, 4-amino-N-(aminocarbonyl)- (sulfaurea); $C_7H_9N_3O_3S$; [547-44-4]
(2) Mannitol; $C_6H_{14}O_6$; [87-78-5]
(3) Methane, trichloro- (chloroform); $CHCl_3$; [67-66-3]
(4) Phosphoric acid, disodium salt; Na_2HPO_4; [7558-94-4]
(5) Phosphoric acid, monopotassium salt; KH_2PO_4; [7778-77-0]
(6) Sodium chloride; NaCl; [7647-14-5]
(7) Water; H_2O; [7732-18-5]

ORIGINAL MEASUREMENTS:

Sonnenberg, H.; Oelert, H.; Baumann, K. *Pflügers Arch. Ges. Physiol.* 1965, *286*, 171-80.

VARIABLES:

pH

PREPARED BY:

R. Piekos

EXPERIMENTAL VALUES:

pH	Relative lipoid solubility determined on the basis of concentration measurements of sulfaurea in perfusates[a,b] before (c_i) and after (c_e) equilibration with chloroform $\left(100 - \frac{100\, c_e}{c_i}\right)$
5[a]	0
8[b]	0

[a] Composition of perfusate: 110 mmol/l NaCl, 35 mmol/l mannitol in a phosphate buffer consisting of 98.8 ml of 0.022M KH_2PO_4 and 1.2 ml of 0.022M Na_2HPO_4.

[b] Composition of perfusate: 68 mmol/l NaCl, 100 mmol/l mannitol in a phosphate buffer consisting of 5.5 ml of 0.022M KH_2PO_4 and 94.5 ml of 0.022M Na_2HPO_4.

AUXILIARY INFORMATION

METHOD/APPARATUS/PROCEDURE:

Lipoid solubilities were detd by shaking equal volumes of the perfusate and chloroform for 20 min and measuring the concn of sulfaurea by the spectrophotometric method of Bratton and Marshall (1) in an aq phase before and after this procedure.

SOURCE AND PURITY OF MATERIALS:

None given.

ESTIMATED ERROR:

None given.

REFERENCES:

1. Bratton, A. C.; Marshall, E. K., Jr. *J. Biol. Chem.* 1939, *128*, 537.

COMPONENTS:

(1) Benzenesulfonamide, 4-amino-N-(aminocarbonyl)- (sulfaurea); $C_7H_9N_3O_3S$; [547-44-4]
(2) Benzene, methyl- (toluene); C_7H_8; [108-88-3]
(3) Mannitol; $C_6H_{14}O_6$; [87-78-5]
(4) Phosphoric acid, disodium salt; Na_2HPO_4; [7558-94-4]
(5) Phosphoric acid, monopotassium salt; KH_2PO_4; [7778-77-0]
(6) Sodium chloride; NaCl; [7647-14-5]
(7) Water; H_2O; [7732-18-5]

ORIGINAL MEASUREMENTS:

Sonnenberg, H.; Oelert, H.; Baumann, K.
Pflügers Arch. Ges. Physiol. 1965, *286*, 171-80.

VARIABLES:

pH

PREPARED BY:

R. Piekos

EXPERIMENTAL VALUES:

pH	Relative lipoid solubility determined on the basis of concentration measurements of sulfaurea in perfusates[a,b] before (c_i) and after (c_e) equilibration with toluene $\left(100 - \frac{100\,c_e}{c_i}\right)$
5[a]	2
8[b]	2

[a] Composition of perfusate: 110 mmol/l NaCl, 35 mmol/l mannitol in a phosphate buffer consisting of 98.8 ml of 0.022 M KH_2PO_4 and 1.2 ml of 0.022 M Na_2HPO_4.

[b] Composition of perfusate: 68 mmol/l NaCl, 100 mmol/l mannitol in a phosphate buffer consisting of 5.5 ml of 0.022 M KH_2PO_4 and 94.5 ml of 0.022 M Na_2HPO_4.

AUXILIARY INFORMATION

METHOD/APPARATUS/PROCEDURE:

Lipoid solubilities were detd by shaking equal volumes of the perfusate and toluene with sulfaurea for 20 min and measuring the concn of sulfaurea by the spectrophotometric method of Bratton and Marshall (1) in an aq phase before and after this procedure.

SOURCE AND PURITY OF MATERIALS:

None given.

ESTIMATED ERROR:

None given.

REFERENCES:

1. Bratton, A. C.; Marshall, E. K., Jr. *J. Biol. Chem.* 1939, *128*, 537.

COMPONENTS:	EVALUATOR:
(1) Benzenesulfonamide, 4-amino-N-[(butylamino)carbonyl]- (carbutamide); $C_{11}H_{17}N_3O_3S$; [339-43-5] (2) Aqueous phosphate buffers	Anthony N. Paruta Department of Pharmaceutics University of Rhode Island Kingston, Rhode Island, USA and Ryszard Piekos Faculty of Pharmacy, University of Gdansk Gdansk, Poland 1986

CRITICAL EVALUATION:

The solubility of this compound was studied by two workers (1,2) at a temperature of 310K and a pH value of 4. Alric and Puech (1) determined the solubility in a McIlvaine type buffer and recorded a value of 1.95 x 10^{-3} mol dm^{-3} as an average of eight determinations. Saffar, Ogata and Ejima (2) used a McIlvaine buffer at a pH value of 4 and also illustrated the equilibrium time for saturation to occur. The value given by these workers (2) coincides very well with the other study and can be given as 1.92 x 10^{-3} mol dm^{-3}. Both workers used at least 48 hours of equilibration and Saffar et al.(2) used an average value at 24, 48 and 72 hours. The recommended value for solubility of carbutamide at pH 4 in McIlvaine's buffer is 1.93 x 10^{-3} mol dm^{-3}.

REFERENCES:

(1) Alric, R.; Puech, R. *J. Pharmacol. (Paris)* 1971, *2(2)*, 141-54.

(2) Saffar, F.; Ogata, H.; Ejima, A. *Chem. Pharm. Bull.* 1982, *30(2)*, 679-83.

COMPONENTS:

(1) Benzenesulfonamide, 4-amino-N-[(butylamino)carbonyl]- (carbutamide); $C_{11}H_{17}N_3O_3S$; [359-43-5]

(2) Hydrochloric acid; HCl; [7647-01-0]

(3) Water; H_2O; [7732-18-5]

ORIGINAL MEASUREMENTS:

Saffar, F.; Ogata, H.; Ejima, A.

Chem. Pharm. Bull. 1982, *30(2)*, 679-83.

VARIABLES:

One temperature: 37°C

PREPARED BY:

R. Piekos

EXPERIMENTAL VALUES:

Solubility of carbutamide in dilute hydrochloric acid of pH 1.2 at 37°C is 1.80 mg/ml (6.63 x 10^{-3} mol dm^{-3}, compiler).

AUXILIARY INFORMATION

METHOD/APPARATUS/PROCEDURE:

A satd soln of carbutamide in dilute HCl of pH 1.2 was shaken at 30 strokes per min at 37°C, and samples were withdrawn for analysis after 48 and 72 h. The samples were taken with a syringe equipped with a membrane filter (1.0 μm), and the absorbances were read after dilution with 0.1 M phosphate buffer (pH 9.2) at 254 nm.

SOURCE AND PURITY OF MATERIALS:

Carbutamide powder was a commercial product from Ono Pharmaceutical Co., Ltd., Osaka, Japan.

Hydrochloric acid was of reagent grade.

ESTIMATED ERROR:

Soly: an average of the detns after 24 h (1.77 mg/ml) and 48 h (1.83 mg/ml) is given (authors).

Temp and pH: not specified.

REFERENCES:

COMPONENTS:

(1) Benzenesulfonamide, 4-amino-N-[(butylamino)carbonyl]- (carbutamide); $C_{11}H_{17}N_3O_3S$; [339-43-5]

(2) Hydrochloric acid; HCl; [7647-01-0]

(3) Potassium chloride; KCl; [7447-40-7]

(4) Water; H_2O; [7732-18-5]

ORIGINAL MEASUREMENTS:

Lippold, B. H.; Sgoll, G. B.

Pharm. Ind. 1978, *40(8)*, 841-8.

VARIABLES:

pH

PREPARED BY:

R. Piekos

EXPERIMENTAL VALUES:

pH	Saturation solubility in HCl - KCl buffer solutions (ionic strength 0.1) at 39±1°C	
	10^3 mol/L	g/L
-0.8	114.94	31.186[a]
0.1	113.0	30.66[a]
0.9	20.2	5.48[a]
1.3	8.79	2.38
1.5	5.37	1.46[a]
2.2	2.9	0.79[a]
2.6	2.7	0.73

[a] Calculated by compiler

AUXILIARY INFORMATION

METHOD/APPARATUS/PROCEDURE:

Satd solns of carbutamide powder in HCl-KCl buffers were prepd under stirring, and carbutamide was assayed by UV spectrophotometry using a DMR 10 (Zeiss, Oberkochen) spectrophotometer. The soly measurements were repeated until const values were obtained.

SOURCE AND PURITY OF MATERIALS:

Neither source nor the purity of the materials was specified.

ESTIMATED ERROR:

Soly and pH: not specified.

Temp: ±1°C (authors).

REFERENCES:

COMPONENTS:	ORIGINAL MEASUREMENTS:
(1) Benzenesulfonamide, 4-amino-N-[(butylamino)carbonyl]- (carbutamide); $C_{11}H_{17}N_3O_3S$; [339-43-5] (2) Acetic acid; $C_2H_4O_2$; [64-19-7] (3) Potassium chloride; KCl; [7447-40-7] (4) Sodium acetate; $C_2H_3NaO_2$; [127-09-3] (5) Water; H_2O; [7732-18-5]	Lippold, B. H.; Sgoll, G. B. *Pharm. Ind.* 1978, *40(8)*, 841-8.
VARIABLES:	PREPARED BY:
pH	R. Piekos

EXPERIMENTAL VALUES:

pH	Saturation solubility at 39±1°C in the Walpole acetate buffer solutions of the ionic strength 0.1 maintained with KCl	
	g/L	10^3 mol/L
3.80	0.604[a]	2.227[b]
3.95	0.604	2.226[a]
5.0	0.69	2.52

[a] Calculated by compiler

[b] Mean of 5 measurements.

AUXILIARY INFORMATION

METHOD/APPARATUS/PROCEDURE:

Satd solns of carbutamide powder in the acetate buffers were prepd under stirring, and carbutamide was assayed by UV spectrophotometry using a DMR 10 (Zeiss, Oberkochen) spectrophotometer. The soly measurements were repeated until const values were obtained.

SOURCE AND PURITY OF MATERIALS:

Neither source nor purity of the materials was specified.

ESTIMATED ERROR:

Soly and pH: not specified.

Temp: ±1°C (authors).

REFERENCES:

COMPONENTS:

(1) Benzenesulfonamide, 4-amino-N-[(butylamino)carbonyl]- (carbutamide); $C_{11}H_{17}N_3O_3S$; [339-43-5]

(2) Phosphoric acid, disodium salt; Na_2HPO_4; [7558-94-4]

(3) Phosphoric acid, monopotassium salt; KH_2PO_4; [7778-77-0]

(4) Potassium chloride; KCl; [7447-40-7]

(5) Water; H_2O; [7732-18-5]

ORIGINAL MEASUREMENTS:

Lippold, B. H.; Sgoll, G. B.

Pharm. Ind. 1978, *40(8)*. 841-8.

VARIABLES:

pH

PREPARED BY:

R. Piekos

EXPERIMENTAL VALUES:

pH	Saturation solubility at 39±1°C in the Sörensen phosphate buffers of ionic strength 0.1 maintained with KCl	
	g/L	10^3 mol/L
6.18	2.1	7.71
6.42	3.04[a]	11.2
6.5	4.75[a]	17.5
6.65	4.94[a]	18.2
6.75	6.43	23.7
7.45	30.0	110.6[a]

[a] Calculated by compiler

AUXILIARY INFORMATION

METHOD/APPARATUS/PROCEDURE:

Satd solns of carbutamide powder in the Sörensen phosphate buffers were prepd under stirring, and carbutamide was assayed by UV spectrophotometry using a DMR 10 (Zeiss, Oberkochen) spectrophotometer. The soly measurements were repeated until const values were obtained.

SOURCE AND PURITY OF MATERIALS:

Neither source nor purity of the materials was specified.

ESTIMATED ERROR:

Soly and pH: not specified.

Temp: ±1°C (authors.

REFERENCES:

COMPONENTS:

(1) Benzenesulfonamide, 4-amino-N-[(butylamino)carbonyl]- (carbutamide); $C_{11}H_{17}N_3O_3S$; [339-43-5]

(2) Potassium chloride; KCl; [7447-40-7]

(3) Sodium hydroxide; NaOH; [1310-73-2]

(4) Water; H_2O; [7732-18-5]

ORIGINAL MEASUREMENTS:

Lippold, B. H.; Sgoll, G. B.

Pharm. Ind. 1978, *40(8)*, 841-8.

VARIABLES:

pH

PREPARED BY:

R. Piekos

EXPERIMENTAL VALUES:

pH	Saturation solubility in NaOH-KCl buffer solutions (ionic strength 0.1) at 39$\pm$1°C	
	g/L	10^3 mol/L
6.7[b]	5.4	19.9[a]
6.75[b]	6.4	23.7
7.0	9.5[a]	35.02
7.2[b]; 7.21[b]	16.7	61.5
7.05[b]	11.4	42.0[a]
7.25	17.5[a]	64.4
7.3	19.3[a]	71.1
7.45	30.0[a]	110.57
8.9	30.7[a]	113.1
10.8	30.6	112.8
12.8	30.61[a]	112.81

[a] Calculated by compiler

[b] Measured in the solubility vessel.

AUXILIARY INFORMATION

METHOD/APPARATUS/PROCEDURE:

Satd solns of carbutamide powder in the NaOH-KCl buffers were prepd under stirring, and carbutamide was assayed by UV spectrophotometry using a DMR 10 (Zeiss, Oberkochen) spectrophotometer. The soly measurements were repeated until const values were obtained.

SOURCE AND PURITY OF MATERIALS:

Neither source nor purity of the materials was specified.

ESTIMATED ERROR:

Soly and pH: not specified.

Temp: $\pm$1°C (authors).

REFERENCES:

COMPONENTS:

(1) Benzenesulfonamide, 4-amino-N-[(butyl-amino)carbonyl]- (carbutamide); $C_{11}H_{17}N_3O_3S$; [339-43-5]

(2) Phosphoric acid, disodium salt; Na_2HPO_4; [7558-94-4]

(3) 1,2,3-Propanetricarboxylic acid, 2-hydroxy- (citric acid); $C_6H_8O_7$; [77-92-9]

(4) Water; H_2O; [7732-18-5]

ORIGINAL MEASUREMENTS:

Alric, R.; Puech, R.

J. Pharmacol. (Paris) 1971, *2(2)*, 141-54.

VARIABLES:

One temperature: 37°C; one pH: 4

PREPARED BY:

R. Piekos

EXPERIMENTAL VALUES:

Intrinsic solubility[a] of carbutamide in a solution 0.025M in Na_2HPO_4 and 0.05M in citric acid, of pH 4, at 37°C is $(19.5 \pm 0.22) \times 10^{-4}$ mol $liter^{-1}$.

[a] Under "intrinsic solubility" a minimum on the solubility - pH curve is meant which corresponds to the limiting concentration of the undissociated form of carbutamide.

AUXILIARY INFORMATION

METHOD/APPARATUS/PROCEDURE:

The soln was equilibrated for 48 h in a thermostat under occasional stirring. Samples were withdrawn through a 1-μ membrane filter, dild with 0.155M NaOH soln to ensure total dissocn of carbutamide, and the sulfonamide was assayed by UV spectrophotometry.

SOURCE AND PURITY OF MATERIALS:

Nothing specified.

ESTIMATED ERROR:

Soly: std error of 8 measurements was $\pm 0.22 \times 10^{-4}$ mol $liter^{-1}$ (authors).

pH: accuracy ± 0.5 pH unit (authors).

Temp: ± 0.1°C (authors).

REFERENCES:

COMPONENTS:

(1) Benzenesulfonamide, 4-amino-N-[(butylamino)carbonyl]- (carbutamide) $C_{11}H_{17}N_3O_3S$; [339-43-5]

(2) Phosphoric acid, disodium salt; Na_2HPO_4; [7558-94-4]

(3) 1,2,3-Propanetricarboxylic acid, 2-hydroxy- (citric acid); $C_6H_8O_7$; [77-92-9]

(4) Water; H_2O; [7732-18-5]

ORIGINAL MEASUREMENTS:

Saffar, F.; Ogata, H.; Ejima, A.

Chem. Pharm. Bull. 1982, *30(2)*, 679-83.

VARIABLES:

pH

PREPARED BY:

R. Piekos

EXPERIMENTAL VALUES:

pH	Solubility (mg/ml) at 37°C in McIlvaine buffer solutions determined after[a] 24 h	48 h	72 h	Average	10^3 mol dm^{-3} [b]
3	0.58	0.57	0.60	0.58	2.14
4	0.55	0.51	0.50	0.52	1.92
5	0.58	0.54	0.52	0.55	2.03
6	1.28	1.14	1.06	1.16	4.27
7.2	7.03	7.11	6.96	7.03	25.9

[a] Numerical values to the graphical data given by the first author (F.S.).

[b] Calculated by compiler.

AUXILIARY INFORMATION

METHOD/APPARATUS/PROCEDURE:

Satd solns of carbutamide in McIlvaine buffer solns of appropriate pH were shaken at 30 strokes/min at 37°C, and samples were withdrawn for analysis after 24, 48 and 72 h. The samples were taken with a syringe equipped with a membrane filter (1.0 μm), and the absorbances were read after diln with 0.1 M phosphate buffer (pH 9.2) at 254 nm.

SOURCE AND PURITY OF MATERIALS:

Carbutamide powder was a commercial product from Ono Pharmaceutical Co., Ltd., Osaka, Japan.

The remaining materials were of reagent grade.

ESTIMATED ERROR:

Soly: accuracy ±0.04; ±0.06; ±0.07; ±0.10; ±0.18 mg/ml, resp. (compiler).

Temp and pH: not specified.

REFERENCES:

COMPONENTS:	ORIGINAL MEASUREMENTS:
(1) Benzenesulfonamide, 4-amino-N-(aminothioxomethyl)- , (Badional); $C_7H_9N_3O_2S_2$; [515-49-1] (2) Ethanol; C_2H_6O; [64-17-5] (3) Water; H_2O; [7732-18-5]	Kuhnert-Brandstätter, M.; Burger, A., *Pharm. Ind.* 1972, *34*, 353-6.
VARIABLES: Temperature	**PREPARED BY:** R. Piekos

EXPERIMENTAL VALUES:

Saturation solubility, c_s, of forms I and II of Badional in 96% ethanol.

Form I (+ + +), form II (● ● ●)

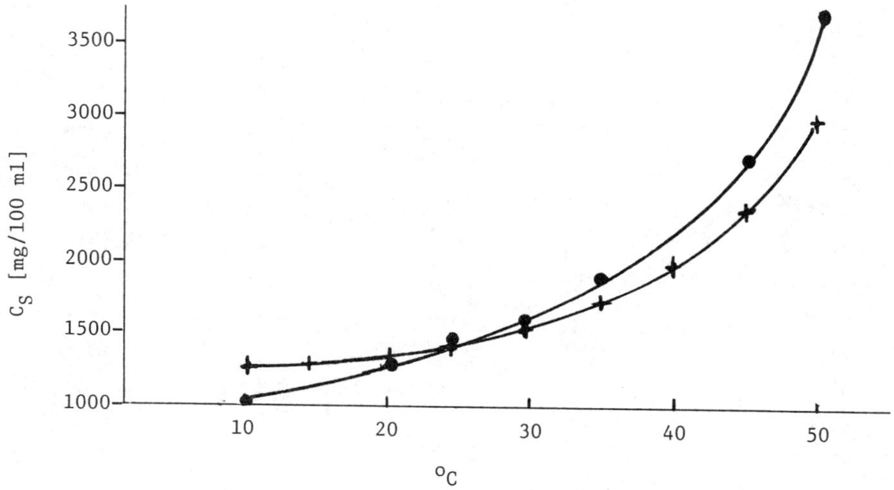

AUXILIARY INFORMATION

METHOD/APPARATUS/PROCEDURE:

An excess of Badional was added to 96% ethanol and stirred until no changes in concn were obsd (2 - 48 h). Aliquots were withdrawn by means of a pipet equipped with a filter, which was preheated to a desired temp, and Badional was assayed spectrophotometrically at 269 nm. The identity of the solid phase was detd by thermomicroscopy and IR spectroscopy.

OURCE AND PURITY OF MATERIALS:

Both forms were obtained by crystn of a compd prep from 96% EtOH;: form I by cooling a hot satd soln to room temp and seeding with crystals of this form, form II by cooling the soln to ∿40°C and scratching the walls.

Purity of the 96% EtOH was not specified.

ESTIMATED ERROR:

Soly: not specified.

Temp: ±0.1°C (authors).

REFERENCES:

COMPONENTS:

(1) Benzenesulfonamide, 4-amino-N-(aminothioxomethyl)-, (Badional); $C_7H_9N_3O_2S_2$; [515-49-1]

(2) Ethanol; C_2H_6O; [64-17-5]

(3) Water; H_2O; [7732-18-5]

ORIGINAL MEASUREMENTS:

Kuhnert-Brandstätter, M.; Burger, A.,

Pharm. Ind. 1972, *34*, 353-6.

VARIABLES:

Temperature

PREPARED BY:

R. Piekos

EXPERIMENTAL VALUES:

$t/^{\circ}C$	Saturation solubility in 96% ethanol of crystalline form I		II	
	g/100 ml	$mol\ dm^{-3}$ [a]	g/100 ml	$mol\ dm^{-3}$ [a]
10.0			1.00	0.0432
10.1	1.06	0.0458		
14.6			1.12	0.0484
15.2	1.18	0.0510		
20.0	1.37	0.0592	1.29	0.0558
25.2			1.54	0.0666
25.4	1.57	0.0679		
30.9			1.81	0.0783
31.0	1.79	0.0774		
37.0			2.21	0.0955
37.1	2.05	0.0886		
41.2	2.30	0.0994	2.52	0.109
46.0	2.68	0.116		
46.2			3.00	0.130
50.8	3.06	0.132		
51.0			3.50	0.151

[a] Calculated by compiler.

AUXILIARY INFORMATION

METHOD/APPARATUS/PROCEDURE:

An excess of Badional was added to 96% ethanol and stirred until no changes in concn were obsd (2-48 h). Aliquots were withdrawn by means of a pipet equipped with a filter, which was preheated to a desired temp, and Badional was assayed spectrophotometrically at 269 nm. The identity of the solid phase was detd by thermomicroscopy and IR spectroscopy.

SOURCE AND PURITY OF MATERIALS:

Both forms of Badional were obtained by crystn of a commprep from 96% EtOH: form I, mp 179-181°C, by cooling a hot satd soln to room temp and seeding with crystals of this form; form II, mp 171°C, by cooling the soln to 40°C and scratching the walls. Purity of the 96% ethanol was not specified.

ESTIMATED ERROR:

Soly: not specified.

Temp: $\pm 0.1^{\circ}C$ (authors).

REFERENCES:

COMPONENTS:

(1) Benzenesulfonamide; 4-amino-N-(aminothioxomethyl)- (sulfathiourea); $C_7H_9N_3O_2S_2$; [515-49-1]

(2) Ethanol; C_2H_6O; [64-17-5]

(3) 1,2,3-Propanetriol; $C_3H_8O_3$; [56-81-5]

(4) Water; H_2O; [7732-18-5]

ORIGINAL MEASUREMENTS:

Dolique, R.; Foucault, J.

Trav. soc. pharm. Montpellier 1952, *12*, 145-53.

VARIABLES:

One temperature: 26-28°C

PREPARED BY:

R. Piekos

EXPERIMENTAL VALUES:

Solubility of sulfathiourea in a mixture of 1,2,3-propanetriol and 95° ethanol (2:1 by wt) at 26-28°C is 7.72% (0.362 mol kg^{-1} solvent, compiler).

AUXILIARY INFORMATION

METHOD/APPARATUS/PROCEDURE:

The sulfathiourea content was detd by diazotization of the amine group in a cold acidified 0.1N KNO_2 soln. An excess of KNO_2 was detected by using iodinated starch.

SOURCE AND PURITY OF MATERIALS:

Nothing specified.

ESTIMATED ERROR:

Nothing specified.

REFERENCES:

COMPONENTS:	ORIGINAL MEASUREMENTS:
(1) Benzenesulfonamide, 4-amino-N-(aminothioxomethyl)- (sulfathiourea); $C_7H_9N_3O_2S_2$; [515-49-1] (2) Ethanol; C_2H_6O; [64-17-5] (3) 1,2,3-Propanetriol; $C_3H_8O_3$; [56-81-5] (4) Urea; CH_4N_2O; [57-13-6] (5) Water; H_2O; [7732-18-5]	Dolique, R.; Foucault, J. *Trav. soc. pharm. Montpellier* 1952, *12*, 145-53.
VARIABLES:	PREPARED BY:
One temperature: 26-28°C	R. Piekos

EXPERIMENTAL VALUES:

Solubility of sulfathiourea at 26-28°C in a saturated solution of urea in a mixture of 1,2,3-propanetriol and 95° ethanol (2:1 by wt), containing 54.5 g of urea per 100 g of the mixture, is 9.47% (0.452 mol kg^{-1} solvent, compiler).

AUXILIARY INFORMATION

METHOD/APPARATUS/PROCEDURE:

The sulfathiourea content was detd by diazotization of the amine group in a cold acidified 0.1N KNO_2 soln. An excess of KNO_2 was detected by using iodinated starch.

SOURCE AND PURITY OF MATERIALS:

Nothing specified.

ESTIMATED ERROR:

Nothing specified.

REFERENCES:

COMPONENTS:	ORIGINAL MEASUREMENTS:
(1) Benzenesulfonamide, 4-amino-N-[imino-(methylthio)methyl]- (sulfamethylisothiourea); $C_8H_{11}N_3O_2S$; [2651-18-5] (2) Water; H_2O; [7732-18-5]	Becher, R.; Leya, S. *Experientia* 1946, *2* 459-60.
VARIABLES: One temperature: 18-19 $^{\circ}$C.	PREPARED BY: R. Piekos

EXPERIMENTAL VALUES:

Solubility of sulfamethylisothiourea in water at room temperature (18-19 $^{\circ}$C) is 19 mg% (8.9×10^{-4} mol dm^{-3}, compiler).

AUXILIARY INFORMATION

METHOD/APPARATUS/PROCEDURE:

After standing for more than two days the soln of sulfamethylisothiourea in water was filtered and the sulfonamide was assayed in the filtrate colorimetrically by the method of Druey and Oesterheld (1).

SOURCE AND PURITY OF MATERIALS:

Nothing specified.

ESTIMATED ERROR:

Nothing specified.

REFERENCES:

1. Druey, J.; Oesterheld, G. *Helv. Chim. Acta* 1942, *25*, 753.

COMPONENTS:	ORIGINAL MEASUREMENTS:
(1) Benzenesulfonamide, 4-amino-N-[imino-(methylthio)methyl]- (sulfamethylisothiourea); $C_8H_{11}N_3O_2S$; [2651-18-5] (2) Sodium chloride; NaCl; [7647-14-5] (3) Water; H_2O; [7732-18-5]	Becher, R.; Leya, S. *Experientia* 1946, *2*, 459-60.
VARIABLES:	**PREPARED BY:**
One temperature: 18-19°C	R. Piekos

EXPERIMENTAL VALUES:

Solubility of sulfamethylisothiourea in a 5% NaCl solution at room temperature (18-19°C) is 21 mg% (9.9×10^{-4} mol dm^{-3}, compiler).

AUXILIARY INFORMATION

METHOD/APPARATUS/PROCEDURE:

After standing for more than two days the soln of sulfamethylisothiourea was filtered and the sulfonamide was assayed in the filtrate colorimetrically by the method of Druey and Oesterheld (1).

SOURCE AND PURITY OF MATERIALS:

Nothing specified.

ESTIMATED ERROR:

Nothing specified.

REFERENCES:

1. Druey, J.; Oesterheld, G. *Helv. Chim. Acta* 1942, *25*, 753.

COMPONENTS:

(1) Benzenesulfonamide, 4-amino-N-[imino-(methylthio)methyl]-; $C_8H_{11}N_3O_2S$; [2651-18-5]

(2) Pectinic acid, sodium salt; $(C_{13}H_{17}NaO_{12})_n$; [9049-37-0]

(3) Water; H_2O; [7732-18-5]

ORIGINAL MEASUREMENTS:

Becher, R.; Leya, S., *Experientia* 1946, *2*, 459-60.

VARIABLES:

One temperature: 18-19°C

PREPARED BY:

R. Piekos

EXPERIMENTAL VALUES:

Solubility of 4-amino-N-[imino(methylthio)methyl]benzenesulfonamide in a 2.6% neutral sodium pectinate solution ([sodium pectinate] = 6.7×10^{-2} mol kg^{-1} (n = 1), compiler) at room temperature (18-19°C) is 32 mg% (1.5×10^{-3} mol dm^{-3}, compiler).

AUXILIARY INFORMATION

METHOD/APPARATUS/PROCEDURE:

The soln was allowed to stand for more than two days at room temp. The soln was then filtered, and the sulfonamide was assayed in the filtrate colorimetrically by the method of Druey and Oesterheld (1).

SOURCE AND PURITY OF MATERIALS:

Nothing specified.

ESTIMATED ERROR:

Nothing specified.

REFERENCES:

1. Druey, J.; Oesterheld, G., *Helv. Chim. Acta* 1942, *25*, 753.

COMPONENTS:

(1) Benzenesulfonamide, 4-amino-N-[imino-(methylthio)methyl]-; $C_8H_{11}N_3O_2S$; [2651-18-5]

(2) Pectin; $(C_{13}H_{18}O_{12})_n$; [9000-69-5]

(3) Water; H_2O [7732-18-5]

ORIGINAL MEASUREMENTS:

Becher, R.; Leya, S., *Experientia* 1946, *2*, 459-60.

VARIABLES:

One temperature: 18-19°C

PREPARED BY:

R. Piekos

EXPERIMENTAL VALUES:

Solubility of 4-amino-N-[imino(methylthio)methyl] benzenesulfonamide in a 2.5% pectin solution ([pectin] = 6.8 x 10^{-2} mol kg^{-1}, compiler), of pH about 2.6, at room temperature (18-19°C) is 28 mg% (1.3 x 10^{-3} mol dm^{-3}, compiler).

AUXILIARY INFORMATION

METHOD/APPARATUS/PROCEDURE:

The soln was allowed to stand for more than 2 days at room temp. The soln was then filtered, and the sulfonamide was assayed in the filtrate colorimetrically by the method of Druey and Oesterheld (1).

SOURCE AND PURITY OF MATERIALS:

A high quality apple pectin was used; the rel viscosity of a 0.5% soln was 6.2, and for neutralization of 1 g of the pectin, 1.67 cm^3 of a 1 mol dm^{-3} NaOH soln was used.

The source and purity of the sulfonamide and water were not specified.

ESTIMATED ERROR:

Nothing specified.

REFERENCES:

1. Druey, J.; Oesterheld, G. *Helv. Chim Acta* 1942, *25*, 753.

COMPONENTS:	ORIGINAL MEASUREMENTS:
(1) Benzenesulfonamide, 4-amino-N-[imino-(methylthio)methyl]- (sulfamethylisothiourea); $C_8H_{11}N_3O_2S_2$; [2651-18-5] (2) D-Glucose; $C_6H_{12}O_6$; [50-99-7] (3) Water; H_2O; [7732-18-5]	Becher, R.; Leya, S. *Experientia* 1946, *2*, 459-60.
VARIABLES:	**PREPARED BY:**
One temperature: 18-19°C	R. Piekos

EXPERIMENTAL VALUES:

Solubility of sulfamethylisothiourea in a 10% D-glucose solution at room temperature (18-19°C) is 23 mg% (9.4×10^{-4} mol dm^{-3}, compiler).

AUXILIARY INFORMATION

METHOD/APPARATUS/PROCEDURE:

After standing for more than two days the soln of sulfamethylisothiourea was filtered and the sulfonamide was assayed in the filtrate colorimetrically by the method of Druey and Oesterheld (1).

SOURCE AND PURITY OF MATERIALS:

Nothing specified.

ESTIMATED ERROR:

Nothing specified.

REFERENCES:

1. Druey, J.; Oesterheld, G. *Helv. Chim. Acta* 1942 , *25*, 753.

COMPONENTS:	EVALUATOR:
(1) Benzenesulfonamide, 4-amino-N-(amino-iminomethyl)- (sulfaguanidine); $C_7H_{10}N_4O_2S$; [57-67-0] (2) Water	Anthony N. Paruta Department of Pharmaceutics University of Rhode Island Kingston, Rhode Island, USA and Ryszard Piekos Faculty of Pharmacy, University of Gdansk Gdansk, Poland 1986

CRITICAL EVALUATION:

The Solubility values are summarized in Table I.

Table I: Solubility of Sulfaguanidine in water 293K and 310K

Reference	10^3 mol dm^{-3} (*indicates mol kg^{-1}) 293K	310K
1	-	8.87
2	35.8*@	9.2*
3	3.0 (291-293K)	-
4	5.191*	-

@ = obvious error in original data

Roblin el al. (1) determined the solubility of sulfaguanidine using an equilibration time of 24 hours, and a colorimetric analytical technique. Sapozhnikova (2) used what appears to be a rather limited length of time for saturation to be reached, but the value reported agrees with that of Roblin et al. (1). The concurrence of these two values therefore allows for an assignments of a tentative value for sulfaguanidine at body temperature of 9.04 x 10^{-3} mol dm^{-3}. Becher and Leya (3) report a value at 291-292K which does not agree with that given by Gerencsér-Németh and Horváth (4). The solubility at 293K as given by Sapozhnikova et al. (2) has apparently a decimal error. It is reasonable to assume that the value should be about 3.575 x 10^{-3} mol kg^{-1}, concurring approximately with other references (3,4). The temperature range in Becher and Leya (3) and the very high value in Gerencsér-Németh and Horváth (4) mitigate against making even a tentative assignment.

REFERENCES:

(1) Roblin, R.O., Jr.; Williams, J.H.; Winnek, P.S.; English, J.P. *J. Am. Chem. Soc.* 1940, *62*, 2002-5.
(2) Sapozhnikova, N.V.; Postovskii, I.Ya. *Zh. Prikl. Khim.* 1944, *17*, 427-34.
(3) Becher, R.; Leya, S. *Experientia* 1946, *2*, 459-60.
(4) Gerencsér-Németh, M.; Horváth, M. *Gyógyszerészet* 1973, *17*, 417-21.

COMPONENTS:

(1) Benzenesulfonamide, 4-amino-N-(amino-iminomethyl)- (sulfaguanidine); $C_7H_{10}N_4O_2S$; [57-67-0]

(2) Water; H_2O; [7732-18-5]

ORIGINAL MEASUREMENTS:

Roblin, R. O., Jr.; Williams, J. H.; Winnek, P. S.; English, J. P.
J. Am. Chem. Soc. 1940, *62*, 2002-5.

VARIABLES:

One temperature: 37°C

PREPARED BY:

R. Piekos

EXPERIMENTAL VALUES:

Solubility of sulfaguanidine in water at 37°C is 190 mg/100 cm^3 solution (8.87×10^{-3} mol dm^{-3}, compiler).

AUXILIARY INFORMATION

METHOD/APPARATUS/PROCEDURE:

Excess sulfonamide in water was heated and stirred on a steam bath for 30 min. The suspension was then agitated for 24 h in a thermostat at 37°C. A sample of the satd soln was withdrawn through a glass filter, dild, and analyzed by the Marshall method (1) using a General Electric recording spectrophotometer for comparing the colors developed with those of the standards.

SOURCE AND PURITY OF MATERIALS:

Sulfaguanidine, mp 189-90°C (dec, cor) was prepd by the authors. Anal: %C 39.2 (calcd 39.3); %H 4.6 (4.7); %N 21.7 (22.4).

Purity of the water was not specified.

ESTIMATED ERROR:

Nothing specified.

REFERENCES:

1. Bratton, A. C.; Marshall, E. K., Jr. *J. Pharmacol.* 1939, *66*, 4.

COMPONENTS:	ORIGINAL MEASUREMENTS:
(1) Benzenesulfonamide, 4-amino-N-(amino-iminomethyl)- (sulfaguanidine); $C_7H_{10}N_4O_2S$; [57-67-0] (2) Water; H_2O; [7732-18-5]	Sapozhnikova, N. V.; Postovskii, I. Ya. *Zh. Prikl. Khim.* 1944, *17*, 427-34.
VARIABLES:	PREPARED BY:
Temperature	R. Piekos

EXPERIMENTAL VALUES:

t/°C	Solubility	
	Weight%	10^2 mol kg^{-1} water[a]
20	0.760	3.58
37	0.196	.92
50	0.430	2.02
75	1.40	6.63
99	3.70	17.93

[a] Calculated by compiler.

AUXILIARY INFORMATION

METHOD/APPARATUS/PROCEDURE:

Sulfaguanidine was dissolved in water to form a satd soln which was occasionally agitated in a glass vessel immersed in a thermostat. The equilibrium was usually attained after 1 h. Five- to 100-cm^3 samples of the satd soln were placed in Pt crucibles or dishes and evapd to dryness at temps lower than 110-115°C. The residue was dried to const wt at 105-110°C and weighed.

SOURCE AND PURITY OF MATERIALS:

Pure, recrystd sulfaguanidine was used. Its mp conformed to that reported in the literature.

Purity of the water was not specified.

ESTIMATED ERROR:

Soly: quite reliable results were obtained over the temp range 20-75°C. At higher temps the accuracy was poor due to evapn of water during sampling (authors).

Temp: ±0.05°C (authors).

REFERENCES:

COMPONENTS:	ORIGINAL MEASUREMENTS:
(1) Benzenesulfonamide, 4-amino-N-(amino-iminomethyl)- (sulfaguanidine); $C_7H_{10}N_4O_2S$; [57-67-0] (2) Water; H_2O; [7732-18-5]	Becher, R.; Leya, S. *Experientia* 1946, *2*, 459-60.
VARIABLES:	**PREPARED BY:**
One temperature: 18-19°C	R. Piekos

EXPERIMENTAL VALUES:

Solubility of sulfaguanidine in water at room temperature (18-19°C) is 65 mg% (3.0×10^{-3} mol dm^{-3}).

AUXILIARY INFORMATION

METHOD/APPARATUS/PROCEDURE:

After standing for more than two days the soln of sulfaguanidine in water was filtered and sulfaguanidine was assayed in the filtrate colorimetrically by the method of Druey and Oesterheld (1).

SOURCE AND PURITY OF MATERIALS:

Nothing specified.

ESTIMATED ERROR:

Nothing specified.

REFERENCES:

1. Druey, J.; Oesterheld, G. *Helv. Chim. Acta* 1942, *25*, 753.

COMPONENTS:	ORIGINAL MEASUREMENTS:
(1) Benzenesulfonamide, 4-amino-N-(amino-iminomethyl)- (sulfaguanidine); $C_7H_{10}N_4O_2S$; [57-67-0] (2) Water; H_2O; [7732-18-5]	Gerencsér-Németh, M.; Horváth, M. *Gyógyszerészet* 1973, *17*, 417-21.
VARIABLES:	PREPARED BY:
One temperature: 20°C	R. Piekos

EXPERIMENTAL VALUES:

Solubility of sulfaguanidine in water at 20°C is 0.1111 g/100 g solution (5.191×10^{-3} mol kg^{-1} water, compiler).

AUXILIARY INFORMATION

METHOD/APPARATUS/PROCEDURE:

A weighed escess of sulfaguanidine in water was shaken in a shaker at 120 rpm for 6 h. The soln was then filtered, the residue was washed with the filtrate and finally with a small amt of distd water, dried and weighed.

SOURCE AND PURITY OF MATERIALS:

Sulfaguanidine (source not specified) was dried at 100°C for 3 h or over conc H_2SO_4 for 72 h. Its mp was 187.5-8.8°C.

Distd water was used.

ESTIMATED ERROR:

Soly: precision ± 0.0047 g/100 g (2 detns) (compiler)

Temp; not specified.

REFERENCES:

COMPONENTS:

(1) Benzenesulfonamide, 4-amino-N-(amino-iminomethyl)- (sulfaguanidine); $C_7H_{10}N_4O_2S$; [57-67-0]

(2) Sodium chloride; NaCl; [7647-14-5]

(3) Water; H_2O; [7732-18-5]

ORIGINAL MEASUREMENTS:

Becher, R.; Leya, S. *Experientia* 1946, *2*, 459-60.

VARIABLES:

One temperature: 18-19°C

PREPARED BY:

R. Piekos

EXPERIMENTAL VALUES:

Solubility of sulfaguanidine in a 5% NaCl solution at room temperature (18-19°C) is 69 mg% (3.2×10^{-3} mol dm^{-3}, compiler).

AUXILIARY INFORMATION

METHOD/APPARATUS/PROCEDURE:

After standing for more than two days the soln of sulfaguanidine was filtered and sulfaguanidine was assayed in the filtrate colorimetrically by the method of Druey and Oesterheld (1).

SOURCE AND PURITY OF MATERIALS:

Nothing specified.

ESTIMATED ERROR:

Nothing specified.

REFERENCES:

1. Druey, J.; Oesterheld, G. *Helv. Chim. Acta* 1942, *25*, 753.

COMPONENTS:

(1) Benzenesulfonamide, 4-amino-N-(amino-iminomethyl)- (sulfaguanidine); $C_7H_{10}N_4O_2S$; [57-67-0]

(2) Sodium chloride, NaCl; [7647-14-5]

(3) Water; H_2O; [7732-18-5]

ORIGINAL MEASUREMENTS:

Avico, U.; Cavazutti, G.; di Francesco, R.; Signoretti Ciranni, E.; Zuccaro, P.

Farmaco, Ed. Pratica 1975, *30(1)*, 40-6.

VARIABLES:

Temperature

PREPARED BY:

R. Piekos

EXPERIMENTAL VALUES:

t/°C	Solubility of amorphous sulfaguanidine in equimolal NaCl solutions	
	g/100 g water	10^3 mol kg^{-1} water[a]
25	0.71	3.3
35	0.84	3.9
40	0.93	4.3

[a] Calculated by compiler.

AUXILIARY INFORMATION

METHOD/APPARATUS/PROCEDURE:

A soln of sulfaguanidine-HCl was added to an NaOH soln contg stoichiometric quantity of the base to neutralize the HCl salt. The neutralization was carried out in a thermostat and the pH of the mixt was maintained close to that of a satd soln of sulfaguanidine in water. The procedure was repeated using various initial concns of the reagents to find the max concn of sulfaguanidine at which no pptn occurred.

SOURCE AND PURITY OF MATERIALS:

Source and purity of sulfaguanidine was not specified. The mp of crystalline sulfaguanidine was 190-3°C.

Purity of the water was not specified.

ESTIMATED ERROR:

Nothing specified.

REFERENCES:

COMPONENTS:	ORIGINAL MEASUREMENTS:
(1) Benzenesulfonamide, 4-amino-N-(amino-iminomethyl)- (sulfaguanidine); $C_7H_{10}N_4O_2S$; [57-67-0] (2) Pectin; $(C_{13}H_{18}O_{12})_n$; [9000-69-5] (3) Water; H_2O; [7732-18-5]	Becher, R.; Leya, S., *Experientia* 1946, *2*, 459-60.
VARIABLES:	**PREPARED BY:**
One temperature: 18-19 ^oC	R. Piekos

EXPERIMENTAL VALUES:

Solubility of sulfaguanidine in a 2.5% pectin solution ([pectin] = 6.8×10^{-2} mol kg^{-1}, compiler), of pH about 2.6, at room temperature (18-19 ^oC) is 111 mg% (5.18×10^{-3} mol dm^{-3}, compiler).

AUXILIARY INFORMATION

METHOD/APPARATUS/PROCEDURE:

The soln was allowed to stand for more than 2 days at room temp. The soln was then filtered, and sulfaguanidine assayed in the filtrate colorimetrically by the method of Druey and Oesterheld (1).

SOURCE AND PURITY OF MATERIALS:

A high quality apple pectin was used: the rel viscosity of a 0.5% soln was 6.2, and for neutralization of 1 g of the pectin, 1.67 cm^3 of a 1 mol dm^{-3} NaOH soln was used. The source and purity of sulfaguanidine and water were not specified.

ESTIMATED ERROR:

Nothing specified.

REFERENCES:

1. Druey, J.; Oesterheld, G., *Helv. Chim. Acta* 1942, *25*, 753.

COMPONENTS:	ORIGINAL MEASUREMENTS:
(1) Benzenesulfonamide, 4-amino-N-(aminoiminomethyl)- (sulfaguanidine); $C_7H_{10}N_4O_2S$; [57-67-0] (2) Pectinic acid, sodium salt; $(C_{13}H_{17}NaO_{12})_n$; [9049-37-0] (3) Water; H_2O; [7732-18-5]	Becher, R.; Leya, S., *Experientia* 1946, *2*, 459-60.
VARIABLES:	**PREPARED BY:**
One temperature: 18-19°C	R. Piekos

EXPERIMENTAL VALUES:

Solubility of sulfaguanidine in a 2.6% neutral sodium pectinate solution ([sodium pectinate] = 6.7×10^{-2} mol kg^{-1} (n = 1), compiler) at room temperature (18-19°C) is 101 mg% (4.71×10^{-3} mol dm^{-3}, compiler).

AUXILIARY INFORMATION

METHOD/APPARATUS/PROCEDURE:

The soln was allowed to stand for more than two days at room temp. The soln was then filtered, and sulfaguanidine assayed in the filtrate colorimetrically by the method of Druey and Oesterheld (1).

SOURCE AND PURITY OF MATERIALS:

Nothing specified.

ESTIMATED ERROR:

Nothing specified.

REFERENCES:

1. Druey, J.; Oesterheld, G., *Helv. Chim. Acta* 1942, *25*, 753.

COMPONENTS:

(1) Benzenesulfonamide, 4-amino-N-(aminoiminomethyl)- (sulfaguanidine); $C_7H_{10}N_4O_2S$; [57-67-0]

(2) Ethanol; C_2H_6O; [64-17-5]

(3) Water; H_2O; [7732-18-5]

ORIGINAL MEASUREMENTS:

Sapozhnikova, N. V.; Postovskii, I. Ya.

Zh. Prikl. Khim. 1944, *17*, 427-34.

VARIABLES:

Concentration of ethanol

PREPARED BY:

R. Piekos

EXPERIMENTAL VALUES:

Concentration of ethanol	Solubility at 75 °C	
Weight%	Weight%	mol kg^{-1} solvent[a]
0	1.40	0.0663
19.2	2.67	0.128
57.6	4.43	0.216
76.4	5.56	0.275
86	4.80	0.235
96	3.53	0.171

[a] Calculated by compiler.

AUXILIARY INFORMATION

METHOD/APPARATUS/PROCEDURE:

Sulfaguanidine was dissolved in EtOH-water mixtures to form satd solns which were occasionally agitated in glass vessels immersed in a thermostat. The equilibrium was usually attained after 1 h. Five- to 100-cm^3 samples of the satd soln were placed in Pt crucibles or dishes and evapd to dryness at temps lower than 110-115 °C. The residue was dried to const at 105-110 °C and weighed.

SOURCE AND PURITY OF MATERIALS:

Pure, recrystd sulfaguanidine was used. Its mp conformed with that reported in the literature.

The purity of ethanol and water was not specified.

ESTIMATED ERROR:

Soly: quite reliable results were obtained (authors).

Temp: ± 0.05 °C (authors).

REFERENCES:

COMPONENTS:

(1) Benzenesulfonamide, 4-amino-N-(amino-iminomethyl)- (sulfaguanidine); $C_7H_{10}N_4O_2S$; [57-67-0]

(2) Ethanol; C_2H_6O; [64-17-5]

(3) 1,2,3-Propanetriol; $C_3H_8O_3$; [56-81-5]

(4) Water; H_2O; [7732-18-5]

ORIGINAL MEASUREMENTS:

Dolique, R.; Foucault, J.

Trav. soc. pharm. Montpellier 1952, *12*, 145-53.

VARIABLES:

One temperature: 26-28°C

PREPARED BY:

R. Piekos

EXPERIMENTAL VALUES:

Solubility of sulfaguanidine in a mixture of 1,2,3-propanetriol and 95° ethanol (2:1 by wt) at 26-28°C is 4% (0.2 mol kg^{-1} solvent, compiler).

AUXILIARY INFORMATION

METHOD/APPARATUS/PROCEDURE:

The sulfaguanidine content was detd by diazotization of the amine group in a cold acidified 0.1N KNO_2 soln. An excess of KNO_2 was detected by using iodinated starch.

SOURCE AND PURITY OF MATERIALS:

Nothing specified.

ESTIMATED ERROR:

Nothing specified.

REFERENCES:

COMPONENTS:	ORIGINAL MEASUREMENTS:
(1) Benzenesulfonamide, 4-amino-N-(amino-iminomethyl)-:(sulfaguanidine); $C_7H_{10}N_4O_2S$; [57-67-0] (2) Ethanol; C_2H_6O; [64-17-5] (3) 1,2,3-Propanetriol; $C_3H_8O_3$; [56-81-5] (4) Urea; CH_4N_2O; [57-13-6] (5) Water; H_2O; [7732-18-5]	Dolique, R.; Foucault, J. *Trav. soc. pharm. Montpellier* 1952, *12*, 145-53.
VARIABLES:	PREPARED BY:
One temperature: 26-28°C	R. Piekos

EXPERIMENTAL VALUES:

Solubility of sulfaguanidine at 26-28°C in a saturated solution of urea in a mixture of 1,2,3-propanetriol and 95° ethanol (2:1 by wt), containing 54.5g of urea per 100 g of the mixture, is 5.77% (0.286 mol kg^{-1} solvent, compiler).

AUXILIARY INFORMATION

METHOD/APPARATUS/PROCEDURE:	SOURCE AND PURITY OF MATERIALS:
The sulfaguanidine content was detd by diazotization of the amine group in a cold acidified 0.1N KNO_2 soln. An excess of KNO_2 was detected by using iodinated starch.	Nothing specified. ESTIMATED ERROR: Nothing specified. REFERENCES:

COMPONENTS:	ORIGINAL MEASUREMENTS:
(1) Benzenesulfonamide, 4-amino-N-(amino-iminomethyl)- (sulfaguanidine) $C_7H_{10}N_4O_2S$; [57-67-0] (2) D-Glucose; $C_6H_{12}O_6$; [50-99-7] (3) Water; H_2O; [7732-18-5]	Becher, R.; Leya, S. *Experientia* 1946, *2*, 459-60.

VARIABLES:	PREPARED BY:
One temperature: 18-19 °C	R. Piekos

EXPERIMENTAL VALUES:

Solubility of sulfaguanidine in a 10% D-glucose solution at room temperature (18-19 °C) is 70 mg% (3.3×10^{-3} mol dm^{-3}, compiler).

AUXILIARY INFORMATION

METHOD/APPARATUS/PROCEDURE:

After standing for more than two days the soln of sulfaguanidine was filtered and the sulfonamide was assayed in the filtrate colorimetrically by the method of Druey and Oesterheld (1).

SOURCE AND PURITY OF MATERIALS:

Nothing specified.

ESTIMATED ERROR:

Nothing specified.

REFERENCES:

1. Druey, J.; Oesterheld, G. *Helv. Chim. Acta* 1942, *25*, 753.

COMPONENTS:

(1) Benzenesulfonamide, 4-amino-N-(aminoiminomethyl)- (sulfaguanidine); $C_7H_{10}N_4O_2S$; [57-67-0]

(2) Sorbitan monooleate, polyoxyethylene derivatives (Tween 80); [9005-65-6]

(3) Water; H_2O; [7732-18-5]

ORIGINAL MEASUREMENTS:

Gerencsér-Németh, M.; Horváth, M.

Gyógyszerészet 1973, *17*, 417-21.

VARIABLES:

Concentration of Tween 80

PREPARED BY:

R. Piekos

EXPERIMENTAL VALUES:

Concentration of Tween 80	Solubility at 20°C	
Weight%	g/100 g soln[a]	10^3 mol kg^{-1} soln[b]
1	0.1131	5.279
	0.1137	5.307
3	0.1449	6.763
	0.1471	6.866
5	0.1636	7.636
	0.1633	7.622
8	0.2078	9.699
	0.2090	9.755

[a] Numerical values supplied by the authors.

[b] Calculated by compiler.

AUXILIARY INFORMATION

METHOD/APPARATUS/PROCEDURE:

An excess of sulfaguanidine in an aq Tween 80 soln was shaken in a lab shaker at 120 rpm for 6 h. The soln was then filtered, the residue was washed first with the filtrate and finally with a small amt of water, dried and weighed.

SOURCE AND PURITY OF MATERIALS:

Sulfaguanidine (source not specified) was dried at 100°C for 3 h or over conc H_2SO_4 for 72 h. Its mp was 187.5-8.8°C.

Source and purity of Tween 80 was not specified.

Distd water was used.

ESTIMATED ERROR:

Nothing specified.

REFERENCES:

COMPONENTS:

(1) Benzenesulfonamide, 4-amino-N-(amino-iminomethyl)- (sulfaguanidine); $C_7H_{10}N_4O_2S$; [57-67-0]

(2) 2-Propanol; C_3H_8O; [67-63-0]

ORIGINAL MEASUREMENTS:

Burlage, H. M.

J. Am. Pharm. Assoc., Sci. Ed. 1948, *37*, 345.

VARIABLES:

One temperature: 25°C

PREPARED BY:

R. Piekos

EXPERIMENTAL VALUES:

Solubility of sulfaguanidine in 2-propanol at 25°C is 0.1770 g/100 cm^3 solution (8.262×10^{-3} mol dm^{-3}, compiler).

AUXILIARY INFORMATION

METHOD/APPARATUS/PROCEDURE:

Satd solns of sulfaguanidine in 2-propanol were prepd at 25°C and definite vols of the solns were measured into tared dishes by means of standard pipets. The alcohol was allowed to evap at room temp and the residue was dried at 105°C. In the case of losses due to apparent decompn, the residue was dried in a desiccator (1).

SOURCE AND PURITY OF MATERIALS:

The sulfaguanidine was manufd by Squibb and was of the U.S.P. purity.

The source and purity of 2-propanol was not specified.

ESTIMATED ERROR:

Nothing specified.

REFERENCES:

1. Burlage, H. M. *J. Am. Pharm. Assoc., Sci. Ed.* 1947, *36(1)*, 16.

COMPONENTS:	ORIGINAL MEASUREMENTS:
(1) Benzenesulfonamide, 4-amino-N-(amino-iminomethyl)- (sulfaguanidine); $C_7H_{10}N_4O_2S$; [57-67-0] (2) 2-Propanone (acetone); C_3H_6O; [67-64-1]	Gutierrez, F. H. *Anales fis. quim. (Madrid)* 1945, *41*, 537-60.
VARIABLES:	**PREPARED BY:**
Temperature	R. Piekos

EXPERIMENTAL VALUES:

$t/^{\circ}C$	G^a	E^b	X_g/l^c	mol/l[d] acetone	mmol/mol acetone	$1:X_g$[e]	$1 + X_{cc}$[f]
0	1.548	1.524	12.610	58.8	4.2	64.60	79.30
5	1.574	1.539	12.730	59.4	4.3	63.53	78.55
10	1.622	1.596	13.025	60.7	4.4	62.89	76.75
15	1.679	1.651	13.383	62.4	4.5	59.56	74.74
20	1.728	1.700	13.672	63.8	4.7	57.87	73.15
25	2.013	1.973	15.806	73.8	5.4	49.73	63.25
30	2.215	2.177	17.261	80.6	6.0	45.15	57.94
35	2.584	2.519	19.982	93.3	7.0	38.70	50.05
40	2.929	2.846	22.847	104.9	7.9	35.14	44.48
45	3.199	3.099	24.357	113.7	8.7	31.26	41.05
50	3.549	3.427	26.813	125.1	9.6	28.18	37.29

[a] $G = \frac{p\ 100}{P - p}$, where p and P are the weights of solute and solution, resp.

[b] $E = \frac{G\ 100}{B + 100}$; [c] g/l acetone; [d] should be mmol/l acetone (compiler);

[e] g of acetone require to dissolve 1 g of solute; [f] volume (cm^3) of acetone required to dissolve 1 g of solute.

AUXILIARY INFORMATION

METHOD/APPARATUS/PROCEDURE:

A special all-glass app was constructed enabling the prepn of satd solns, agitation by bubbling a stream of acetone-satd N, filtration, and distn off the solvent without contact with air. Two exchangeable dissoln vessels of 15 and 8 cm^3 working capacity were used depending on the soly of solute. The app was immersed in a thermostat. The vols of acetone used were 15 or 5 cm^3, and the equilibration time was 2-2.5 h. The satd solns were filtered, weighed, the solvent was distd off, the residues were dried at 105°C, weighed, and examd for the presence of solvated acetone.

SOURCE AND PURITY OF MATERIALS:

The source of the materials was not specified. Pure, anhyd acetone was used. The absence of impurities and water was confirmed by procedures of the German Pharmacopeia VI and Spanish Pharmacopeia VIII.

The purity of sulfaguanidine was not specified.

ESTIMATED ERROR:

Soly: measurements were repeated until 2 values not differing in the second decimal were obtained (author).

Temp: ±0.1°C (author).

REFERENCES:

COMPONENTS:	ORIGINAL MEASUREMENTS:
(1) Benzenesulfonamide, 4-amino-N-(amino-iminomethyl)- monohydrate (sulfaguanidine monohydrate); $C_7H_{10}N_4O_2S \cdot H_2O$ [6190-55-2] (2) Ethanol; C_2H_6O; [64-17-5] (3) 1,2,3-Propanetriol; $C_3H_8O_3$; [56-81-5] (4) Urea; CH_4N_2O; [57-13-6] (5) Water; H_2O; [7732-18-5]	Dolique, R.; Foucault, J. *Trav. soc. pharm. Montpellier* 1952, *12*, 145-53
VARIABLES:	PREPARED BY:
One temperature: 26-28°C	R. Piekos

EXPERIMENTAL VALUES:

Solubility of sulfaguanidine monohydrate at 26-28°C in a saturated solution of urea in a mixture of 1,2,3-propanetriol and 95° ethanol (2:1 by wt), containing 54.5 g of urea per 100 g of the mixture, is 6.26% (0.287 mol kg^{-1} solvent, compiler).

AUXILIARY INFORMATION

METHOD/APPARATUS/PROCEDURE:

The sulfaguanidine monohydrate was detd by diazotization of the amine group in a cold acidified 0.1N KNO_2 soln. An excess of KNO_2 was detected by using iodinated starch.

SOURCE AND PURITY OF MATERIALS:

Nothing specified.

ESTIMATED ERROR:

Nothing specified.

REFERENCES:

COMPONENTS:

(1) Benzenesulfonamide, 4-amino-N-(aminoiminomethyl)- monohydrate (sulfaguanidine monohydrate); $C_7H_{10}N_4O_2S \cdot H_2O$; [6190-55-2]

(2) Ethanol; C_2H_6O; [64-17-5]

(3) 1,2,3-Propanetriol; $C_3H_8O_3$; [56-81-5]

(4) Water; H_2O; [7732-18-5]

ORIGINAL MEASUREMENTS:

Dolique, R.; Foucault, J.

Trav. soc. pharm. Montpellier 1952, *12*, 145-53.

VARIABLES:

One temperature: 26-28°C

PREPARED BY:

R. Piekos

EXPERIMENTAL VALUES:

Solubility of sulfaguanidine monohydrate in a mixture of 1,2,3-propanetriol and 95° ethanol (2:1 by wt) at 26-28°C is 4.33% (0.195 mol kg^{-1} solvent, compiler).

AUXILIARY INFORMATION

METHOD/APPARATUS/PROCEDURE:

The sulfaguanidine monohydrate content was detd by diazotization of the amine group in a cold acidified 0.1N KNO_2 soln. An excess of KNO_2 was detected by using iodinated starch.

SOURCE AND PURITY OF MATERIALS:

Nothing specified.

ESTIMATED ERROR:

Nothing specified.

REFERENCES:

COMPONENTS:

(1) Acetamide, N-[4-[[(iminomethyl)amino] sulfonyl]phenyl]-; $C_9H_{12}N_4O_3S$; [19077-97-5]

(2) Water; H_2O [7732-18-5]

ORIGINAL MEASUREMENTS:

Sapozhnikova, N. V.; Postovskii, I. Ya.

Zh. Prikl. Khim. 1944, *17*, 427-34.

VARIABLES:

Temperature

PREPARED BY:

R. Piekos

EXPERIMENTAL VALUES:

$t/^{\circ}C$	Solubility	
	Weight %	10^3 mol kg^{-1} water[a]
20	0.0154	0.601
37	0.039	1.52
50	0.100	3.91
75	0.320	12.5
99	0.868[b]	34.2

[a] Calculated by compiler.

[b] Calculated from the heat of dissolution (10,667 cal mol^{-1})

AUXILIARY INFORMATION

METHOD/APPARATUS/PROCEDURE:

The sulfonamide was dissolved in water to form a satd soln which was occasionally agitated in a glass vessel immersed in a thermostat. The equilibrium was usually attained after 1 h. Five- to 100-cm^3 samples of the satd soln were placed in Pt crucibles or dishes and evapd to dryness at temps lower than 110-115°C. The residue was dried to const wt of 105-110°C and weighed.

SOURCE AND PURITY OF MATERIALS:

Pure, recrystd sulfonamide was used. Its mp conformed to that reported in the literature.

Purity of the water was not specified.

ESTIMATED ERROR:

Soly: quite reliable results were obtained over the temp range 20-75°C. At higher temps the accuracy was poor due to evapn of water during sampling (authors).

Temp: $\pm 0.05^{\circ}$C (authors).

REFERENCES:

COMPONENTS:

(1) Acetamide, N-[4-[[(aminoiminomethyl)amino]sulfonyl]phenyl]-; $C_9H_{12}N_4O_3S$; [19077-97-5]

(2) Ethanol; C_2H_6O; [64-17-5]

(3) Water; H_2O; [7732-18-5]

ORIGINAL MEASUREMENTS:

Sapozhnikova, N. V.; Postovskii, I. Ya.

Zh. Prikl. Khim. 1944, *17*, 427-34.

VARIABLES:

Concentration of ethanol

PREPARED BY:

R. Piekos

EXPERIMENTAL VALUES:

Concentration of ethanol	Solubility at 75°C	
Weight%	Weight%	mol kg^{-1} solvent[a]
0	0.32	0.012
19.2	0.77	0.030
38.3	1.31	0.052
57.6	3.24	0.131
76.4	3.78	0.153
96	2.50	0.100

[a] Calculated by compiler.

AUXILIARY INFORMATION

METHOD/APPARATUS/PROCEDURE:

The sulfonamide was dissolved in EtOH-water mixts to form satd solns which were occasionally agitated in glass vessels immersed in a thermostat. The equilibrium was usually attained after 1 h. Five- to 100-cm^3 samples of the satd soln were placed in Pt crucibles or dishes and evapd to dryness at temps lower than 110-115°C. The residue was dried to const wt at 105-110°C and weighed.

SOURCE AND PURITY OF MATERIALS:

Pure, recrystd sulfonamide was used. Its mp conformed to that reported in the literature.

The purity of ethanol and water was not specified.

ESTIMATED ERROR:

Soly: quite reliable results were obtained (authors).

Temp: $\pm$0.05°C (authors).

REFERENCES:

COMPONENTS:	ORIGINAL MEASUREMENTS:
(1) Benzenesulfonamide, 4-(aminomethyl)- (sulfamyd); $C_7H_{10}N_2O_2S$; [138-39-6] (2) 2-Propanol; C_3H_8O; [67-63-0]	Burlage, H. M. *J. Am. Pharm. Assoc., Sci. Ed.* 1948, *37*, 345.
VARIABLES:	PREPARED BY:
One temperature: 25°C	R. Piekos

EXPERIMENTAL VALUES:

Solubility of sulfamyd in 2-propanol at 25°C is 5.6650 g/100 cm^3 solution (0.3042 mol dm^{-3}, compiler).

AUXILIARY INFORMATION

METHOD/APPARATUS/PROCEDURE:

Satd solns of sulfamyd in 2-propanol were prepd at 25°C and definite vols of the solns were measured into tared dishes by means of standard pipets. The alcohol was allowed to evap at room temp and the residue was dried at 105°C. In the case of losses due to apparent decompn, the residue was dried in a desiccator (1).

SOURCE AND PURITY OF MATERIALS:

The sulfamyd was manufd by Schering Corp (purity not specified). The source and purity of 2-propanol was not reported.

ESTIMATED ERROR:

Nothing specified.

REFERENCES:

1. Burlage, H. M. *J. Am. Pharm. Assoc., Sci. Ed.* 1947, *36(1)*, 16.

COMPONENTS:	ORIGINAL MEASUREMENTS:
(1) Benzenesulfonamide, 4-(aminomethyl)-, monosodium salt (sodium sulfamyd); $C_7H_{10}N_2O_2S{\cdot}Na$; [60758-21-6] (2) 2-Propanol; C_3H_8O; [67-63-0]	Burlage, H. M. *J. Am. Pharm. Assoc., Sci. Ed.* 1948, *37*, 345.
VARIABLES:	**PREPARED BY:**
One temperature: 25°C	R. Piekos

EXPERIMENTAL VALUES:

Solubility of sodium sulfamyd in 2-propanol at 25°C is 0.6290 g/100 cm^3 solution (3.021×10^{-2} mol dm^{-3}, compiler).

AUXILIARY INFORMATION

METHOD/APPARATUS/PROCEDURE:

Satd solns of sodium sulfamyd in 2-propanol were prepd at 25°C and definite vols of the solns were measured into tared dishes by means of standard pipets. The alcohol was allowed to evap at room temp and the residue was dried at 105°C. In the case of losses due to apparent decompn, the residue was dried in a dessicator (1).

SOURCE AND PURITY OF MATERIALS:

The sodium sulfamyd was manufd by Schering Corp., (purity not specified).

The source and purity of 2-propanol was not specified.

ESTIMATED ERROR:

Nothing specified.

REFERENCES:

1. Burlage, H. M. *J. Am. Pharm. Assoc., Sci. Ed.* 1947, *36(1)*, 16.

SYSTEM INDEX

Page numbers preceded by E refer to evaluation texts whereas page numbers not preceded by E refer to compiled tables.

REGISTRY NUMBER INDEX

Page numbers preceded by E refer to evaluation texts whereas page numbers not preceded by E refer to compiled tables.

AUTHOR INDEX

Page numbers preceded by E refer to evaluation texts whereas page numbers not preceded by E refer to compiled tables.

SOLUBILITY DATA SERIES

Volume 1	H. L. Clever, *Helium and Neon*
Volume 2	H. L. Clever, *Krpton, Xenon and Radon*
Volume 3	M. Salomon, *Silver Azide, Cyanide, Cyanamides, Cyanate, Selenocyanate and Thiocyanate*
Volume 4	H. L. Clever, *Argon*
Volume 5/6	C. L. Young, *Hydrogen and Deuterium*
Volume 7	R. Battino, *Oxygen and Ozone*
Volume 8	C. L. Young, *Oxides of Nitrogen*
Volume 9	W. Hayduk, *Ethane*
Volume 10	R. Battino, *Nitrogen and Air*
Volume 11	B. Scrosati and C. A. Vincent, *Alkali Metal, Alkaline Earth Metal and Ammonium Halides. Amide Solvents*
Volume 12	C. L. Young, *Sulfur Dioxide, Chlorine, Fluorine and Chlorine Oxides*
Volume 13	S. Siekierski, T. Mioduski and M. Salomon, *Scandium, Yttrium, Lanthanum and Lanthanide Nitrates*
Volume 14	H. Miyamoto, M. Salomon and H. L. Clever, *Alkaline Earth Metal Halates*
Volume 15	A. F. M. Barton, *Alcohols with Water*
Volume 16/17	E. Tomlinson and A. Regosz, *Antibiotics: I. ß-Lactam Antibiotics*
Volume 18	O. Popovych, *Tetraphenylborates*
Volume 19	C. L. Young, *Cumulative Index: Volumes 1–18*
Volume 20	A. L. Horvath and F. W. Getzen, *Halogenated Benzenes, Toluenes and Phenols with Water*
Volume 21	C. L. and Young P. G. T. Fogg, *Ammonia, Amines, Phosphine, Arsine, Stibine, Silane, Germane and Stannane in Organic Solvents*
Volume 22	T. Mioduski and M. Salomon, *Scandium, Yttrium, Lanthanum and Lanthanide Halides in Nonaqueous Solvents*
Volume 23	T. P. Dirkse, *Copper, Silver, Gold, and Zinc, Cadmium, Mercury Oxides and Hydroxides*
Volume 24	W. Hayduk, *Propane, Butane and 2-Methylpropane*
Volume 25	C. Hiraymana, Z. Galus and C. Guminski, *Metals in Mercury*
Volume 26	M. R. Masson, H. D. Lutz and B. Engelen, *Sulfites, Selenites and Tellurites*
Volume 27/28	H. L. Clever and C. L. Young, *Methane*
Volume 29	H. L. Clever, *Mercury in Liquids, Compressed Gases, Molten Salts and Other Elements*
Volume 30	H. Miyamoto and M. Salomon, *Alkali Metal Halates, Ammonium Iodate and Iodic Acid*
Volume 31	J. Eysseltová and T. P. Dirkse, *Alkali Metal Orthophosphates*
Volume 32	P. G. T. Fogg and C. L. Young, *Hydrogen Sulfide, Deuterium Sulfide and Hydrogen Selenide*
Volume 33	P. Franzosini, *Molten Alkali Metal Alkanoates*
Volume 34	A. N. Paruta and R. Piekos, *4-Aminobenzenesulfonamides. Part I: Non-cyclic Substituents*
Volume 35	A. N. Paruta and R. Piekos, *4-Aminobenzenesulfonamides. Part II: 5-Membered Heterocyclic Substituents*